T0122552

Human–Computer Interaction Series

The Human–Computer Interaction Series, launched in 2004, publishes books that advance the science and technology of developing systems which are effective and satisfying for people in a wide variety of contexts. Titles focus on theoretical perspectives (such as formal approaches drawn from a variety of behavioural sciences), practical approaches (such as techniques for effectively integrating user needs in system development), and social issues (such as the determinants of utility, usability and acceptability).

HCI is a multidisciplinary field and focuses on the human aspects in the development of computer technology. As technology becomes increasingly more pervasive the need to take a human-centred approach in the design and development of computer-based systems becomes ever more important.

Titles published within the Human–Computer Interaction Series are included in Thomson Reuters' Book Citation Index, The DBLP Computer Science Bibliography and The HCI Bibliography.

More information about this series at http://www.springer.com/series/6033

Gerrit Meixner
Editor

Smart Automotive Mobility

Reliable Technology for the Mobile Human

Editor
Gerrit Meixner
UniTyLab
Heilbronn University
Heilbronn, Baden-Württemberg, Germany

ISSN 1571-5035 ISSN 2524-4477 (electronic)
Human–Computer Interaction Series
ISBN 978-3-030-45133-2 ISBN 978-3-030-45131-8 (eBook)
https://doi.org/10.1007/978-3-030-45131-8

This Springer imprint is published by the registered company Springer Nature Switzerland AG
The registered company address is: Gewerbestrasse 11, 6330 Cham, Switzerland

Preface

Mobility is a defining element of our modern way of life and is of crucial importance for securing basic needs as well as for the satisfactory organization of everyday life. Against the background of progressive urbanization, demographic change and the structural changes that go hand in hand with it, ensuring mobility is mentioned by most people as a decisive factor for an active and self-determined life.

Personalized mobility is the design of mobility means and solutions that are tailored to the individual needs of the individual person, thus increasing personal safety, flexibility and comfort. Such human-friendly mobility for all generations is a fundamental prerequisite for achieving the overarching goal of intelligent and smart mobility. At the heart of this approach are always the abilities and preferences of the individual person. Technology should adapt to the individual, adapt to human needs and support us without overburdening or even controlling us. Accordingly, technical solutions must consider the cognitive and physiological abilities and intentions of the human users in the situational context. Thanks to current technological developments, the possibilities and conditions for mobility are changing. In both the private and public sphere, information and communication technologies have the potential to create individually tailored, easily accessible offers and services as well as new types of technical functions. Today, means of transport can be provided according to individual needs and used according to individual possibilities. People can be guided, warned, supported or brought into contact with each other according to their individual needs or preferences. Networked technologies and mobility concepts that span different means of transport can ensure that people are transported reliably, quickly, safely and comfortably.

This book reports on five German research projects and their latest results in the field of intelligent and smart mobility focusing automobile. The projects have been funded by the Federal Ministry of Education and Research (BMBF) in Germany. Chapter 1 reports on the **PAKoS** project (personalized, adaptive cooperative systems for highly automated cars). This chapter discusses how a human-centered control transition can be designed. Another aspect the chapter discusses is how to adapt the driving of highly automated cars to the passengers in the normal full autonomous driving scenario. The project proposes a novel concept of a human-centered highly automated car. Chapter 2 reports on the **KomfoPilot** project. This chapter focusses on investigating, assessing and enhancing comfort during automated driving by two driving simulator studies and a test track study. Sensors such as wearable devices, eye tracking, face tracking and motion tracking allowed for the integration of driver state data with information about the vehicle and the surroundings. Various driving styles as well as display solutions were evaluated for reducing discomfort. In addition, privacy issues were continuously monitored for all aspects over the project lifetime. Chapter 3 reports on the **KoFFI** project (cooperative driver–vehicle interaction). This chapter describes how KoFFI supports the driver in typical traffic situations during a drive. The vehicle acts as a cooperative partner for the driver. On the one hand, there is the so-called Guardian Angel function, which helps the driver to survive critical traffic situations but also offers some convenient features during manual driving. On the other hand, KoFFI can assist the driver at system boundaries and vice versa in various cooperative driving scenarios. The results of the project show that the cooperative assistant KoFFI is able to ensure a convenient, pleasant and safe drive in either manual or automated driving mode. Ethical recommendations from the KoFFI project are described in Chap. 4. Chapter 5 reports on the **Vorreiter** project. This chapter addresses the control transition between the driver and the automated vehicle. It has been inspired by the metaphor of a rider and a horse to provide intuitive steering gestures on the steering wheel which initiate maneuvers to be executed by the automation, and to be supervised by, influenced or interrupted by the driver. The gestures are built up in a design for all which helps both average drivers, beginners and drivers with a handicap. Chapter 6 reports on the **KOLA** project (cooperative laser beam). This chapter describes how KOLA supports a better communication between the different road participants, e.g., cars, bicyclists and pedestrians and potentially optimizes road safety. The project designed prototypical messages and evaluated them in two driving simulator studies using light which was projected from a car as the medium of the messages. They found that expressing one's intention in a specific situation furthers understanding and prosocial behavior of others, which leads to positive emotions on both sides. Thus, light-based communication is an excellent way to further communication, cooperation and positive emotions in traffic.

This book could not be completed without the help of many people. I would like to thank all the authors for their contribution to the book. Furthermore, I would like to thank Helen Desmond as Editor from Springer for her valuable guidance in editing the manuscript and Prof. Jean Vanderdonckt as Editor-in-Chief for his feedback in the initial stage of this book project.

Heilbronn, Germany
September 2020

Gerrit Meixner

Contents

Editor and Contributors

About the Editor

Gerrit Meixner got his diploma and his master's degree in computer science from the University of Applied Sciences Trier and his doctoral degree in Mechanical Engineering focusing on human–machine interaction from the Technical University of Kaiserslautern.

From 2009 to 2013, he was head of the human–machine interaction group at the Innovative Factory Systems department of the German Research Center for Artificial Intelligence (DFKI) in Kaiserslautern. Since March 2013, he is full professor for human–computer interaction in the faculty of computer science of Heilbronn University and managing director of the Usability and Interaction Technology Laboratory (UniTyLab). Since December 2019, he is Affiliated Professor at KTH Royal Institute of Technology Stockholm, and since March 2020, he got a research professorship from the Heilbronn University.

Currently, he is a member of the executive committee of the section "Interactive Systems" and "Software Ergonomics" at the German Informatics Society (GI) as well as Chairman of the Technical Committee 5.31 "User-Centered Development of Industrial User Interfaces" at the Association of German Engineers (VDI). From 2011 to 2014, he was Chairman of the working group "Model-based User Interfaces" at W3C.

He has published more than 130 papers in different human–computer interaction-related conferences, journals and books. In 2015, he got the Research Transfer Award from the Chamber of Industry and Commerce Heilbronn-Franken. His main research interests are in the area of innovative interaction technologies, usability engineering and model-based user interface development. In 2020, he acts as general chair for the 8th IEEE International Conference on Healthcare Informatics (ICHI).

Weblinks/Profiles:

- https://hs-heilbronn.de/gerrit.meixner
- https://www.unitylab.de
- ORCID: https://orcid.org/0000-0001-9550-7418
- LinkedIn: https://www.linkedin.com/in/gerritmeixner/

Contributors

Lenne Ahrens Robert Bosch GmbH, Heilbronn, Germany

Eugen Altendorf Institute of Industrial Engineering and Ergonomics IAW, RWTH Aachen University, Aachen, Germany

Clemens Arzt Berlin School of Economics and Law (HWR Berlin), Berlin Institute for Safety and Security Research (FÖPS Berlin), Berlin, Germany

Emre Aydin Institute of Human Factors and Technology Management IAT, Stuttgart, Germany

Ralph Baier Institute of Industrial Engineering and Ergonomics IAW, RWTH Aachen University, Aachen, Germany

Martin Baumann Universität Ulm, Ulm, Germany

Matthias Beggiato Cognitive and Engineering Psychology, Chemnitz University of Technology, Chemnitz, Germany

Sven Bischoff Institute of Human Factors and Technology Management IAT, Stuttgart, Germany

Anja Valeria Bopp-Bertenbreiter Institute of Human Factors and Technology Management IAT, Stuttgart, Germany

Evin Bozbayir Valeo Schalter und Sensoren GmbH, Bietigheim-Bissingen, Germany

Angelika C. Bullinger Ergonomics and Innovation, Chemnitz University of Technology, Chemnitz, Germany

André Dettmann Ergonomics and Innovation, Chemnitz University of Technology, Chemnitz, Germany

Frederik Diederichs Fraunhofer Institute for Industrial Engineering IAO, Stuttgart, Germany

Daniel Diers Institute of Human Factors and Technology Management IAT, Stuttgart, Germany

Bernhard Doentgen Valeo Schalter und Sensoren GmbH, Bietigheim-Bissingen, Germany

Ute Ehrlich Daimler AG, Ulm, Germany

Stephan Enhuber Communications Engineering, Chemnitz University of Technology, Chemnitz, Germany

Rainer Erbach Robert Bosch GmbH, Renningen, Germany

Volker Fischer EML European Media Laboratory GmbH, Heidelberg, Germany

Michael Flad Karlsruhe Institute of Technology, Karlsruhe, Germany

Boris Flecken mVISE AG, Düsseldorf, Germany

Frank Flemisch Institute of Industrial Engineering and Ergonomics IAW, RWTH Aachen University, Aachen, Germany

Dagmar Gesmann-Nuissl Private Law and Intellectual Property Rights, Chemnitz University of Technology, Chemnitz, Germany

Petra Grimm Hochschule der Medien, Stuttgart, Germany

Franziska Hartwich Cognitive and Engineering Psychology, Chemnitz University of Technology, Chemnitz, Germany

Nicolas Herzberger Institute of Industrial Engineering and Ergonomics IAW, RWTH Aachen University, Aachen, Germany

Sören Hohmann Karlsruhe Institute of Technology, Karlsruhe, Germany

Katharina Hottelart Valeo Schalter und Sensoren GmbH, Bietigheim-Bissingen, Germany

Vera Kaim Institute of Industrial Engineering and Ergonomics IAW, RWTH Aachen University, Aachen, Germany

Frank Kaiser Valeo Schalter und Sensoren GmbH, Bietigheim-Bissingen, Germany

Luis Kalb Technical University of Munich, Garching, Germany

Burak Karakaya Technical University of Munich, Garching, Germany

Philipp Karg Karlsruhe Institute of Technology, Karlsruhe, Germany

Verena Kaschub Fraunhofer Institute for Industrial Engineering IAO, Stuttgart, Germany

Tobias Keber Hochschule der Medien, Stuttgart, Germany

Erdi Kenar Robert Bosch GmbH, Heilbronn, Germany

Tobias Kiefer Valeo Schalter und Sensoren GmbH, Bietigheim-Bissingen, Germany

Judith Klink-Straub Hochschule der Medien, Stuttgart, Germany

Marius Koller UniTyLab, Heilbronn University, Heilbronn, Germany

Josef Krems Cognitive and Engineering Psychology, Chemnitz University of Technology, Chemnitz, Germany

Jakob Landesberger Daimler AG, Ulm, Germany

Carolin Lange Spiegel Institut Mannheim GmbH & Co. KG, Mannheim, Germany

Julian Ludwig Karlsruhe Institute of Technology, Karlsruhe, Germany

Manuel Martin Fraunhofer IOSB, Karlsruhe, Germany

Steffen Maurer Robert Bosch GmbH, Renningen, Germany

Marcus Mazewitsch Spiegel Institut Mannheim GmbH & Co. KG, Mannheim, Germany

Gerrit Meixner UniTyLab, Heilbronn University, Heilbronn, Germany

Ronald Meyer Institute of Industrial Engineering and Ergonomics IAW, RWTH Aachen University, Aachen, Germany

Julia Maria Mönig Hochschule der Medien, Stuttgart, Germany

Klaus Mößner Communications Engineering, Chemnitz University of Technology, Chemnitz, Germany

Timo Pech Communications Engineering, Chemnitz University of Technology, Chemnitz, Germany

Matthias Powelleit Technische Universität Braunschweig, Lehrstuhl für Ingenieur- und Verkehrspsychologie, Braunschweig, Germany

Achim Pruksch Robert Bosch GmbH, Heilbronn, Germany

Patrice Reilhac Valeo Schalter und Sensoren GmbH, Bietigheim-Bissingen, Germany

Alina Roitberg Karlsruhe Institute of Technology, Karlsruhe, Germany

Patrick Roßner Ergonomics and Innovation, Chemnitz University of Technology, Chemnitz, Germany

Simone Ruth-Schumacher Berlin School of Economics and Law (HWR Berlin), Berlin Institute for Safety and Security Research (FÖPS Berlin), Berlin, Germany

Anna Sommer Institute of Human Factors and Technology Management IAT, Stuttgart, Germany

Julia Spies Institute of Industrial Engineering and Ergonomics IAW, RWTH Aachen University, Aachen, Germany

Rainer Stiefelhagen Karlsruhe Institute of Technology, Karlsruhe, Germany

Marcel Usai Institute of Industrial Engineering and Ergonomics IAW, RWTH Aachen University, Aachen, Germany

Mark Vollrath Technische Universität Braunschweig, Lehrstuhl für Ingenieur- und Verkehrspsychologie, Braunschweig, Germany

Marcel Walch Universität Ulm, Ulm, Germany

Michael Weber Universität Ulm, Ulm, Germany

Reto Wechner Institute of Human Factors and Technology Management IAT, Stuttgart, Germany

Gina Weßel Institute of Industrial Engineering and Ergonomics IAW, RWTH Aachen University, Aachen, Germany

Harald Widlroither Fraunhofer Institute for Industrial Engineering IAO, Stuttgart, Germany

Susann Winkler Technische Universität Braunschweig, Lehrstuhl für Ingenieur- und Verkehrspsychologie, Braunschweig, Germany

Marcel Woide Universität Ulm, Ulm, Germany

Chapter 1
Personalisation and Control Transition Between Automation and Driver in Highly Automated Cars

Michael Flad, Philipp Karg, Alina Roitberg, Manuel Martin, Marcus Mazewitsch, Carolin Lange, Erdi Kenar, Lenne Ahrens, Boris Flecken, Luis Kalb, Burak Karakaya, Julian Ludwig, Achim Pruksch, Rainer Stiefelhagen, and Sören Hohmann

Abstract The goal for the distant future is fully autonomous vehicles, which can handle every possible situation and change the role of the driver to that of a passenger. On the way to that goal, drivers of highly automated cars will still have to take over the control of the vehicle at the boundaries of the operational domain of the automation. However, several studies show that the driver's ability to take over depends on his current status and activities. As for highly automated cars more activities beyond the actual driving task, like reading and writing emails, will be allowed, the driver might need of assistance from the vehicle automation during the control transition phase. Ironically, the fact that higher levels of automation result in decreased driver responsibilities makes considering the driver in the automation algorithms even more essential. Therefore, this chapter discusses how a human-centred control transition can be designed. Another aspect we also want to discuss is how to adapt the driving of highly automated cars to the passengers in the normal full autonomous driving scenario. Hence, we propose a novel concept of a human-centred highly automated car. First, we present an approach for monitoring and interpreting the driver in the

---

M. Flad (✉) · P. Karg · A. Roitberg · J. Ludwig · R. Stiefelhagen · S. Hohmann
Karlsruhe Institute of Technology, Karlsruhe, Germany
e-mail: Michael.Flad@gmx.de

M. Martin
Fraunhofer IOSB, Karlsruhe, Germany

M. Mazewitsch · C. Lange
Spiegel Institut Mannheim GmbH & Co. KG, Mannheim, Germany

E. Kenar · L. Ahrens · A. Pruksch
Robert Bosch GmbH, Heilbronn, Germany

B. Flecken
mVISE AG, Düsseldorf, Germany

L. Kalb · B. Karakaya
Technical University of Munich, Garching, Germany

G. Meixner (ed.), *Smart Automotive Mobility*,
Human–Computer Interaction Series,
https://doi.org/10.1007/978-3-030-45131-8_1

interior and a systematic design of a cooperative control transfer. We further point out the crucial aspects of the implementation of such kind of adaption concepts and present results from various driving studies.

1.1 Introduction

Nowadays, all commercially available vehicles, which are permitted for the general traffic environment, still require a human driver. Even features that are advertised by the manufacturers as "automated" rely on the supervision of the driver, also due to legal constraints. The driver has to monitor the correct functioning of the assisting feature continuously and is not allowed to digress from the driving task legally. In order to describe the term "automated vehicle" more precisely, current series cars can be classified with automation level 2 according to the definition of the Society of Automotive Engineers (SAE) (SAE 2018). These levels rate the influence of the automation on the driving task and how strongly the human driver will be supported by the assistance system. The six levels extend from 0, no automation active, to 5, fully autonomous driving with no need for supervision at all.

Naturally, the vision is vehicles with automation and assistance systems that will not only support their drivers instead they will also take over the full control in some situations and, even more important, the responsibility for the driving task. At a first glance, this trend seems to reduce the need of focusing on the human driver during vehicle development, but in contrast, this results in new challenging interactions between the driver and the vehicle automation. For example, the driver of a SAE level 3 car (SAE 2018) has not to supervise the automated vehicle continuously. However, the automation system can transfer the vehicle control back to the driver in an adequate period, if it is not able to handle upcoming situations safely and autonomously. This could be the case, e.g. in scenarios with an unconventional temporal track guiding during a road construction site or in the degeneration case, when sensor failures occur. Even in context of vehicles of the higher SAE level 4, a control transition back to the driver is still required at the boundaries of the automation operational domain like for example the end of the motorway. At least in the near future, fully autonomous and thus driverless vehicles will only be capable to handle very tight operational domains like people mover scenarios with a predefined environment. The transition to SAE level 5 would demand that a system could handle all situations in all possibly occurring driving scenarios.

In contrast to current systems on the market, which only provide assisting features, also due to legal constraints, the driver of a highly automated car (SAE level 3 or higher) will be allowed to disengage from the driving task completely. Several studies point out the challenge for drivers to resume manually controlling the vehicle after a long period of automated driving (Merat et al. 2014; Brandenburg and Skottke 2014; Damböck et al. 2012; Gold et al. 2013; Endsley and Kris 1995). For a commuter that had to handle every day the same exception situation at the same place, it would most likely be possible to perform an adequate takeover but for the driver of a minivan, which he has rented for family holiday and drives in an unfamiliar environment, this would probably not be the case.

Hence, automated cars with higher automation levels need to provide a concept to support the driver by resuming the driving task, and additionally, it has to act adaptive and individualised with respect to the current state of the user and his capabilities. Furthermore, such an adaption and personalisation is not only beneficial in case of the described transition scenarios, where the driver is supported by taking over the control of the vehicle. Adjusting the behaviour of the vehicle automation in purely automated driving scenarios based on the needs of the passengers as well as an evaluation of the current situation inside the car can increase the overall driving comfort.

Therefore, the vehicle automation should ideally not only adapt itself to the individual preferences of the user (e.g. driving faster or slower, considering the preferred lateral dynamics), it should also consider the current situation inside the car (e.g. is the driver drinking coffee, is he reading a newspaper). In the end, including such an adaption procedure in the automation concept has the potential to increase the acceptance of automated cars. This especially refers to people, who are less affine to technology. Without a proper personalisation concept for the automation the current "designed for all" approach will not work in future mobility scenarios. In addition, due to the decreasing diversification potential in autonomous cars, such a feature will most likely become a unique selling point for a car manufacturer.

In order to summarise the discussion above, there should be a cooperation between the automation and the driver. This is essential for the control transition task as well as for the "normal" automated driving mode to increase user acceptance and the driving comfort for the passengers.

Therefore, our goal for this chapter is to introduce the idea of a human-centred highly automated car. We want to discuss how a human-centred control transition can be designed and how the driving behaviour of an automated car can be adapted to the needs and individual preferences of its passengers in general. The last-mentioned aspect refers to the transition phase as well as the purely automated drive itself.

To build such kind of system, several technical modules are required, and various questions need to be answered. At first, an ideal technical system with the aim to adapt itself towards a better cooperation with a human partner would have the ability to understand the human being itself. Since this general statement is a quite challenging aspect for an artificial system, technology may possibly never master it. More realistically, for the scenarios addressed in the project, it would be beneficial, in our opinion essential, that the automation can at least understand the driving situation, and additionally, the status of the driver and the other passengers. In a vehicle of SAE level 3 or higher, the driver can be currently inactive but could receive a request for taking over the control at any time. Whereas grasping the driving situation is a basic aspect of all highly automated cars, we focus on the human in this research project and design a software module that can identify the activities the occupants perform by using a camera-based driver monitoring system. By understanding, the current activity (e.g. is the driver sleeping, reading a book or focused on the driving situation), the automation can reason about his current takeover capability. This knowledge is used by the automation for the design of the takeover. In addition, the current human activities also greatly affect the desires of the passengers regarding the driving behaviour of the automated vehicle. Therefore, this knowledge is the relevant input for the adaption algorithm altering the purely automated driving behaviour, as well.

In a second step, a concept to support the driver in the control transition is needed. We suggest to divide the control transition into two phases. An information and preparation phase we call priming and the actual transition phase. In the first phase, since a level 3 car allows the driver to perform activities that are not related to the driving task, the driver is normally not physically connected to the vehicle control interfaces like the steering wheel and the pedals. Thus, we focus on informing the driver about the current driving situation in order to prepare him for the driving task, i.e. manually controlling the vehicle. For this, a multimodal communication concept is required. After the hands on (and with respect to the pedals the feet on) the driver can influence the vehicle dynamics. However, during the priming phase, he will most likely not have obtained full situational awareness and he is prone to errors (e.g. see Brandenburg and Skottke 2014). Therefore, in contrast to assistance systems currently on the market where the control is shifted more or less binary, we present an automation system, which slowly transfers the control authority to the driver. The system is intended to adapt itself to the current driver, his situation and personal skill. During this cooperative driving phase where driver and automation jointly operate the vehicle for a short time, the driver can build up his full situational awareness.

To realise this cooperative transition phase but also to adapt the autonomous driving style to the drivers needs information about his personal preferences and capabilities are required. We call this set of data driver profile. Although some data are obtained and processed during the actual interaction of the users with the system, an essential question for implementing the proposed concept is how to manage the personal data of the driver and how to integrate them into the electric and digital architecture of a vehicle. As automotive mobility becomes more and more flexible (e.g. car sharing), the data should be transferable, which means that the user should be able to access his personalisation and his transition profiles, not only in one single car but also use them in every other vehicle supported. This is necessary because these profiles are unique for one individual person and the automation can only provide the best solution by exploiting these data. Ideally, this should even be possible between cars of different manufacturers. Thus, a transportable data storage is required, and we have decided to realise a mobile phone application for this purpose. A modular user profile concept is suggested as candidate for a manufacturer-independent standard. In order to guarantee the data sovereignty for the user, we have implemented a Face ID system for the purpose of authentication.

Our concept of a human-centred highly automated car is summarised in Fig. 1.1. As described above, to realise this vision, a driver observation module, a concept for the control transition, an approach to adapt the automated driving behaviour and a module that handles the user data in a secure way are required. The driver observation data act as the basis for the individualisation and personalisation of the

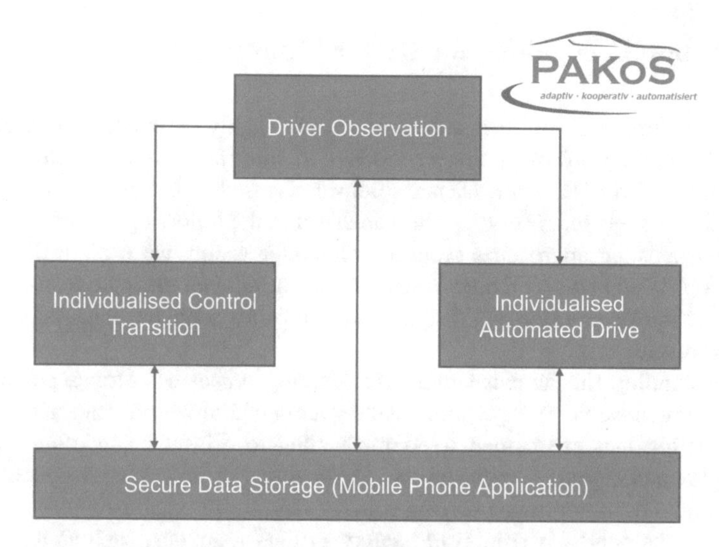

Fig. 1.1 Concept of a human-centred highly automated car

control transition as well as the purely automated driving mode. The mobile phone application builds a secure data storage to manage the personal data and makes it transferable between different vehicles.

Finally, after the implementation of the proposed concept, its measurable benefit needs to be proven using several driving studies. These studies were performed on a driving simulator as well as with an actual test car. The results highlight the improvement of objective measures like the control performance and subjective measures like the user acceptance.

The ideas presented in this chapter primarily base on the work of the Personalised, Adaptive Cooperative Systems for highly automated cars (PAKoS) that was funded by the Federal Ministry of Education and Research.

The chapter is structured as follows: in the first section *Action Recognition for Driver Monitoring* the system for identifying and interpreting the activities of the driver is presented. The section *Human–Machine Interface and Control Transition Concept* introduces the interaction concept between human and automation as well as the control algorithms for the vehicle control transition back to the human driver. The overall system architecture along with other technical implementation details are given in the section *Implementation Aspects.* Finally, the section *Study Results* concludes the chapter by presenting the evaluation results and benefits of the proposed concept.

1.2 Action Recognition for Driver Monitoring

One of the first steps towards a human-centred highly automated car according to Fig. 1.1 is the implementation of a driver monitoring system, which is able to recognise the driver's activities. Therefore, we discuss the different facets of action recognition models for observing a human driver in the following section first. After evaluating existing approaches based on computer vision, we explain the dataset *Drive&Act*, which acts as the basis to train our developed machine learning algorithms for driver monitoring. The section concludes with the discussion of the achieved results.

Understanding the human behind the steering wheel has strong potential to improve human–vehicle cooperation, making it more intuitive and safe. Rising levels of automation increase human freedom, leading to drivers being often engaged in complex and distractive behaviours. Applications of driver activity recognition models range from improving driving comfort (e.g. by automatically turning on the light when the person is reading) to safety–critical functions, such as identifying distraction. We recognise four major use cases for applications of driver activity recognition models in practice.

- **Increased safety through assessing the level of alertness**. The current activity directly affects the human cognitive workload (Wolf et al. 2018). An important application of driver activity recognition models inside vehicles of SAE levels 0–3 is quantifying the distraction level of the driver and reacting accordingly by, for example, issuing a warning signal or adapting the takeover concept (see Fig. 1.2).
- **Increased comfort**. Vehicle controls, such as driving dynamics, lighting or music volume, can be automatically adjusted depending on the detected activity. For example, in context of SAE levels 4 and 5, the car might move smoother if the person is sleeping or drinking coffee.
- **Intuitive communication**. Gestures have been identified as a natural way for human–machine interaction in many fields and might enhance the existing communication interfaces inside the vehicle.
- **Prediction of potentially dangerous manoeuvres**. The majority of traffic fatalities is caused by human errors (Singh 2015; Lloyd et al. 2014). An important application of activity recognition models in context of manual driving is predicting, what the driver is doing next. Such an information would allow Advanced Driver Assistance Systems (ADAS) to identify that the person intends to induce a dangerous lane change, and thus, it is able to prevent the accident.

1.2.1 Impact of Driver Activities on Takeover Manoeuvres

In this subsection, we focus on the specific use case of takeover manoeuvres. As long as cars are not fully automated (SAE level 5), the driver still has some responsibilities

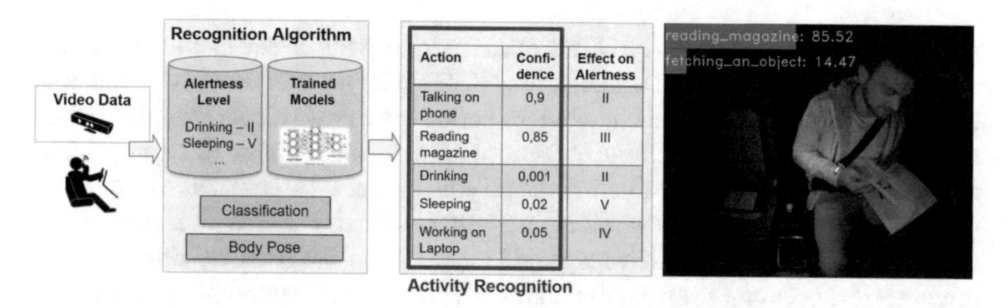

Fig. 1.2 A major safety–critical application of activity recognition models is identifying driver's level of alertness through his current behaviour

in the driving process. Depending on thc degree of automation, he is required to supervise and intervene if necessary (SAE level 2), take over on short notice (SAE level 3) or take over on system boundaries (SAE level 4). Apart from environmental factors such as the complexity (Damböck et al. 2012) or criticality (Gold et al. 2013) of the traffic situation, the driver's activities before a takeover request and his behaviour while taking over have a big impact on the safety of the control transfer itself and the driver's ability to drive manually afterwards.

At the time of writing, there are strong indications that future automated cars are legally required to monitor the driver in some way to determine his readiness to take over.[1] However, the method and extend are not yet clear. Current commercial systems rely mostly on sensors in the steering wheel. However, there are methods to fool these sensors,[2] and as soon as the driver is allowed to remove his hands from the controls, this method is rendered ineffective. Research and development of driver monitoring systems will therefore focus on these takeover scenarios for the near future. The following section summarises some studies, which examine the impact of driver activities on the takeover process.

Even in current commercial vehicles with adaptive cruise control or at most level 2 automation research indicates that due to over reliance and comfort some drivers already disengage from the main driving task taking their hands and feet from the control interfaces for a prolonged time or even engaging in secondary activities (Banks et al. 2018).

On higher automation levels, the driver is not required to monitor the car constantly. Instead, he can perform other activities, which means that the ability to take over the control of the vehicle can vary greatly. The effect of such non-driving activities on different phases of the takeover process have already been investigated in several studies. It has been shown that the reaction time for the first driving action after a takeover request significantly increases with increasing load (Petermann-Stock et al. 2013; Radlmayr et al. 2014). Further studies point out that drivers who

---

[1] https://www.unece.org/fileadmin/DAM/trans/doc/2019/wp29/WP29-177-19e.pdf, accessed 2020.02.03.

[2] https://www.nhtsa.gov/press-releases/consumer-advisory-nhtsa-deems-autopilot-buddy-product-unsafe, accessed 2020.02.03.

are distracted by secondary activities make more driving mistakes in the first few seconds after the takeover (Damböck et al. 2012) and drive less safely (Gold et al. 2013). In addition, an analysis of a longer period after the takeover exhibits that the manual driving behaviour returns to its normal performance after 40 s compared to drivers who are not distracted before the takeover request (Merat et al. 2014).

The presented studies show mainly two effects. First, the driver's condition and activity prior to a takeover have a major influence on the takeover time. Second, driving errors occur more often after taking over the vehicle control from the automation if the human drivers were distracted beforehand. Hence, the given overview highlights the importance of a driver monitoring system and an action recognition system in the context of a human-centred automated car, which supports the driver in the control transition phase.

1.2.2 Action Recognition in Computer Vision and Its Application in the Interior of Cars

Until now, the extraordinary progress in the field of vision-based activity recognition has had a rather slow effect on the field of driver observation. Since machine learning models are heavily driven by data, this is caused by the lack of large-scale datasets for training such models, presumably. Existing algorithms focus on a rather coarse classification of very limited set of driver behaviours (e.g. Ohn-Bar et al. 2014; Abouelnaga et al. 2018). For example, (Yan et al. 2016) detect three states: *operating the shift gear*, *eating/smoking* (combined together) and *talking on the phone*. The proposed method, achieving 99.78% accuracy, might draw an artificially idealistic picture as the three posture types are highly discriminative and easy to distinguish. Similarly, (Ohn-Bar et al. 2014) evaluate their method on only three behaviours linked to the interior region: activities of the instrument cluster region, gear region and steering wheel region. Due to increasing automation reinforcing driver engagement in secondary tasks and diversifying possible activities behind the steering wheel, fine-grained semantic analysis becomes more important. Such detailed understanding of the driver state has been largely overlooked in the past and was first enabled by (Martin et al. 2019) by means of the first large-scale dataset for driver activity recognition covering both, autonomous and manual driving together with over 80 activity classes.

In terms of activity recognition algorithms, proposed approaches can be categorised into (1) methods based on manually designed features and (2) end-to-end methods based on convolutional neural networks (CNNs). In the feature-based approaches, the feature is often computed from body pose (Martin et al. 2018; Ohn-Bar et al. 2014), eye gaze (Ohn-Bar et al. 2014), head pose (Jain et al. 2015) or vehicle dynamics (Lefèvre et al. 2015). The resulting feature vector is further passed to a machine learning framework based on, e.g. support vector machines, hidden Markov models (Jain et al. 2015) or recurrent neural networks (Martin et al. 2018, Jain et al.

2015). End-to-end methods operate directly on the input video so that the intermediate representation is not defined but learned through convolution filters, which are optimised for classification and have been successfully applied in fields such as driver intention recognition (Gebert et al. 2019), fine-grained activity recognition (Martin et al. 2019) and driver posture classification (Abouelnaga et al. 2018). Such approaches are often based on 3D CNNs (Tran et al. 2015) for capturing spatiotemporal structures and require large amount of training data. In general, the field of driver activity recognition has been largely driven by progress in computer vision research. Despite being slowed by the lack of detailed annotated datasets, the recent emergence of large-scale benchmarks, such as *Drive&Act* and progress in transfer learning, opens new possibilities for intelligent driver monitoring systems.

1.2.3 **Drive&Act** *Dataset for Fine-Grained Driver Activity Recognition*

To address the lack of large-scale datasets for driver activity recognition, (Martin et al. 2019) have introduced the *Drive&Act* benchmark, which is described in the following subsections in detail. Therefore, the data collection method is explained as well as how the available actions in the dataset are selected and finally how the dataset is structured and annotated.

1.2.3.1 Recording Environment

Even with prototype vehicles for automated driving, it is not safe to perform highly distracting activities in road traffic. The driver is responsible for monitoring the vehicle in order not to endanger himself or other road users.

For this reason, the collection of the dataset took place in a static driving simulator. To keep the scenario as realistic as possible the driving simulator is equipped with a converted Audi A3. A polygonal screen in front of the vehicle and several other displays serve to show the simulated environment. The simulation software used is SILAB from WIVW. Manual and automated driving runs, including takeovers as well, were part of the overall simulated scenario.

A complex camera system was used to record the test participants. Six cameras were distributed in the vehicle cabin in total. All cameras were calibrated to each other and the vehicles interior. The focus of the used camera technology was their viability to work under realistic driving conditions both in daylight and at night in a prototype vehicle. To this end, near infrared cameras with active lighting were installed at five positions in the interior to cover the front part of the cabin. In addition, a Microsoft Kinect for Xbox One was used to record depth and colour images. Figure 1.3 shows an overview of the camera positions and views.

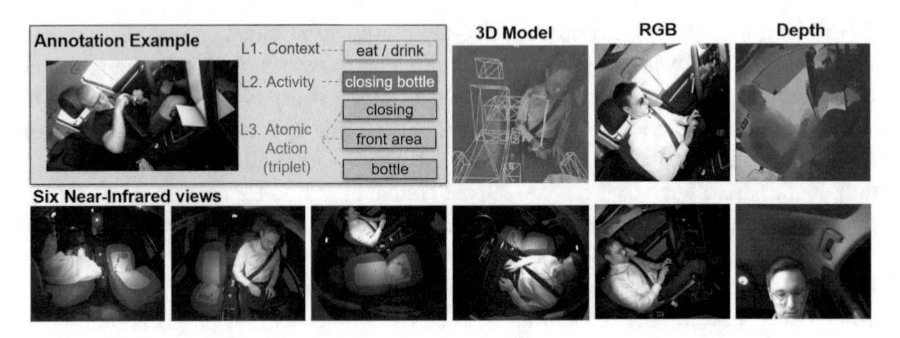

Fig. 1.3 Overview of the Drive&Act dataset for driver activity recognition

1.2.3.2 Survey of Driver Activities

The first question to ask when building an activity recognition system is *what do we want to recognise*? To answer this question, we identify the behaviours which (1) will occur frequently inside the car and (2) will be used by the manufacturers, e.g. due to their impact on accident odds or interesting multimedia applications. We examined these aspects and derived a catalogue of desired driving behaviours serving as the foundation for *Drive&Act*. The design of the activity catalogue followed three principles:

Literature research on driver engagement in secondary tasks In order to represent the situations, that occur while driving, adequately, we have conducted a throughout literature review on driver engagement in manual driving based on three kinds of sources: driver interviews, police reviews of accident studies as well as naturalistic car studies. The first list of activities was produced by revising previously published work on secondary activities during manual driving and their impact on attention/accident risk. Frequently listed secondary activities from different studies were compiled in an activity list and divided into ten activity categories: *food and drink*, *clothing and hygiene*, *interaction with in-vehicle devices*, *interaction with external devices and objects*, *interaction with the environment outside the vehicle*, *smoking*, *self-related activities*, *interaction with persons and pets*, *critical situations* and *driving-related activities*.

Evaluation of the activities by interviewing application-oriented project partners Five partners from research and industry rated the activities on a numerical scale, gave suggestions for further behaviours and additional feedback. Statistical evaluation of the results (whether the median rating of the activity is above the chosen threshold value) resulted in the first version of the activity set.

Feasibility analysis of the activities Subsequently, a feasibility analysis of the activities (e.g. the implementation of "smoking" is difficult) was carried out together with the Fraunhofer IOSB and the final catalogue of objectives for the activities was drawn up. A total of 64 activities were identified, 30 of which are classified as technically feasible with high priority and 13 with lower priority (partner survey), 9 of which were classified as a special case (e.g. sleeping), which is not excluded but considered separately, and 12 activities were excluded due to their difficult implementation regarding the data collection (e.g. smoking).

After the three steps described above, we designed a data collection protocol aimed at both, encouraging diverse behaviour, which captures the desired action classes, and giving the driver certain freedom to reassure a realistic environment. Each driver completed two sessions, starting with entering the car, making adjustments, beginning to drive manually and switching to the autonomous driving mode after several minutes. In each session, the driver was given instructions to solve twelve different tasks, which were displayed on a tablet (e.g. look up the current weather forecast with the laptop and report it via SMS). The order of the tasks was randomised in every session. Note, that while the high-level tasks were given to the driver as instructions, he or she could freely decide how to approach them. Following the data collection, we have designed a hierarchical annotation scheme, labelling each video frame on three levels of abstraction. The categories cover (1) the high-level *context tasks* (instructions, which were currently given to the user), (2) *fine-grained activities* (which are presumably most interesting for the recognition) and (3) *atomic action units* (which are primitive interactions with the environment and are described in the next subsection in detail).

1.2.3.3 Dataset and Hierarchical Activity Labels

Drive&Act is the first large-scale dataset for driver activity recognition which covers both, autonomous and manual driving scenarios (Martin et al. 2019). The dataset features colour-, depth-, infrared- and body pose data streams and is densely annotated on three levels of abstraction (see Fig. 1.3). The hierarchical annotation scheme groups behaviours according to their granularity and comprises the following categories:

- 12 **Context Tasks** the user was asked to complete each session in a random order (e.g. read a magazine to answer a question).
- 34 **Fine-grained Activities** (e.g. closing bottle), which alternate freely in one session, as the participants got no instructions *how* to solve the high-level task.
- **Atomic Action Units** defined as a triplet of ***Action*** (5 in total), ***Object*** (17 in total), ***Location*** (14 in total) depict primitive human interactions with the environment.

Figure 1.4 illustrates different annotation examples on the three abstraction levels. The context task annotations (level 1) are linked to the instructions the driver was given and are therefore long-term behaviours, which can be decomposed in smaller activities. Such fine-grained activities are annotated at level 2 and are further decomposed in the atomic action units.

The behaviours are highly diverse in their complexity and duration, even if we consider the fine-grained activity classes only. For example, an activity segment of

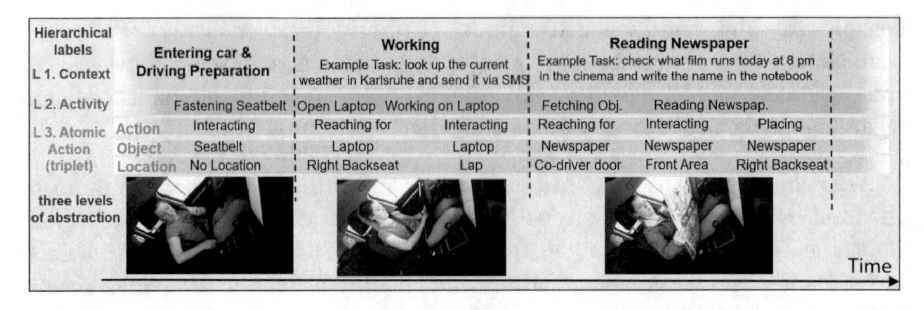

Fig. 1.4 Examples of the hierarchical annotation snapshots in Drive&Act (simplified)

a person reading a magazine often lasts for almost a minute, while other activities, such as pressing the automation button, on average, lasts only one second. Categories are also unevenly distributed in the dataset in terms of their frequency, which is characteristic for real-time applications, where data are scarce and certain events are rare. An overview of how often an action has occurred in the dataset and its average duration is presented in Fig. 1.5.

High level of detail is an additional challenge *Drive&Act* poses for the action recognition models. While previous works tackle highly discriminative actions, such as *eating* versus *talking on the phone* (e.g. Yan et al. 2016), the fine-grained annotations of our benchmark require models to distinguish between *opening* and *closing a bottle* or between *eating* and *preparing food*.

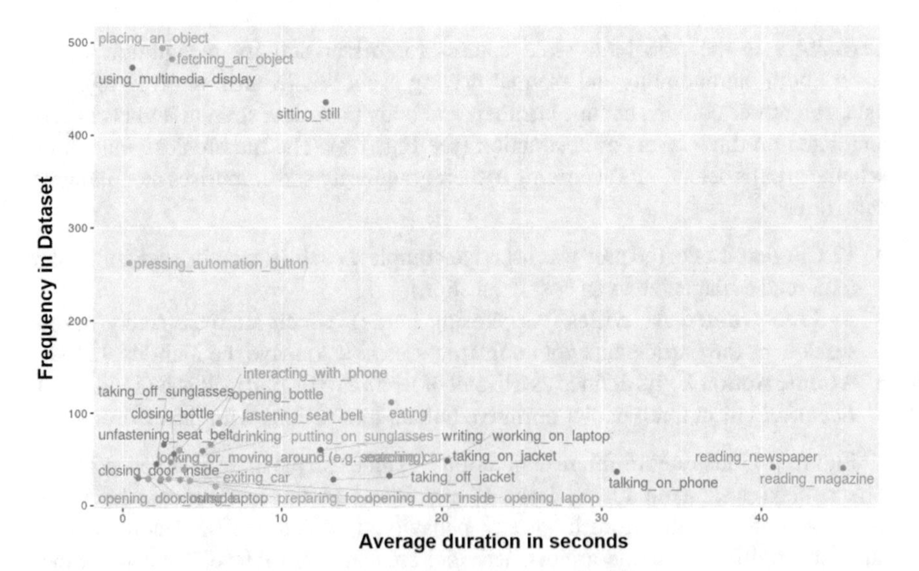

Fig. 1.5 Average duration statistics of fine-grained activities (X-axis) and their frequency (Y-axis) in the Drive&Act dataset. Recorded activities are diverse in their duration, ranging from fractions of a second (e.g. pressing the automation button) to almost a minute on average (e.g. reading a magazine)

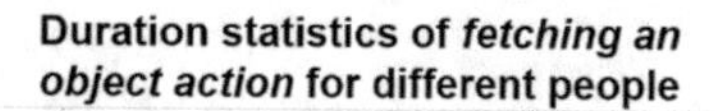

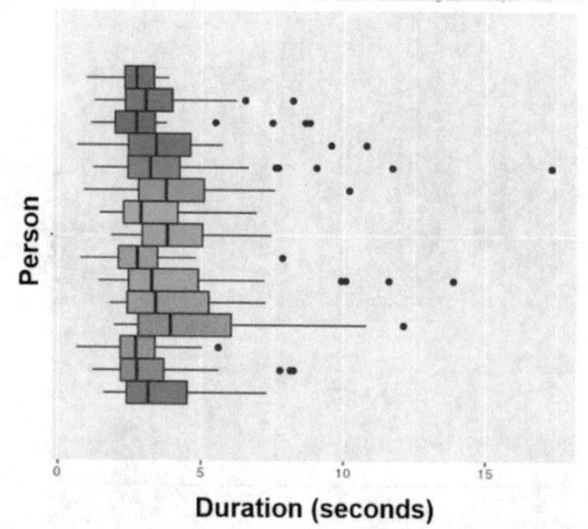

Fig. 1.6 People execute actions in different ways. The boxplot illustrates how long it took, to **fetch an object**, grouped by person (blue points depict outliers). While for some people, fetching an object only takes a couple of seconds, other drivers have frequent outliers with very long durations (over 15 s). Personalisation of recognition models and their online adjustment to the characteristic movement patterns of the user is an important future research direction, which might make the recognition model more robust

Furthermore, we have observed that people execute similar tasks in different ways. Some people move slower than others move or perform different steps to achieve the same goal, i.e. to accomplish the same high-level task. Studying such inter-personal differences, depicted in Fig. 1.6 for *fetching an object* as an example, acts as an important starting point for future work, as it leads to the development of personalised action recognition models. For example, an action recognition model trained on *Drive&Act* might dynamically be adjusted to different people with individual movement characteristics by improving the accuracy through learning on the fly.

For evaluation purposes, we divide the video recordings into chunks of three seconds. These video samples with their corresponding annotation are used for quantify the model performance. Our task is therefore to assign the correct labels on all hierarchy levels to a 3-s video chunk. Figure 1.7 illustrates the distribution of samples for each possible label in our dataset. The clearly unbalanced distribution of different classes is typical for real-world applications, as certain actions (e.g. taking laptop from the backpack) are rare while others (e.g. sitting still) are very common.

1.2.4 Recognition Approaches

As described in previous sections, there are different approaches to realise an action recognition model with machine learning methods.

Often camera images are directly used as input of neural networks to infer activities. The advantage of this approach is that the neural network can learn which

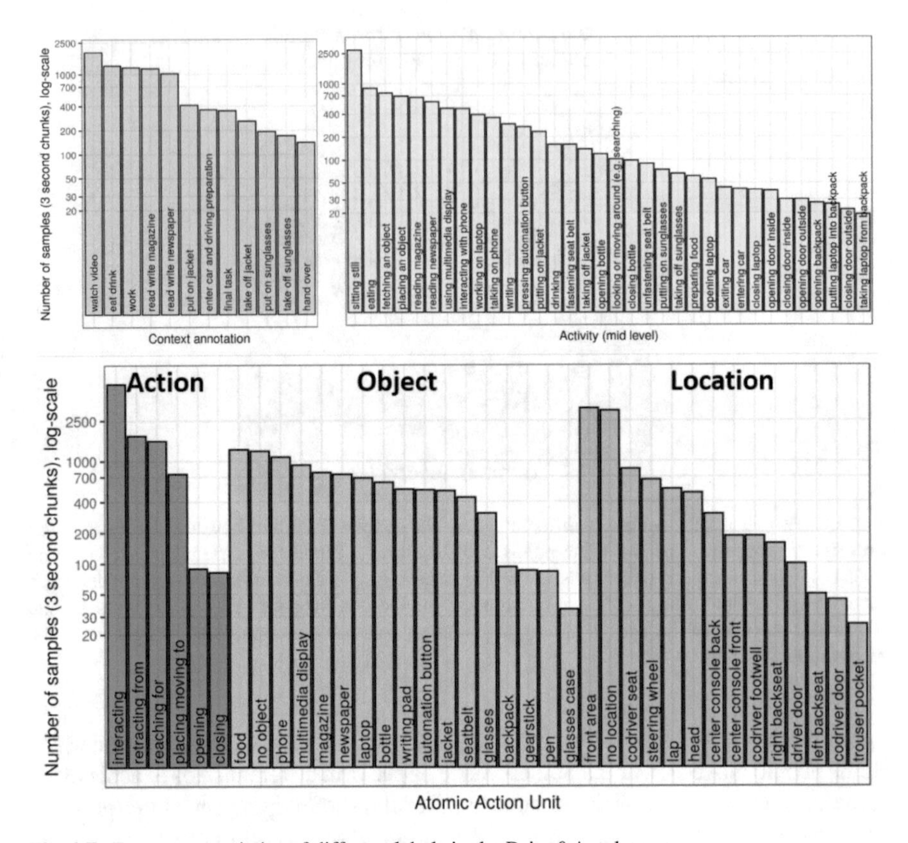

Fig. 1.7 Frequency statistics of different labels in the Drive&Act dataset

features of the image are best suited for recognition. However, the disadvantage of this approach is the black box character of such models, which makes it difficult to draw conclusions in case of errors. Furthermore, these methods are also dependent on the used cameras as well as their viewing angles and require extensive data augmentation. If these parameters change, it might be necessary to collect training data again.

Another way of performing action detection uses an intermediate representation. This decouples the action recognition system to a large extent from the used cameras and allows an easier interpretation in case of an error. The disadvantage of this procedure is the loss of information, since not all information contained in the camera images is retained in the intermediary representation.

The following sections present methods for both approaches and explore their strengths and weaknesses in more detail based on results achieved on the presented *Drive&Act* dataset. Figure 1.8 shows an overview of both ways to perform action recognition. Formally, given a 3-s video chunk, our goal is to assign it the correct activity labels for each hierarchy level (five labels in total: one label for the context task and the fine-grained activity and three labels for the atomic action unit triplet).

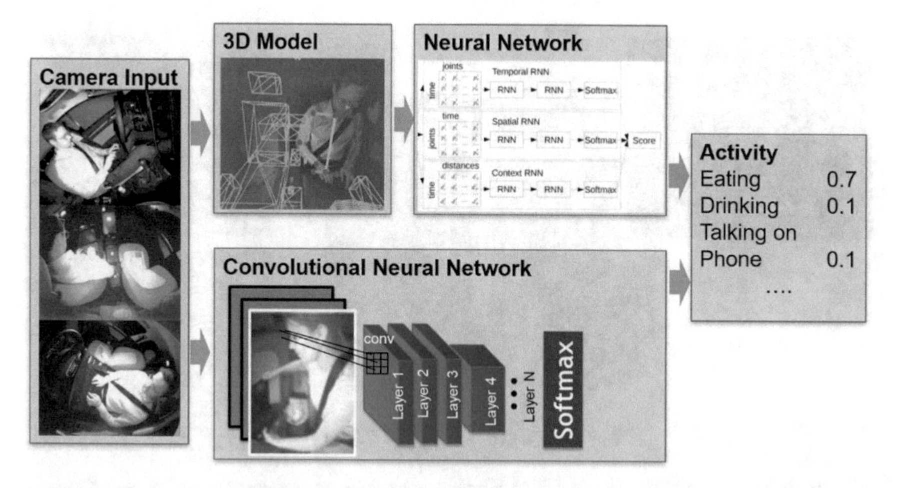

Fig. 1.8 We categorise action recognition algorithms in two groups: (1) methods which first calculate the body pose and 3D features and then classify them, e.g. via a recurrent neural network (top); (2) end-to-end methods based on convolutional neural networks, which calculate the prediction directly from the video (bottom)

Note that we use separate models for every hierarchy level, as training a single model with multiple outputs has led to a decline in recognition rate.

1.2.5 Approach Based on Body Pose and 3D Model

The following subsection presents a method for action recognition using an intermediary representation. At first, the intermediary representation and its creation is explained. Afterwards, it is shown how this representation can be used for action recognition of the driver.

1.2.5.1 Interior Model Creation

The goal of the intermediate representation is to retain all necessary information for action recognition from the camera images while at the same time keeping the dependence on the parameters of the sensor system as low as possible. One way to gain invariance to the camera position and orientation is to use a 3D representation referenced to the vehicle instead of the camera. This representation must at least show the movements of the driver. A possible representation is his 3D upper body skeleton and its movement over time. To determine this skeleton the method OpenPose (Cao et al. 2018) is used. It uses a neural network to determine the 2D coordinates of the driver's body joints from camera images. 2D joint positions of multiple cameras are then used to triangulate the 3D pose of the driver using the intrinsic and extrinsic parameters

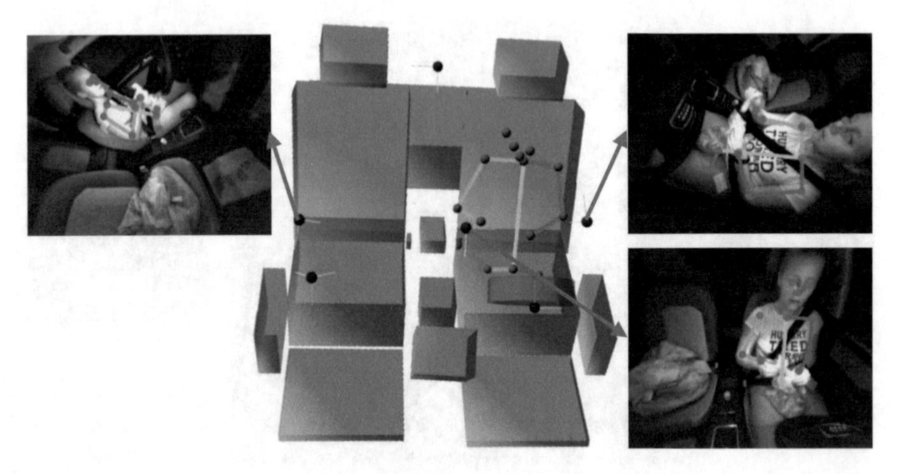

Fig. 1.9 Overview of the intermediary representation used to estimate the driver's actions. The pose is triangulated based on the depicted camera images. Geometric primitives positioned by using an a priori calibration method define the interior

of the camera system. Based on this abstract representation, activity recognition can already be performed. However, since the vehicle interior is a very static environment, additional 3D information describing the driver's environment can be included in the model. For this purpose, individual elements of the vehicle interior, such as the steering wheel or the passenger seat, are represented as cylinders and cubes in 3D space. Through this additional context, a method for action recognition should be able to learn the relationship of certain actions with the interior of the car (Fig. 1.9).

1.2.5.2 Interior Model Interpretation

One way to interpret the presented intermediate representation are recurrent neural networks. In general, they are designed to model time series data. The presented approach uses multiple neural networks to focus on different parts of the intermediary model. Overall, three views of the representation are analysed by three separate models:

Pose: All joint nodes of the skeleton are merged into one vector and used as input for the first neural network in each time step. This allows the network to learn temporal relationships between joints.

Spatial: The skeleton is transformed into a sequence of joint nodes using depth-first search. This sequence is used as input for a second recurrent network. This representation is suitable to analyse the spatial relationship between adjacent joints.

Interior: The distance between the different primitives of the interior model and the wrist and head joints are determined with the third neural network for each time step. This method is suitable to learn the spatiotemporal interactions between the driver and the different parts of the interior.

All three models use a recurrent neural network with two Long Short-Term Memory Units (Hochreiter and Schmidhuber 1997) followed by a fully connected layer with Softmax activation. A weighted average is used to combine the results of multiple models. Wang and Wang (2017) investigated the fusion of the pose and spatial model. In the following, this model is called "2-stream" (Martin et al. 2018) extended this method to the "3-stream" model by adding the model focusing on the interior.

1.2.6 CNN-Based Models

In CNN-based models, intermediate representations are not implemented explicitly, but *learned* together with the final classifier. Multiple convolution filters are simultaneously applied on image patches computing a weighted sum of the input channels (e.g. RGB channels in the first layer). The weights of the convolution filters are optimised during training to learn features, which are useful for the specific application. Although such models need a huge dataset, recent progress in *transfer learning* allows network pre-training on large available datasets from other domains and then re-using the knowledge and optimising it for the specific application (Roitberg et al. 2018). Consequently, such end-to-end models have been successfully applied for driver monitoring (Martin et al. 2019; Gebert et al. 2019). Still, the sensitivity of neural networks to changes in the data distribution (e.g. illumination, unknown behaviours of the driver) poses a challenge in realistic environments. Recent research in the area of uncertainty-aware models for activity recognition (Roitberg et al. 2018) shows an encouraging evidence that deep CNNs enhanced with probabilistic inference are effective in handling uncertain observations of an open world.

In the following, we describe multiple CNN-based architectures, which have been successfully adopted for driver activity recognition.

C3D The first neural network used for learning features from videos was the C3D architecture proposed by (Tran et al. 2015). C3D contains eight blocks of $3 \times 3 \times 3$ convolution kernels followed by $2 \times 2 \times 2$ pooling operations. Besides being the first framework for generic spatiotemporal feature extraction from videos, it is compact and efficient through the small kernel sizes. It takes a 112×112 video snippet of 16 frames as input and produces a 4096-dimensional video feature, which is then classified with a fully connected layer.

Inflated 3D ConvNet The inflated 3D CNN (I3D) is the off-the-shelf activity recognition architecture initially proposed by (Carreira and Zisserman 2017) for conventional action recognition. This neural network architecture takes a video snippet with 64 frames and 224×224 resolution as input and deals with spatial and temporal dimensions through 3D convolutions as well as pooling operations. I3D gains in efficiency by stacking nine characteristic *Inception* modules: small sub-networks, which execute $5 \times 5 \times 5$, $3 \times 3 \times 3$ and $1 \times 1 \times 1$ convolution operations in parallel and

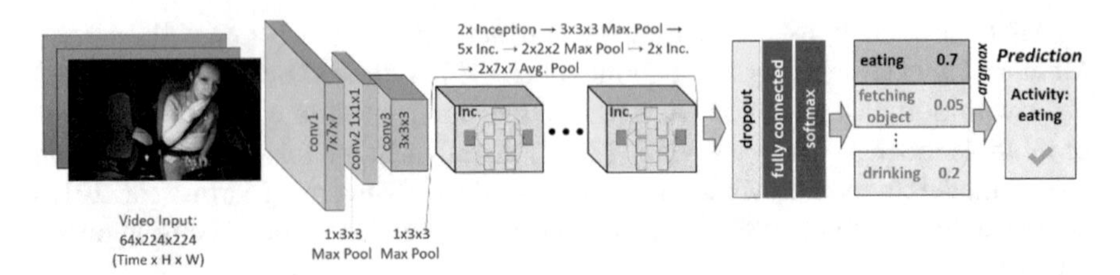

Fig. 1.10 Overview of the I3D ConvNet, which performs best on the Drive&Act benchmark. I3D leverages 3D convolution to learn intermediate video representations

concatenate the output. A key feature of the *Inception* module are multiple $1 \times 1 \times 1$ convolutions, which reduce the dimensionality and keep the number of dimensions low. The complete I3D network consists of 27 layers with an overview provided in Fig. 1.10 (note, that the nine *Inception* modules are themselves two layers deep). A useful quality of the *inflated* 3D network is its ability for *knowledge transfer* from 2D datasets. First, a two-dimensional version of I3D is trained on an image recognition dataset, such as ImageNet (Deng et al. 2009). Then, the convolution kernels are *inflated* to become 3D-kernels, with the learned weights being copied along the time dimension. According to this procedure, one could recycle the knowledge learned from still images using large available datasets and fine-tune the model for the application-specific video data.

Pseudo-3D ResNet Besides the 2D to 3D inflation, another way to reuse the available 2D CNNs on spatiotemporal data is to first convolve only the spatial domain, for which the pre-trained weights from 2D models can be transferred, and then apply the convolution in the time dimension only. Formally, we "simulate" spatiotemporal convolution by first applying a $1 \times 3 \times 3$ filter and then convolving the time-dimensions with a $3 \times 1 \times 1$ filter. Following this paradigm, Pseudo-3D ResNet (Qiu et al. 2017) resembles a very deep residual network (ResNet). Remarkable ResNet depth (152 layers) allows learning of complex hierarchical structures, but leads to the *vanishing gradient* problem when training with backpropagation. The network weights are modified proportional to the partial derivative of the error function computed using the chain rule. Consequently, repeated multiplication of small numbers lets the gradient decrease exponentially with the depth of the network, which hinders the learning process. To alleviate this shortcoming, residual networks use *skip connections* to allow the gradient flow by "skipping" multiple layers and adding the values to a later layer. Pseudo-3D ResNet leverages the skip connections to allow effective training of a very deep model with impressive results in action recognition.

CNN-Based Multimodal Fusion Multimodality is an important concept in driver observation systems, as different sensors have their individual limitations. For example, depth sensors are less susceptible to changes in illumination, while colour-based models can benefit from pre-training on large-scale computer vision datasets, which are often based on colour images. The most popular strategy for multimodal

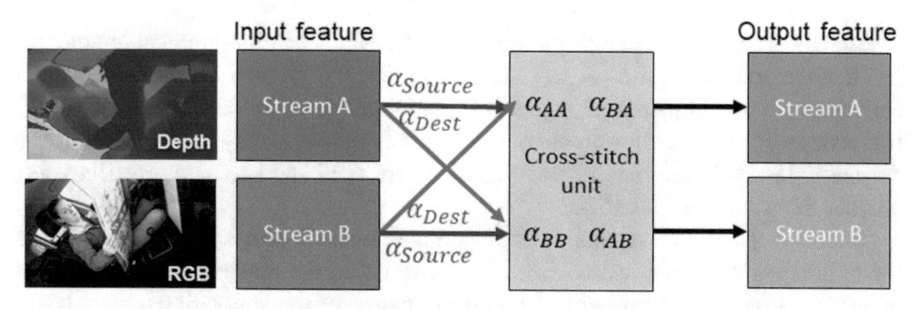

Fig. 1.11 Multimodal fusion at intermediate layers: cross-stitch units allow information sharing at any network depth by learning a weighted linear combination of the features

recognition with CNNs are 2-stream networks with late fusion, i.e. training individual networks for each modality, which are then linked via averaging the resulting scores at the last layer (Simonyan and Zissermann 2014). While being simple and surprisingly effective, such a late fusion dismisses high correlation of the input data at early stages.

Multiple strategies have been proposed to learn how to share the information between the modalities at earlier network stages. One way for such a fusion is to use separate streams at early layers and then fuse them into a joint model in a later stage via $1 \times 1 \times 1$ convolutions. In this case, however, the point of fusion happens at a single network layer and must be chosen manually. A better way to achieve the desired multimodal fusion is to exchange knowledge at multiple layers simultaneously. This can be achieved via *cross stitch units*, which learn how to combine the activations of both networks, learning a weight parameter for each input feature map (see Fig. 1.11). Initially designed for multi-task learning (Misra et al. 2016), *cross-stitch units* linking two action recognition networks (Roitberg et al. 2019) enable knowledge exchange at multiple layers simultaneously leading to better recognition results.

1.3 Evaluation and Discussion

The *Drive&Act* dataset is used for training and evaluation of the presented methods. The videos are divided into segments of 3s, which are then used for training and testing. We randomly split the dataset person-wise into training (10 people), validation (2 people) and test set (3 people). Following the standard practice, we use the validation data to compute intermediate results, e.g. for selecting model hyper parameters, such as learning rate, while the test set is used during the final evaluation only. We use the average accuracy over all classes as our evaluation metrics. This is computed by averaging the Top 1 detection rate for each category. As baseline for the evaluation, the accuracy of selecting the correct class by chance is provided. The baseline differs for each annotation level and varies between 0.31 and 16.67%.

Models are trained separately for each method and each hierarchical annotation level: 12 instructed high-level tasks (Level 1), 34 fine-grained activities (Level 2) and atomic action units with 372 possible combinations of [action, object, location] triplets (Level 3). Since the number of triplet combinations is very high, the results for correctly classified activities, objects and locations (6, 17 and 14 classes, respectively) are also determined separately.

Table 1.1 shows the results on the instructed tasks (Level 1). Overall, they are lower than on the other levels. However, pose-based approaches perform slightly better. Because of the abstract and long-term nature of the annotations, the models would most likely perform better on sections longer than 3s.

Table 1.2 shows the results on the fine-grained semantic activities annotation (Level 2). This level achieves the overall best scores. The end-to-end model I3D achieves the best results both compared to the pose-based methods and compared to other end-to-end models. The results of the pose-based approaches clearly shows that the fusion of the different views on the intermediary representation improves the performance. With the 3-stream method achieving the best result in this group.

Table 1.3 shows the results regarding the atomic action units (Level 3). The performance on whole triplets is still a big challenge due to the large number of possible combinations. The separate results on activity, object and location show much better performance. End-to-end methods still perform best. However, the pose-based approach performs better on the classification of the location because this information is modelled in the intermediary representation. On the other hand, it performs worse on object classification because this information is not contained in the intermediary representation and can only be inferred by the pose of the driver.

Overall, the results show that a detailed analysis of the interior of the car is still a challenge. Collecting a suitable dataset is a lot of effort. However, more data would likely improve results further. In addition, the collected data show that performed actions vary strongly both in frequency and in length which is still difficult to handle with state-of-the-art machine learning approaches. The evaluation shows that end-to-end models still work better overall. Designing a suitable intermediary interpretation without too much loss of information is difficult. However, if the right data is well modelled, these approaches can work better.

In Fig. 1.12, we provide qualitative results of the I3D model predictions, highlighting the image location with high impact on the decision. The analysis shows the significance of driver hands due to common object manipulations (Tables 1.1, 1.2, 1.3).

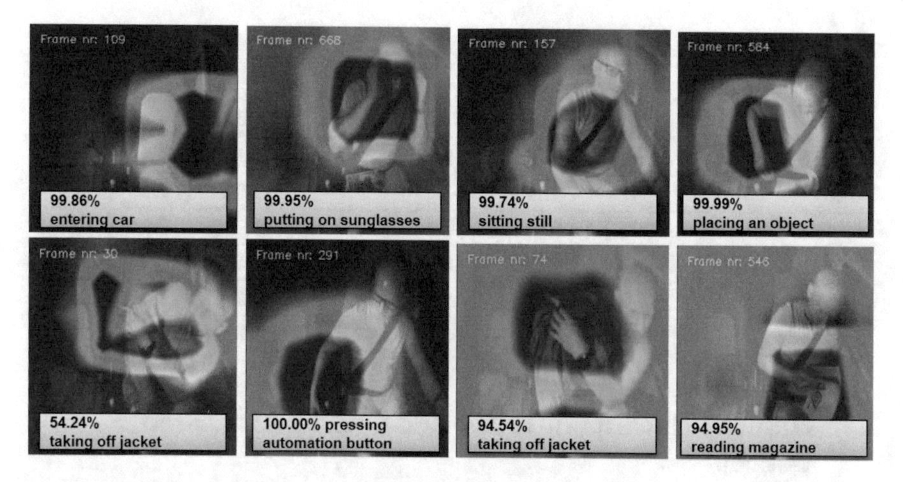

Fig. 1.12 Examples of prediction results using a CNN and visualisation of image regions with highest neural network activations, i.e. where the neural network was looking at, when it made the prediction. Networks attention is often focused on the hands

Table 1.1 Results for instructed tasks in percent

Type	Model	Validation	Test
Baseline	Random	8.33	8.33
Pose	Interior	35.76	29.75
	Pose	37.18	32.96
	2-Stream	39.37	34.81
	3-Stream	41.70	**35.45**
End-to-end	I3D Net	**44.66**	31.80

Table 1.2 Results for fine grained actions in percent

Type	Model	Validation	Test
Baseline	Random	2.94	2.94
Pose	Interior	45.23	40.30
	Pose	53.17	44.36
	2-Stream	53.76	45.39
	3-Stream	53.67	46.95
End-to-end	C3D	49.54	43.41
	P3D ResNet	55.04	45.32
	I3D Net	**69.57**	**63.64**

Table 1.3 Results for atomic action estimation in percent

Model	Activity		Object		Location		All	
	Val	Test	Val	Test	Val	Test	Val	Test
Random	16.6	16.6	5.8	5.88	7.1	7.1	0.4	0.3
Pose	57.6	47.7	51.4	41.7	53.3	52.6	9.2	7.0
Interior	54.2	49.0	50.0	40.7	53.8	53.3	8.8	6.9
2-Stream	57.9	48.8	52.7	42.8	54.0	54.7	10.3	7.11
3-Stream	59.3	50.7	55.6	45.3	**59.5**	**56.5**	11.6	8.1
I3D Net	**62.8**	**56.1**	**61.8**	**56.2**	47.7	51.1	**15.6**	**12.1**

1.4 Human–Machine Interface and Control Transition Concept

An essential part of the concept of a human-centred highly automated car (see Fig. 1.1) is the design of a proper transition concept. In this context, most important is the transition from the purely automated driving mode back to the human driver in case of a fallback situation. The driver needs to be informed and supported suitably through the human–machine interfaces. Therefore, the following section explains how this transition phase is addressed in the described research project. This includes the information of the driver as well as the design of a novel control algorithm to support the driver until he reaches his full mental readiness.

Several studies have investigated the transition of the driving task from an automation system to the human driver with different focuses (see for example Feldhütter et al.2019; Körber 2018; Radlmayr et al. 2019). These studies stuck to the concept of a transition described according to the SAE standard J3016. The automation system informs the driver of his duty to take over the driving task and shuts off completely at a given time after the so-called Request to Intervene (RtI) (SAE 2018). In accordance with other publications, this is also referred to as a takeover request (TOR).

Within the project PAKoS, we follow a different approach, as mentioned in the introduction and further explained by Karakaya et al. (2018). The driver gets primed in a first transition phase in order to prepare him cognitively for the cooperation in the second transition phase, where the driver is supported by the automation in such a way that the human and the assistance system jointly control the vehicle. Priming is a concept where one stimulus affects how a following stimulus is received and processed. In this case, the first stimulus is a multimodal information about the steering intentions of the automation in phase two. The second stimulus is the proprioceptively perceived torque applied on the steering wheel by the automation in the second phase. The two phases are separated by the point in time where the driver puts his hands on the steering wheel (often called the hands on) and thus is ready to perceive the torque of the assistance system (cf. Fig. 1.13).

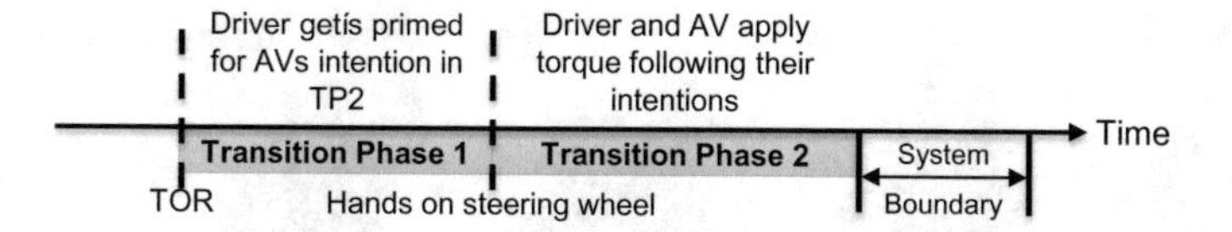

Fig. 1.13 Transition concept in the project PAKoS. After the TOR, the driver gets primed for the automated vehicles (AVs) intention in the second transition phase. In transition phase 2 (starting with the hands on), the driver is supported by the automation in order to solely control the vehicle at the system boundary, which describes the end of the operational domain of the automated vehicle (e.g. fallback or end of the motorway)

1.4.1 Transition Phase 1—Priming the Driver

1.4.1.1 Literature Review

Other experiments in the literature have tested different ways to provide the driver more detailed information than a simple RtI. The information used most widely was a suggestion, in which direction the driver could solve the transition scenario. Petermeijer et al. (2017a) and Petermeijer et al. (2017b) used vibrotactile and auditory stimuli on the drivers back to convey directional information (cf. Figs. 1.13, 1.14 and 1.15).

Fig. 1.14 Vibration mat used in the experiments by Petermeijer et al. (2017a). The red dots indicate the positions of the vibration motors. Image taken from (Petermeijer et al. 2017a)

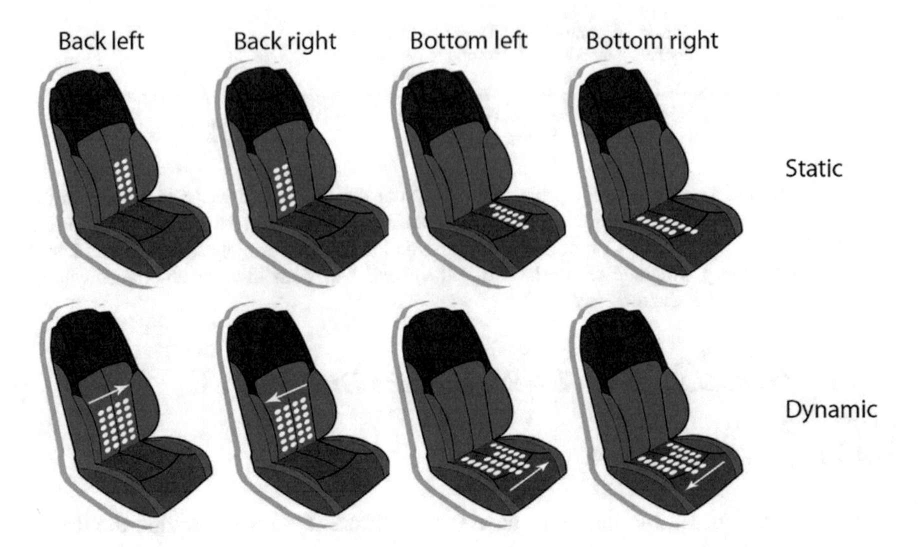

Fig. 1.15 Vibration patterns used by Petermeijer et al. (2017b). Image taken from (Petermeijer et al. 2017b)

Lorenz et al. (2014) presented a visual aid in the form of green and red carpets projected on the road via a head-up display (HUD) to direct the driver around an obstacle in a transition scenario (see Fig. 1.16).

Eriksson et al. (2017) conducted a similar study. The authors additionally used symbols, such as arrows to improve the directional characteristics (see Fig. 1.17).

Fig. 1.16 Visual aids used by Lorenz et al. (2014). The red carpet shows the direction the driver should avoid, whereas the green carpet shows the direction the driver should follow. Image taken from (Lorenz et al. 2014)

Fig. 1.17 Visual aids used by Eriksson et al. (2017). The red line shows where the driver should not switch lanes, whereas the green carpet shows the recommended lane for the driver to use. The green arrow shows the recommended maneuver to switch lanes, and the red triangle shows the recommended maneuver to brake. Image taken from (Eriksson et al. 2017)

1.4.1.2 Pre-Study and Design of the Priming Phase

In order to evaluate the above-mentioned literature approaches and to develop the design of our priming phase, a pre-study was conducted in the project PAKoS. Several subjective metrics were recorded from the participants during a driving simulator study. The goal was to identify the most suiting priming setup for transition phase one. The complete study is described in more in detail in (Kalb et al. 2018a, b). A short summary is given below followed by the key facts that were used for the design of the final simulator studies.

The pre-study took place in the static driving simulator of the Chair of Ergonomics at the Technical University of Munich (cf. Fig. 1.18). The participants drove in a full BMW E64 Mock-Up using the simulation software SILAB (WIVW). They got a takeover request six seconds before a static obstacle by means of the information to swerve to the left. Three groups completed four drives each. The groups differed in their modalities for conveying directional information to the driver in a transition scenario. This should simulate the later use case when the driver would be informed about the intention of the automation to steer in a certain direction. The first group

Fig. 1.18 Static driving simulator of the Chair of Ergonomics at the Technical University of Munich. Image taken from (TUM 2020)

received the RtI by the haptic seat (cf. Figs. 1.14 and 1.15) and the directional information by means of a green arrow shown in a HUD (cf. Fig. 1.19). The second group received the RtI by the haptic seat as well but the directional information by an auditory announcement. The information the third group received was a combination of those of both other groups. The haptic seat was augmented by visual and auditory directional information.

The pre-study revealed that drivers preferred the audio announcement for the directional information. The evaluation of glance behaviour showed less drivers than expected looked over their shoulder before changing lanes. Hence, the audio message for the next simulator studies was changed from "*left*" to "*if possible, swerve left*" (translation of the authors, the original messages were in German). This procedure was used to (1) include the suggestive nature of the information rather than a command ("*if possible*") and to (2) clarify that left was the direction of the solution, not the direction of the hazard ("*swerve left*"). The feedback from the drivers also showed that the multimodal priming increases the workload in an already time-critical situation. The application of a cooperative torque on the steering wheel in transition phase two would most likely have increased it even further. In line with this conclusion, we have decided to limit the priming to the auditory announcement and removed the haptic seat and the HUD.

Fig. 1.19 Green arrow displayed in a HUD as used in the visual condition. Image taken from (Streit 2018)

1.4.2 Transition Phase 2—Supporting the Driver to Take Over the Control

In comparison with transition phase 1, there are far less publications that focus on transition phase 2. Two questions arise in context of the realisation of this phase. First, it has to be answered how to model the interaction between the human and the automation in such a way that the chosen model framework can describe the transition of controlling the vehicle between both partners. As mentioned at the beginning of this section, phase 2 starts with the hands on of the driver on the steering wheel (the proposed concept applies for the longitudinal control of the vehicle as well). Since the driver achieves its normal performance on the driving task only successively, he has to be supported by the automation. We aim at designing a cooperative transition phase where both partners (human and automation) apply torque on the steering wheel and thus control the vehicle. In order to design the automation in this scenario a description is needed to model the mutual operation. In addition, it should be possible to characterise the influence on the actual control task, so how strongly is one partner involved into controlling the vehicle. By means of the parameters used for the last-mentioned aspect, it is possible to describe the shifting of the control authority, i.e. the transition, between the automation and the human driver. As soon as such a framework is found, the final question is how to choose the specific courses of these transitions and how to adapt them to the human and hence reaching an individualisation.

Before explaining our concept in detail, we introduce the basic idea of haptic shared control, which acts as our design paradigm for the cooperative transition phase. Abbink and Mulder (2010) define haptic shared control as methods of human–machine interaction, which allows both partners (human and automation) to exert forces on an actuator by which the system is controlled. Abbink et al. (2012) give a literature overview of various applications of haptic shared control in the control of robots and vehicles. In line with this, the concept is a promising way to support people in controlling a technical system. Due to the physical connection between the human and the automation through the same interface (i.e. applying torque on the steering wheel), an implicit and intuitive communication between both partners is possible by considering the feedback of the partner's action. Such an action-based cooperation (Flad 2016) forgoes any kind of explicit communication channels like displays. This reduces the human cognitive workload and makes the concept applicable in time-critical applications. In addition, through the feedback of the human at the interface, the automation can reason about his situational awareness and thus is able to assist him in building up his full mental readiness. By performing the driving task jointly, the human is slowly led to the sole fulfilment of the task by the automation. To guide the human and to determine the amount of control, which can be handed over to him, the direct feedback of the partner's action at the steering wheel acts as the starting point for the control and adaption algorithm of the automation.

The described situation of the mutual operation of the vehicle by the human driver and the automation is illustrated through the control structure in Fig. 1.20, which introduces our used modelling approach as well. In order to realise the cooperative driving based on haptic shared control, we use a game theoretic approach in accordance with several other authors (Tamaddoni et al. 2011; Na and Cole 2013; Flad et al. 2014).

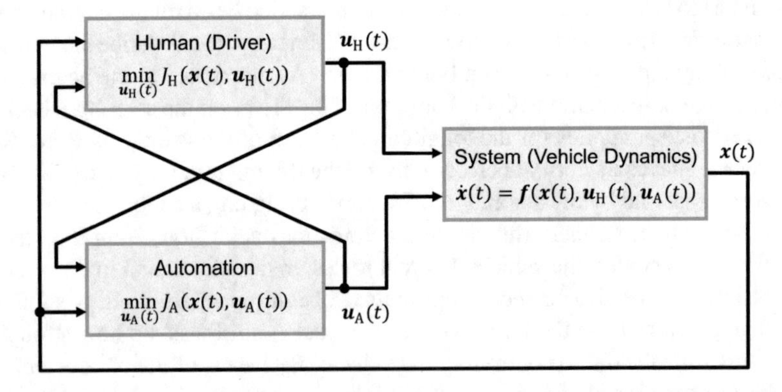

Fig. 1.20 Modelling of the human–machine interaction by means of a game theoretic approach

They describe the interaction between both partners (driver and automation) and the vehicle based on a differential game. The cooperation partners are regarded as players, who pursue individual goals. Hence, the human driver aims at achieving his personal goal, which describes the driving task and can mathematically be expressed by an objective function J_{H}. In context of the lateral control of the vehicle, J_{H} includes, among others, the deviation of the vehicle to the desired trajectory as well as the control effort of the human, i.e. his control signal $\boldsymbol{u}_{\mathrm{H}}(t)$. By minimising this objective function, the driver chooses the control signal $\boldsymbol{u}_{\mathrm{H}}(t)$ in such a way that the driving task is accomplished under a simultaneous minimisation of the energy expenditure. Modelling the human as such an optimal controller acts as a state-of-the-art method in describing human movements (Todorov and Jordan 2002). In addition, designing an automation by minimising an objective function represents a well-known control approach. Thus, J_{A} can be formulated and explained in a very similar manner. In Fig. 1.20 $\boldsymbol{u}_{\mathrm{A}}(t)$ denotes the control signal applied by the automation, which results from the minimisation of J_{A}.

However, since both partners interact with each other and are coupled through the same actuator as well as the system respectively vehicle dynamic $\dot{\boldsymbol{x}}(t) = \boldsymbol{f}(\boldsymbol{x}(t), \boldsymbol{u}_{\mathrm{H}}(t), \boldsymbol{u}_{\mathrm{A}}(t))$, where $\boldsymbol{x}(t)$ denotes the state of the system, a coupled dynamic optimisation problem follows, also known as differential game. The players need to agree on their control variables to fulfil their goals as well as possible. Normally, this arrangement respectively game ends in an equilibrium, in which no player deviates from the chosen strategy. In general, this equilibrium strategy combination differs from the globally best combination that would achieve minimal costs for each player without worsen other players due to the individual optimisation of each player.

1.4.2.1 Modelling Human–Machine Interaction

The above-mentioned control structure based on a differential game is used in the following to model the human–machine interaction. Like examined by, e.g. Flad (2016) this approach performs well in case of ADAS.

To provide the ability to model changes in the behaviour of the corresponding partners, we have extended the framework with time-variant weighting parameters as introduced in (Ludwig et al. 2017) and (Ludwig et al. 2018a, b). This allows designing transitions between the human partner (index H) and his automation counterpart (index A). The process of transition phase 2 starts with the automation fully controlling the vehicle while the human partner just starts to regain control. With increasing situational awareness, the performance of the human also increases and he gets better in accomplishing the driving task by his own. This can be modelled in the objective function J_{H} of the driver by increasing the weighting of the tracking error (deviation to the desired vehicle trajectory) by means of a transition parameter $\alpha_{\mathrm{H}}(t)$. Its values range from 0 (no control) to 1 (full control). To achieve an ideal transition, the weighting factor of the automation $\alpha_{\mathrm{A}}(t)$ decreases only to a complementary degree. Mathematically, an ideal transition is then given by

$$\alpha_\mathrm{A}(t) + \alpha_\mathrm{H}(t) = 1.$$

More formally, we define an ideal transition as a transition of a control task between two partners, if this transfer process does not affect the controlled system states. This is equivalent to the requirement that the joint influence on the system during the transition does not differ from the way in which one single partner would perform the task. The prerequisite for such a transition is that both partners are able to carry out the task alone. For the given problem of the control transition in a car of SAE level 3, this is fulfilled. Otherwise, the actions, which the partners consider as optimal at the beginning and the end of the transition, would be different.

In order to summarise the preceding discussion and to explain the realisation of our concept in more concrete terms, the mathematical starting points are highlighted in the following. We assume quadratic objective functions for the human and the automation (note: for ease of notation, the time dependencies are omitted):

$$J_\mathrm{H} = \frac{1}{2}(\mathbf{x}_T - \overline{\mathbf{x}}_T)^T \alpha_{\mathrm{H},T} \mathbf{S}^* (\mathbf{x}_T - \overline{\mathbf{x}}_T) + \frac{1}{2}\int_0^T (\mathbf{x} - \overline{\mathbf{x}})^T \alpha_\mathrm{H} \mathbf{Q}^* (\mathbf{x} - \overline{\mathbf{x}}) + \mathbf{u}_\mathrm{H}^T \mathbf{R} \mathbf{u}_\mathrm{H} \mathrm{d}t$$

$$J_\mathrm{A} = \frac{1}{2}(\mathbf{x}_T - \overline{\mathbf{x}}_T)^T \alpha_{\mathrm{A},T} \mathbf{S}^* (\mathbf{x}_T - \overline{\mathbf{x}}_T) + \frac{1}{2}\int_0^T (\mathbf{x} - \overline{\mathbf{x}})^T \alpha_\mathrm{A} \mathbf{Q}^* (\mathbf{x} - \overline{\mathbf{x}}) + \mathbf{u}_\mathrm{A}^T \mathbf{R} \mathbf{u}_\mathrm{A} \mathrm{d}t.$$

On one side, the cost functions sanction the deviation of the actual system states $\boldsymbol{x}$ from the desired state respectively vehicle trajectory $\overline{\boldsymbol{x}}$. This refers to the quadratic error function $1/2(\boldsymbol{x} - \overline{\boldsymbol{x}})^T \alpha_{(H/A)} \boldsymbol{Q}^* (\boldsymbol{x} - \overline{\boldsymbol{x}})$, which is integrated over the considered time horizon T, as well as the error function $1/2(\boldsymbol{x}_T - \overline{\boldsymbol{x}}_T)^T \alpha_{\mathrm{H/A},T} \mathbf{S}^* (\boldsymbol{x}_T - \overline{\boldsymbol{x}}_T)$ at the final state $\boldsymbol{x}_T = \boldsymbol{x}(T)$. The matrices $\mathbf{Q}^*$ and $\mathbf{S}^*$ denote the corresponding weighting matrices. They as well as the reference trajectory $\overline{\boldsymbol{x}}$ are assumed identical for the human and the automation. The transition parameters α_A and α_H provide an additional weighting factor of the tracking error and how already explained, they describe how strongly one partner is involved into controlling the vehicle and thus accomplishing the driving task. On the other side, the objective functions sanction the overall control effort of the corresponding partner. This is done by the integration of the error function $1/2\mathbf{u}_{\mathrm{H/A}}^T \mathbf{R} \mathbf{u}_{\mathrm{H/A}}$, where $\mathbf{R}$ represents a weighting matrix.

Both objective functions J_H and J_A are minimised under the constraint of the system dynamics

$$\dot{\mathbf{x}} = \mathbf{A}\mathbf{x} + \mathbf{B}\mathbf{u}_\mathrm{H} + \mathbf{B}\mathbf{u}_\mathrm{A}$$
$$\mathbf{x}(0) = \mathbf{x}_0,$$

where $\mathbf{A}$ and $\mathbf{B}$ denote the matrices of the linearised vehicle dynamics. Finally, a finite linear-quadratic differential game defined for a two player tracking problem follows.

Now, the shift of the control authority is modelled by the transition parameters α_A and α_H. In order to fulfil the prerequisite of an ideal transition, the following must apply for the automation part of the cooperation:

$$\alpha_A(t) = 1 - \alpha_H(t).$$

By determining $\alpha_H(t)$, which corresponds with the mental readiness of the driver, this relation yields the appropriate influence of the automation represented by the course $\alpha_A(t)$. The estimation of $\alpha_H(t)$ based on the current actions of the human driver is the most promising way to achieve the best individualisation and performance. This procedure is explained in the next subsection in detail. In contrast to such an adaption process, one can also postulate a course for $\alpha_H(t)$ in advance, e.g. a simple linear increase after the hands on. Such an example is illustrated in Fig. 1.21 together with the resulting transition parameter of the automation.

With the given formulation of the differential game, its solution directly leads to the optimal reaction of the automation to the human partner for the cooperative control transition phase. Therefore, we want to outline this aspect shortly. Both players select their respective actions to minimise their individual objective function. Since the players are linked via the state equations, the action of one player influences the optimal strategy of the other one. Thus, the resulting set of optimal strategies $(\mathbf{u}_H^*, \mathbf{u}_A^*)$ must have the property that none of the partners can achieve an improvement by unilaterally choosing a different strategy. Such a set of strategies is called a Nash equilibrium and a standard dynamic game problem (e.g. see Engwerda 2005). Ludwig et al. (2017) address the solution of the proposed special case.

Solving this dynamic game problem requires, compared to other functions needed for a SAE level 3 car, less computing power and can hence be implemented on a control unit.

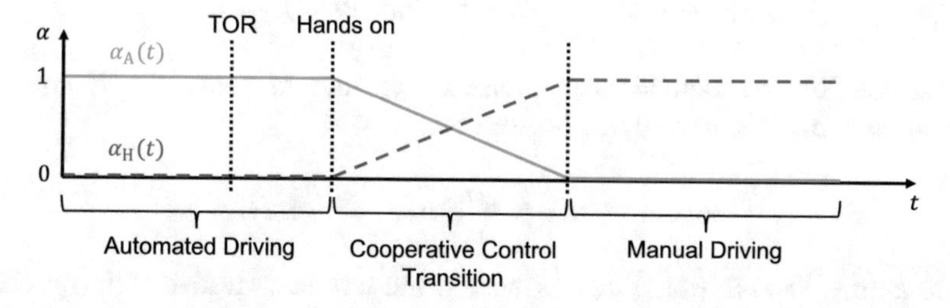

Fig. 1.21 Modelling the transition of the vehicle control by means of the transition parameters α_H and α_A. The parameter of the automation results from the relation of an ideal transition under the assumed linear increase of the transition parameter of the human

1.4.2.2 Action-Based Model Adaption

The model framework introduced in the previous subsection is used to describe the interaction between both partners (human and automation) and the vehicle, by choosing them as players of the described differential game. However, the question of how to design a specific transition, i.e. the course of the parameter $\alpha_{\mathrm{A}}(t)$, remains open. There are two options for designing the transition between both players. The first option, that was already introduced in the previous subsection, is to specify the transition process by using a predefined trajectory $\alpha_{\mathrm{A}}(t)$, which corresponds with the postulation of a course $\alpha_{\mathrm{H}}(t)$, since both parameters are connected through the relation of an ideal transition. From a practical point of view, the course of $\alpha_{\mathrm{A}}(t)$ defines how fast the automation delegates the control to the driver. The takeover time can be adapted to the current driving situation, the driver's actions before the takeover request or learned empirical values. However, in this case, all parameters are set in advance to the cooperative control transition. No online adaption will take place. Different transition courses have been investigated in the study of Ludwig et al. (2018a).

The other option is to adapt the control transition online to the current situation of the human driver, i.e. its mental readiness. Here, it is also possible to use a driver monitoring system as described in the previous section of this chapter but also to estimate the driver's state directly based on his control actions. The ability to take over the manual control of the vehicle can be evaluated by estimating the transition parameter α_{H}. This identification procedure is based on the current actions of the driver $\mathbf{u_H}(t_0)$ and $\mathbf{u}^*_{\mathrm{H}}(\alpha_{\mathrm{H}}, t_0)$, the action the driver should ideally apply according to the calculated solution of the mentioned differential game. The error of the currently used model is computed via the quadratic error function

$$E(\alpha_{\mathrm{H}}) = \frac{1}{2}\left(\mathbf{u_H}(t_0) - \mathbf{u}^*_{\mathrm{H}}(\alpha_{\mathrm{H}}, t_0)\right)^2,$$

we use the Newton method to minimise it and thus to iteratively improve the estimation of α_{H}. The update rule is given by

$$\alpha_{\mathrm{H},n+1} = \alpha_{\mathrm{H},n} - E''\left(\alpha_{\mathrm{H},n}\right)^{-1} E'\left(\alpha_{\mathrm{H},n}\right),$$

where $E'\left(\alpha_{\mathrm{H},n}\right)$ and $E''\left(\alpha_{\mathrm{H},n}\right)$ denote the first and second derivative of the quadratic error function with respect to α_{H} and n is the iteration index of the Newton method. Finally, from the estimation, the formula of an ideal transition and the solution of the differential game the optimal reaction of the automation to the human driver is calculated.

1.5 Implementation Aspects

In addition to the special aspects of the realisation and individualisation of the control transition phases, a concept is needed to adapt the vehicle automation to the users' needs during the purely automated driving phases. This along with the data management system (mobile phone application) completes our introduced concept of a human-centred highly automated car (see Fig. 1.1). Both facets are explained in the following section.

As stated in the introduction of this chapter, conditional automation (SAE level 3) lets the driver carry out non-driving-related tasks while in automated driving mode. Acting according to this automation level, the driver's hands must not be on the steering wheel and supervision of the vehicle environment by the driver is not necessary. However, the automation can still request the driver to react in certain situations wherein the automated system reaches its limits. Therefore, a completely mind-off situation, such as sleeping, is not allowed during this automation level. The conditional automation (SAE level 3) provides the required framework (e.g. driver interaction) for our development and evaluation of an adaptive, individualised automated driving functionality.

An automated car is capable of sensing its environment and moving in this environment safely without human (driver) input. Self-driving cars combine a variety of sensors to perceive their surroundings, such as radar, video and inertial measurement units. Advanced control systems interpret sensory information to identify appropriate navigation paths, as well as obstacles and relevant signs. Our concept is to design a human-centred highly automated car. First, we investigate the possible scenarios for personalisation in such a system.

1.5.1 Driving Scenarios

The possible scenarios for a personalised and adaptive highly automated driving within the given automation boundary of SAE level 3 and higher can be categorised into five basic situations:

(a) Intended takeover by the driver,
(b) Unintended (accidental) takeover by the driver,
(c) Strategic (planned) takeover by the driver before reaching designed system boundaries (system request),
(d) Critical fallback to the driver like described in SAE Level 3 (system request),
(e) Adaption of the driving style and the strategy of the automated system.

In context of PAKoS, we address the scenarios (c)–(e). Since the takeover situations (c) and (d) correspond with the design and personalisation of the control transition between the automation and the human driver, they are addressed in the preceding section *Human–Machine Interface and Control Transition Concept* in detail. Scenario (e) will be outlined in course of the following section focusing on the implementation aspects.

1.5.2 Concept Vision

The concept vision for a personalised highly automated driving is described with the help of a user journey where the defined scenarios are illustrated in their given scopes (see Fig. 1.22).

At the beginning of the user journey, the driver gets into the car and is identified by the driver monitoring system. For the purpose of a secure authentication, a special camera together with a Face ID application completes the driver monitoring system developed for action recognition, which is explained in the section *Action Recognition for Driver Monitoring*. The driver's identification is then used to load his or her specific user profile, which can, for example, additionally be used to release the car to the specific driver (such as in the car sharing use case). Not only at the beginning of a drive, but also while driving, the driver's identity has to be checked regularly or triggered by a specific action to ensure that the appropriate user profile is selected all the time (e.g. after a driver change).

Based on the selected user profile, previous personalisation aspects, like adjusting the position of the seat, settings of the infotainment system or configuring the dynamics of the automated driving function, can be applied.

After importing the driving profile and applying the necessary adjustments in the car, the driver starts driving manually, e.g. to the motorway, where the automated driving function is available. During the manual drive, the driving behaviour is analysed continuously and compared to the user profile. If new user specific characteristics or changes of the driving behaviour arise, the user profile and thus the behaviour

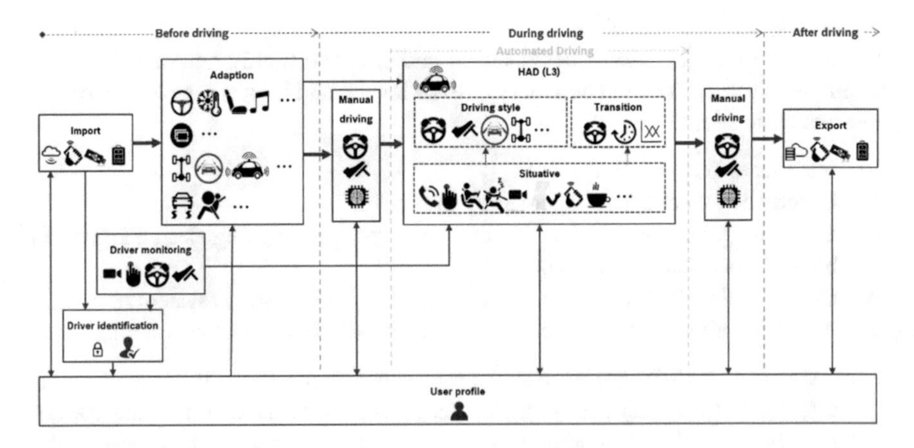

Fig. 1.22 Concept vision of the augmented automation system

of the vehicle are adjusted accordingly, considering legal and safety aspects as well. All profile suggestions can be disabled, adjusted or postponed by the user.

In addition to the adjustments mentioned in context of the manual driving, the interior scenario, i.e. the status of the driver, is observed via cameras as described in section *Action Recognition for Driver Monitoring*. This information can then be used for the situational adaptation of the automated driving function.

Once the driver carries out a non-driving-elated task during the highly automated drive, the behaviour of the vehicle automation is adapted based on this contextual understanding (e.g. driving slower when the driver is making a call or driving with lower lateral dynamics when the driver is working on the laptop, etc.).

After the automated drive on the motorway, the end of the operational domain of the SAE level 3 car approaches and the automated driving function hands the driving task over to the human driver. The transition behaviour of the system is also adapted to driver and the current situation inside the vehicle. For example, the system can give elder drivers more time to take over the control of the vehicle or if the driver was engaged in a secondary task, such as working on the laptop, an earlier takeover request is triggered to provide a sufficiently long preparation time for the driver. In case of a critical takeover, in line with SAE level 3, the transition behaviour can be adapted similarly; however, the system degeneration is a limiting aspect here.

After the successful handover, the driver manually drives the remaining route to his destination. The driving behaviour is again analysed continuously and compared to the user profile. At the destination, the user profile is exported to a mobile device (e.g. a smartphone) by means of a mobile phone application, which acts as a secure and offline available data storage. Thus, the user profile can be managed offline via the portable device and transferred to different cars of other manufacturers (OEM's).

1.5.3 System Architecture

The system architecture, which is derived from the introduced concept, is illustrated in Fig. 1.23. In order to keep the complexity manageable, we have summarised the vehicle related subsystems (e.g. braking system, steering system) as the so-called local systems. The realisation of them highly depends on the overall vehicle

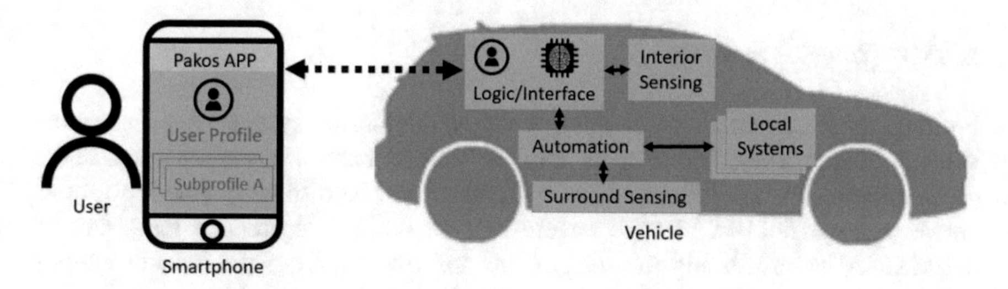

Fig. 1.23 System architecture

architecture of the actual manufacturer. Since these local systems are not the focus of our research project, we explain the other subsystems of the given structure in this section in detail. This refers also to the user profile, which is the key feature of the mobile phone application (APP). This APP represents the focus of the next section.

1.5.3.1 Logic/Interface

The fundamental part of the overall system in Fig. 1.23 is the *Logic/Interface* block. This block enables the communication between all subsystems and coordinates the whole system. The communication between the vehicle subsystems is mainly based on standardised interfaces, such as CAN or WLAN, to achieve portability and integrity of the system in various vehicles. This software block can be implemented on any electronic control unit (ECU) as long as it has the needed interfaces and is connected with all the subsystems.

Information from the *User Profile* and the *Interior Sensing* provide the knowledge needed for the personalisation and adaptation of the *Vehicle Automation*.

1.5.3.2 Surround Sensing and Vehicle Automation

As mentioned before, the *Vehicle Automation* uses the data processed by the *Surround Sensing* to control the vehicle accordingly.

For the *Surround Sensing*, multiple radars (4 at the corners and 1 at the front of the car) and a video camera (mono-video camera) are available. All the surrounding information are processed by means of a sensor fusion to an environment model. This is utilised to navigate the automated car in given boundaries and system constraints, respectively.

These boundaries can be universal for various automation systems as well as different levels of automation. Hence, the personalisation and situational adaptation of the automated driving is realised at this point. Depending on the specific driving situation, the information about the boundaries (e.g. limits for lateral and longitudinal acceleration, set speed, etc.) can be adapted to adjust the driving style of the vehicle automation to the needs of the passengers.

1.5.3.3 Interior Sensing

For the advanced situational analysis, a driver monitoring system is needed in the vehicle. This is solved by using the camera- and AI-based system presented in the section *Action Recognition for Driver Monitoring*. The different camera positions enable a large field of view to ensure that all non-driving-related tasks can be detected. After processing the images, the information from the vehicle interior such as the current activity of the driver, like telephoning or working on the laptop, is given to the *Logic/Interface* block to calculate the situational adaption of the automated driving behaviour.

1.5.3.4 User Profile

To develop a flexible and portable user profile, different degradation levels have to be considered. These allow interpreting the user profile knowledge by various vehicle types and manufacturers. In order to map the conceptual user profile, three different levels have been defined. The structural composition of them is shown in Fig. 1.24 and will be explained in the following subsection in detail.

Persona

The first level of the user profile contains the personal data and, therefore, represents the OEM-independent *Persona* level. These are largely demographic in nature. Safety assistance systems, for example could, automatically configure themselves according to the respective user with values such as size and weight. Generally, most of the personal data is optional.

User Needs

The second abstraction level contains information about driving comfort-related preferences of the user, such as different driving styles (e.g. defensive, normal and sporty). This knowledge can be addressed by suitable parameters of the *Product Applications* level of the user profile or used for further non-suitable vehicle/manufacturer architectures.

Product Applications

All user relevant settings for the actual vehicle are saved in the vehicle settings, which represent the OEM-dependent data of the *Product Applications* level of the user

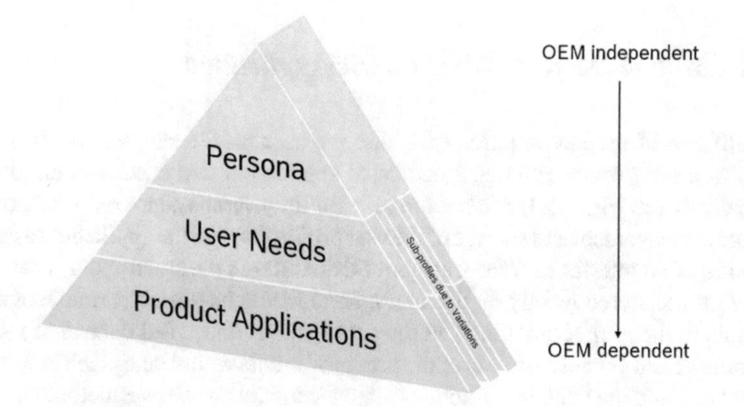

Fig. 1.24 User profile architecture

profile. Thus, these settings form a combination of explicit low-level user requirements and OEM-dependent application parameters. This includes settings that are relevant for manual driving, such as the position of the mirrors, seat and steering wheel, as well as settings for the driver assistance systems, like takeover parameters or settings of the air condition and infotainment system. In general, any quantity that can be defined by digital values and has a benefit by being individualised can be saved on this level of the user profile.

In order to save specific vehicle settings in a meaningful and usable manner, they should be saved individually for each vehicle model, in so-called vehicle profiles. Corresponding data structures, that match the values of vehicle models or calculate them based on the OEM-independent parameters, could be adopted from manufacturers or worked out with them. There is the advantage that components and systems that are not integrated in every vehicle can also be captured. Using vehicle profiles would ensure a personalised and satisfactory vehicle setup for the user.

Subprofiles

Driving profiles capture the desired driving behaviour and further needs of the user. These profiles are used when the vehicle is driving autonomously. The overall driving behaviour results from the combination of the user requirements (*User Needs*) and the corresponding OEM-dependent application parameters.

The user profile always contains a "basic" driving profile, which can be configured individually by the user and thus represents the desired standard driving behaviour. Optionally, the user can add and configure additional driving profiles; the so-called *Subprofiles* (cf. Fig. 1.23). In these, the driving behaviour deviates from the basic profile to fulfil specific user needs depending on the actual specific situations the user is faced with, like family trip or drive to work. This solution gives the user the opportunity to change the driving behaviour quickly and comfortably. In the application, all profiles are managed in an overview, from where the profiles can be selected and further configured.

1.5.4 Mobile Device/Mobile Phone Application

To fulfil one of the key requirements, that the user profile can be used in varying vehicles, a smartphone running a dedicated application was chosen as an interface to the driver (cf. Fig. 1.23). This setup also fulfils several system requirements like creation, management and storage of the user profile through the application as well as provision of an interface for the transfer of data through the smartphone's hardware. A user profile stored locally on the smartphone would be portable, transferable and available to the user at any time and the data remains under full control of the user. The transfer and synchronisation of the user profile data would take place between the smartphone and the vehicle. Various common standards, such as Bluetooth or WIFI, can be used as an interface for data transmission. Mandatory offline functionality would also be given directly.

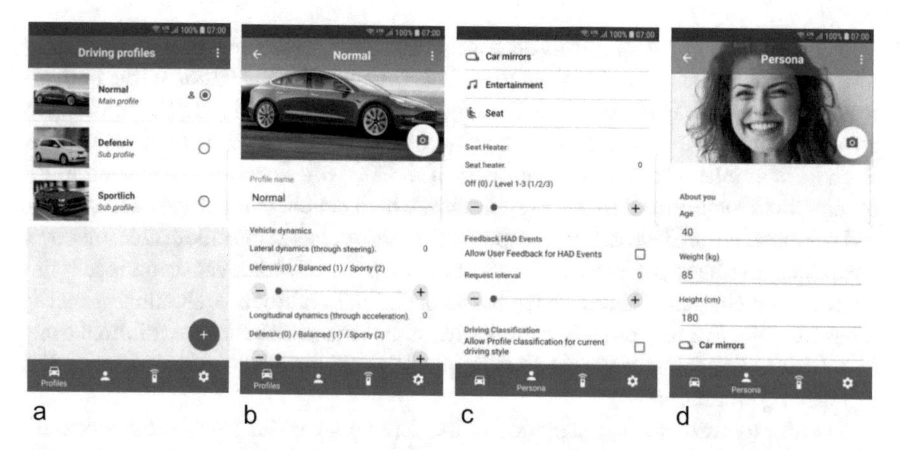

Fig. 1.25 Mobile phone application for managing the user profiles

The benefit using this digital user profile is as followed: Initially, the user enters the vehicle, starts the application on the smartphone and transmits his user profile to the vehicle where it is stored in internal memory. After that, the vehicle starts automatically the setup for manual driving periphery, safety and comfort systems using the data from vehicle settings and persona. Once the vehicle is in the automated driving mode, the driving profiles become relevant. The user can choose freely between his profiles to set a driving behaviour, which fits his current needs. The active profile can further be adjusted, using feedback dialogues during the automated driving. At the end of the trip, all data are synchronised with the data in the application, due to changes in driving profiles or vehicle settings that were made during the driving period. Finally, the user leaves the vehicle with the updated user profile on the smartphone.

As an example, Fig. 1.25 shows details of the prototype application that was developed during the research project. Figure 1.25a depicts the overview of the driving profiles, with the "Normal" profile selected as active. Figure 1.25b shows details of a profile, in form of configurable driving parameters. Several vehicle settings can be seen in Fig. 1.25c, whereas Fig. 1.25d shows some input values for the personal data.

The overall system according to Fig. 1.23 has been developed and integrated in a prototype vehicle. In addition, the proof of concept is carried out by multiple driving studies, which are presented in the next section.

1.6 Study Results

The evaluation of the proposed concept of a human-centred highly automated car took place in two simulator studies and one real-life driving study.

The developed control transition concept for SAE level 3 has been evaluated by investigating critical takeover situations. This includes the evaluation of the human–machine interface (HMI) to prime the driver as well as the evaluation of the cooperative control algorithm to support the driver in taking over the control of the vehicle. Such critical takeover situations are part of the *Driving Scenarios* identified in the section *Implementation Aspects* as situations, which benefit from an individualisation and personalisation. Due to the more pretentious time constraints, the critical takeover situations have been preferred over the strategic (planned) takeover scenarios. In this context, two simulator studies have been carried out and their evaluation primarily focus on objective respectively quantitative safety measures, like the minimal time-to-collision. To guarantee the safety of the participants, we have conducted the studies in a driving simulator.

In order to evaluate the adaptation of the driving style and the general personalisation of the automated vehicle, a real-life driving study was performed using a test vehicle on a proving ground. In contrast to the other studies, we focus on the subjective user perception in this case. As described later, the use of the driving simulator would have affected the perception and evaluation of the driving style negatively.

1.6.1 Preparations

Since the essential parts of our vision of a human-centred highly automated car are the personalisation, the associated cooperation and the adaptation of the machine to the human, a detailed preliminary examination of possible users was useful. For that purpose, extensive desk research and qualitative in-depth interviews were used to analyse driver types regarding partially and highly automated vehicles. Based on the available empirical information, driver types were formed within the group by means of clustering similar characteristics that differ clearly from the characteristics of the other driver types. In sum, three different driver types have to be distinguished:

- **Self-confident drivers**: For this group, the self-determination and the acting out of their individual driving style is very important. Therefore, these persons have problems with the idea of a completely autonomously driving vehicle. Support in critical driving situations or when needed, however, is somewhat conceivable to them.
- **Balanced drivers**: Their anticipatory and quiet driving style comes closest to the presumed driving style of autonomous vehicles; so, it is easy for them to imagine driving an autonomous vehicle.
- **Cautious drivers**: In particular, they desire support in critical driving situations or in situations that cause uncertainty.

In a further step, these driver types were described in classes, enabling all project participants to develop an idea of the future users of the system, their characteristics, motivation and goals.

1.6.2 Simulator Studies

In the following, the results of the driving simulator studies are presented. They focus on the novel concept for the control transition from automated driving mode to manual driving as part of the degeneration strategy of a SAE level 3 car.

Setup of the Simulator Studies.

Two simulator studies were conducted within the project PAKoS. Both studies used the driving simulator shown in Figs. 1.18 and 1.26. The fixed base driving simulator used six projectors to generate a 180° front view as well as all mirror views. The simulation software was SILAB 5 (WIVW). The steering wheel was connected to the electric motor SENSO-Wheel SD-LC (SENSODRIVE).

The participants drove on the same three lane German highway in both studies for approx. 15 km before encountering a broken-down vehicle in the middle lane, which triggered an RtI. The obstacle was 200 m away at the time of the RtI, which equals a time-to-collision (TTC) of six seconds. The takeover scenario always occurred on a straight part of the road. The automated vehicle drove at a speed of 120 km/h and switched lanes automatically several times before staying on the middle lane for the takeover scenario. Since the right lane was blocked, the participants as well as the automation had to avoid the obstacle on the left.

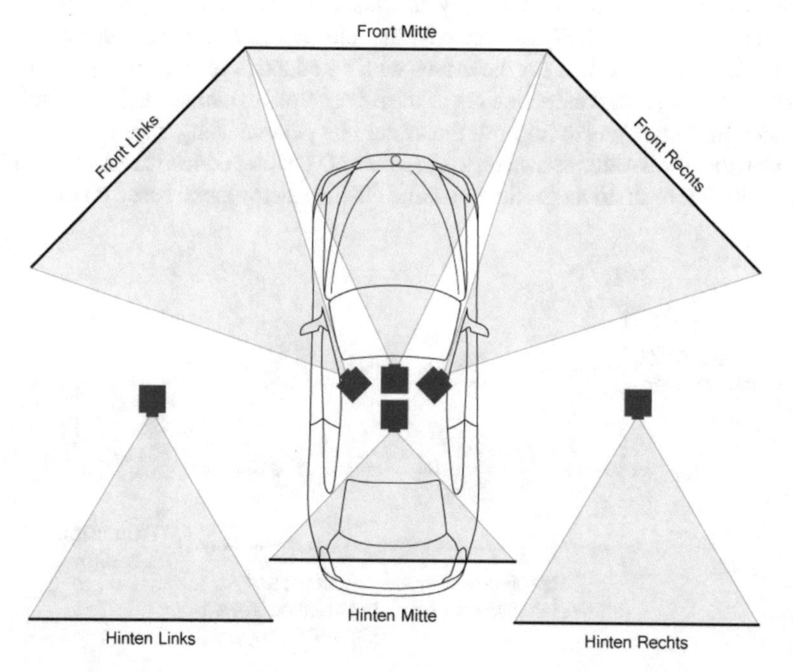

Fig. 1.26 Schematic image of the driving simulator of the Chair of Ergonomics (TUM). Picture taken from (TUM 2020)

Study 1: Evaluation of HMI Concepts and Standard Transition Concept

The first study included four experimental drives for each participant. One of these four drives was intentionally different from the other three, both in street design and length of the track. It was always included as the second drive and intended to minimise learning effects of the drivers. The other three drives differed in the visual and auditory HMI used as well as in the configuration used to support the driver in transition phase 2.

The longitudinal control was switched off with the RtI in every drive. The support to steer around the obstacle was either activated or not. More precise, in drives with support of the automation system the system started performing a lane change manoeuvre as soon as the driver's hands were detected at the steering wheel (start of transition phase 2). Until this point in time, the automation only aimed at keeping the lane. Apart from some lane change manoeuvres, the automated vehicle has performed the lane keeping manoeuvre on the 15 km drive before as well. Furthermore, the automation has decreased its steering support gradually during transition phase 2 and thus handed the control over to the driver. In drives without support of the automation, all assistance was switched off completely with the RtI.

An overview of the four experimental drives and their respective HMI and lateral support is given in Fig. 1.27. How mentioned before, the second drive corresponds with the baseline HMI and transition design for every participant, whereas the other numbers of the drives in Fig. 1.27 vary between them. For example, the drive marked with number 1 in Fig. 1.27 could also be the third or fourth drive of one participant.

The design of the HMI for the drives with an adaptive respectively personalised HMI was based on the categorisation of the driver types explained at the beginning of this section. The aim is to improve the system by personalising it for each driver. In contrast, the information shown on the baseline HMI were reduced to a minimum (see Fig. 1.28). For each driver group, a specific HMI was developed based on their needs

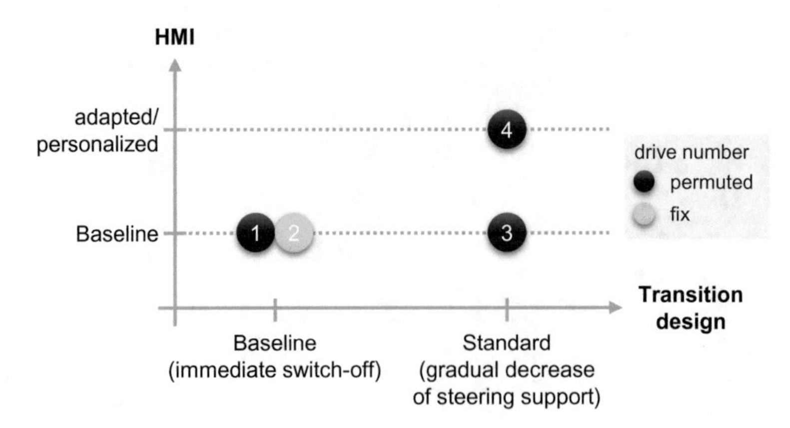

Fig. 1.27 Overview of the experimental drives and their respective HMI and transition design for study 1

Fig. 1.28 HMI for the baseline group representing the system status when **a** the automation is active and **b** the automation triggers an RtI

and preferences (adaptive HMI); function-oriented (FO), usage-oriented (UO) or service-oriented (SO). A function-oriented (FO) HMI contains the most information about the system status (see Fig. 1.29).

Self-confident drivers have been assigned to the FO group, which corresponds to their relatively high technical interest in driving. Balanced and cautious drivers have been characterised as not very eager to engage more in the technical aspects of driving. These drivers were the least confident and used cars only as a mean of transport. In accordance with the needs of these two driver types, the balanced drivers have been assigned to the group that saw the user-oriented (UO) HMI and cautious drivers to the group with a service-oriented (SO) HMI. The SO group had the same HMI as the UO group but with the additional possibility of speech commands. Figure 1.30 shows the instrument cluster for both groups. The study investigator realised the speech commands by (de-)activating the desired functions manually per remote. More details about the prototyping development and the speech commands can be found in the publication of (Kalb et al. 2020).

The baseline transition concept used in the second transition phase is to immediately switch-off the steering torques that the autonomous steering controller applies on the steering rack. The other configuration (standard transition) applies the transition concept proposed in section *Human–Machine Interface and Control Transition Concept*. The transition parameter α_A is thereby not adapted to the human. Instead, it is linearly decreased starting from 1 at the start of transition phase 2 towards 0 at

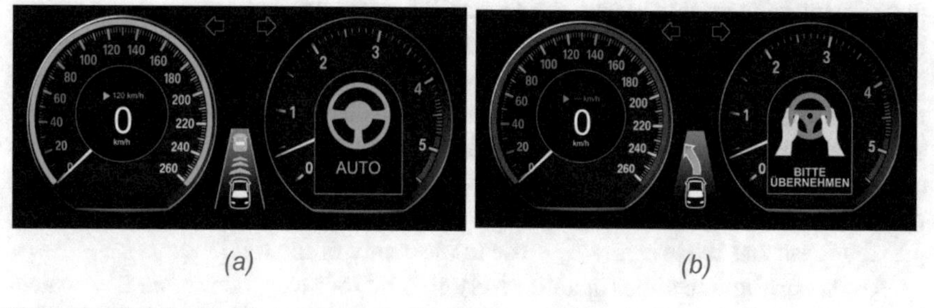

Fig. 1.29 HMI for the FO group representing the system status when **a** the automation is active and **b** the automation triggers an RtI

Fig. 1.30 HMI for the UO and SO group representing the system status when **a** the automation is active and **b** the automation triggers an RtI

the end of the phase. Thus, a situation similar to Fig. 1.21 results. More information about possible transition concepts can be found in (Ludwig et al. 2018a).

Each participant rated several statements after a critical situation to measure the acceptance of the adaptive HMI. The statements refer to the visual and auditory aspects of the HMI, e.g. "the amount of information given (display and output) by the system was appropriate" or "the message from the system (display and output) was easy to understand".

We recorded valid datasets from 31 drivers in the first study. Figure 1.31 shows the gender and age distribution. There was a wide age range in study 1 ($M = 32.9$ years, $SD = 11.7$) because the personalisation of the HMIs based on the three presented groups was of interest.

Results of Study 1

The presentation of the results of the preceding introduced study is divided into two parts. First, we describe the results achieved with the different designs of the driver's steering support in transition phase 2. Afterwards, the subsection concludes with a comparison of the results regarding the variations of the HMI concept.

Before introducing and evaluating our quantitative safety measures, we want to investigate the achieved vehicle trajectories from a subjective perspective. Hence, the trajectories of all study participants, which result from using the baseline transition concept (immediate switch-off of the automation at the RtI respectively takeover request), are illustrated in Fig. 1.32. The trajectories show a wide dispersion. Many participants have trouble to perform a smooth lane change and require many corrective interventions to stabilise the vehicle on the target lane. In this way, some drivers get dangerously close to the obstacle or guardrail. In contrast to this, the standard transition concept, i.e. designing a cooperative transition phase where the automation starts performing a lane change manoeuvre and decreases its support linearly, yields significantly better results. This can already be deduced by a subjective evaluation of the vehicle trajectories in Fig. 1.33. The trajectories are more focused than in the previous case and the drivers stay close to the centre of the left lane.

To compare both methods quantitatively, four metrics were determined. We calculated the minimum time-to-collision K_{t}, the minimum lateral distance to another object (obstacle or guardrail) K_{d}, the lateral distance between the trajectory point

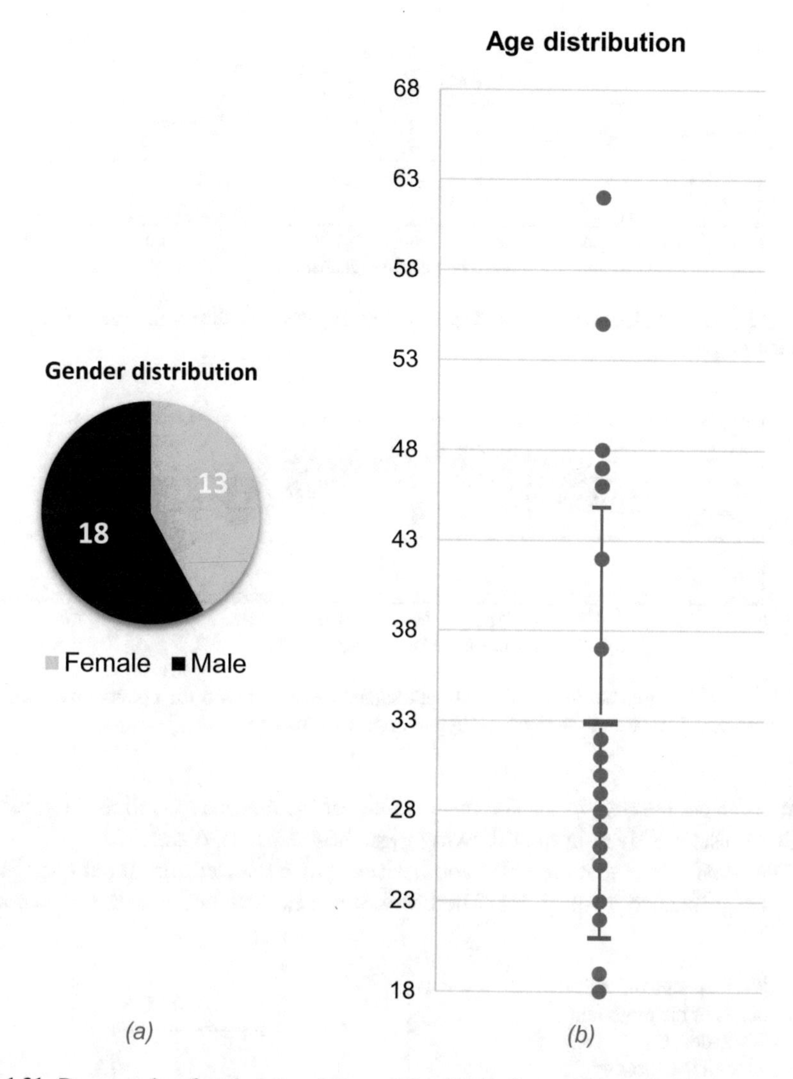

Fig. 1.31 Demography of study 1 ($n = 31$): **a** gender distribution and **b** age distribution

nearest to the obstacle and the point nearest to the guardrail K_D and the maximum lateral acceleration K_a. The metric K_D can be interpreted as the utilisation of the left lane of the highway. In order to evaluate the transition concept, we aim at higher values for K_t and K_d, since they correspond with a lower criticality of the lane change and thus improve safety. So is with lower values for K_D. In addition, the described improvement of the values correlate with a higher driving comfort as well. Such an improvement of the driving comfort is also highlighted by lower values for K_a.

Several *t*-tests were conducted to check the null hypothesis that the mean values of the baseline runs are better; i.e. K_t and K_d are larger and K_D and K_a are smaller

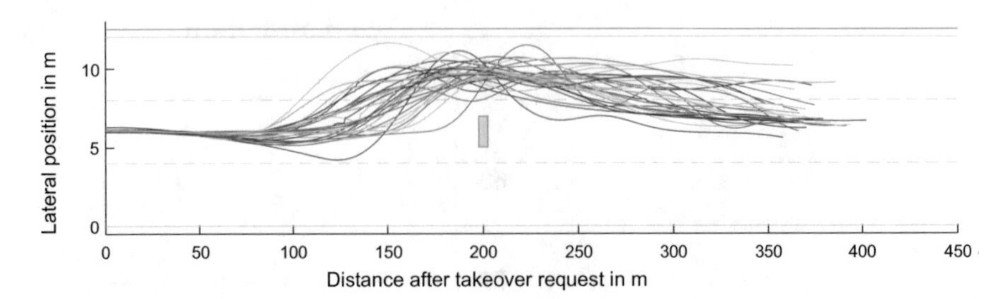

Fig. 1.32 Vehicle trajectories of all study participants achieved with the baseline transition concept (immediate switch-off)

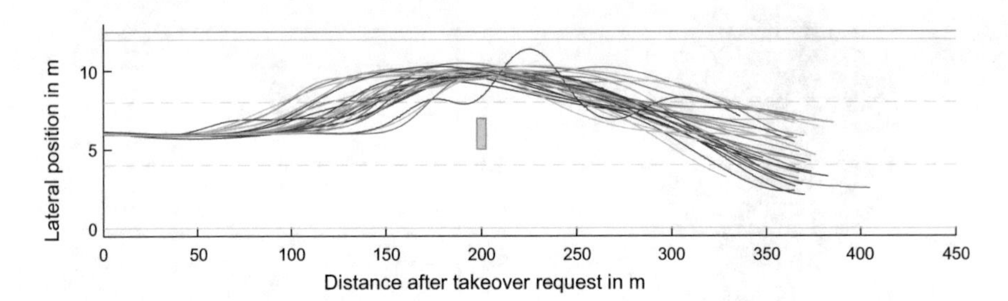

Fig. 1.33 Vehicle trajectories of all study participants achieved with the cooperative transition concept (linear decrease of the steering support after the hands on)

than in the cooperative case. The mean values of the metrics as well as the p-values of these tests are given in the following Figs. 1.34, 1.35, 1.36 and 1.37.

The t-tests show that the null hypothesises can be rejected for all but one metric with a significance level of 5%. Therefore, it can be assumed that the cooperative

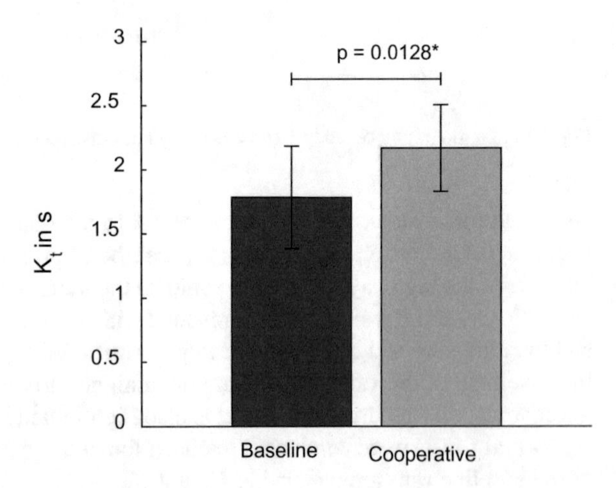

Fig. 1.34 Comparison of the mean values of the minimum time-to-collision K_{D} achieved with the baseline respectively cooperative transition concept. In addition, the standard deviation and the p-value of the t-test are illustrated

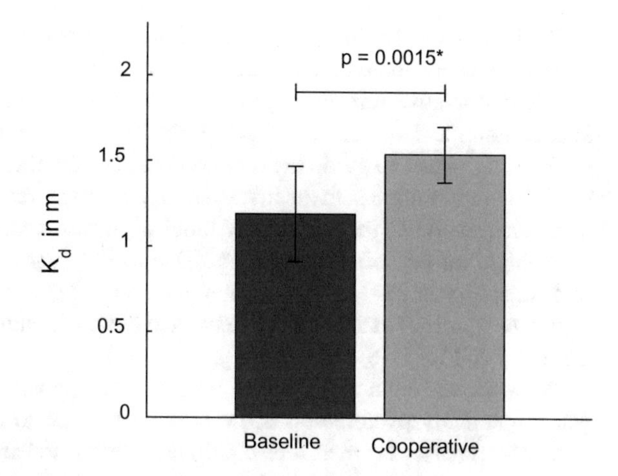

Fig. 1.35 Comparison of the mean values of the minimum lateral distance to an object K_d achieved with the baseline respectively cooperative transition concept. Standard deviation and *p*-value of the *t*-test are illustrated as well

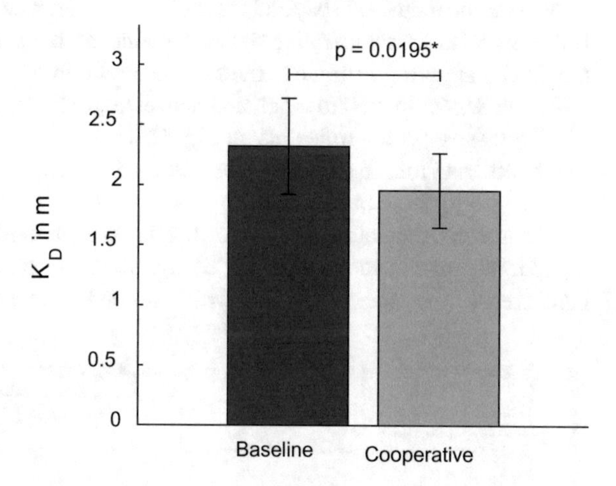

Fig. 1.36 Comparison of the mean values of the lane utilisation K_D achieved with the baseline respectively cooperative transition concept. Standard deviation and *p*-value of the *t*-test are illustrated as well

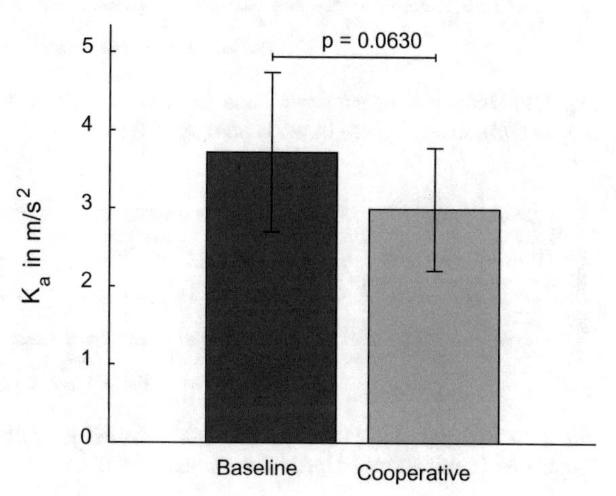

Fig. 1.37 Comparison of the mean values of the maximum lateral acceleration K_a achieved with the baseline respectively cooperative transition concept. Standard deviation and *p*-value of the *t*-test are illustrated as well

method increases the time-to-collision as well as the distance to other objects for the given evasive manoeuvre. No difference for the lateral acceleration can be found.

The cooperative transition improves the driving behaviour by compensating the lack of individual situational awareness in the given scenario by applying a torque at the steering wheel to guide the human to the correct direction. This is illustrated in the following examples. In the first example, the driver reacted to the takeover request by a quick swerve to the left lane. Without any assistance, he had troubles to stabilise the vehicle on the left lane and got close to the guardrail as well as the obstacle (red trajectory in Fig. 1.38). The new method mitigated the evasive manoeuvre by a counter torque. Thus, the driver executed a smooth and stable lane change (green trajectory in Fig. 1.38).

In the second example, the driver showed the opposite reaction. Instead of acting fast respectively overambitious, he took more time to decide what to do. In the baseline scenario, he therefore continued driving towards the obstacle, evaded it rather late and then drove beyond the centre of the left lane (red trajectory in Fig. 1.39). If driving with the cooperative transition concept, the system supports the decision-making by applying a torque to the evasive direction on the steering wheel. This way, the driver was animated to react, and hence, he performed the manoeuvre earlier and less aggressive (green trajectory in Fig. 1.39).

The steering torque applied by the automation during transition phase 2 is depicted in Fig. 1.40 for both examples.

The results indicate that the novel takeover method via a cooperative control transition is able to reduce the risk of collision and to increase the vehicle stability compared to the state of the art, which immediately switches the automation off.

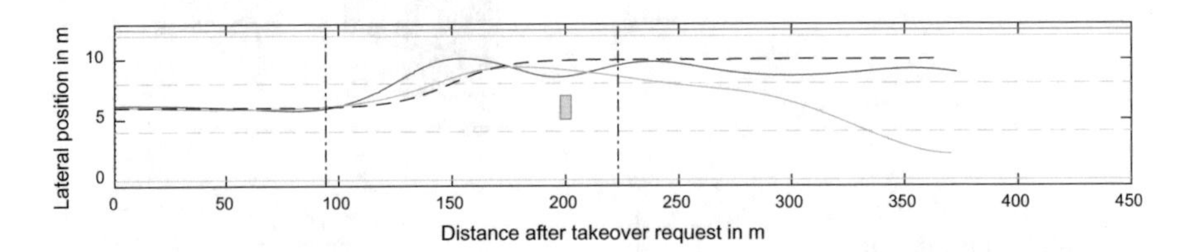

Fig. 1.38 Example trajectories of one study participant with support of the automation in a cooperative transition phase (green) and without (red)

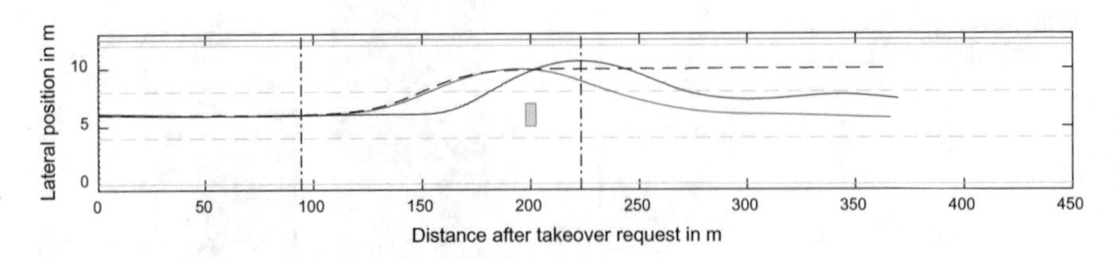

Fig. 1.39 Example trajectories of one study participant with support of the automation in a cooperative transition phase (green) and without (red)

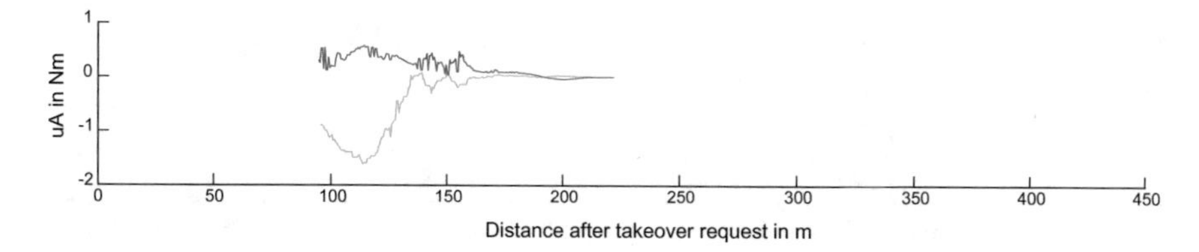

Fig. 1.40 Applied steering torque of the automation in the first (light green) and the second example (dark green)

Now, we present the results of the interviews, which were carried out to evaluate the different HMI concepts.

The evaluation of the HMI concepts revealed an overall acceptance of the HMIs that where investigated. The self-confident drivers in the FO group seem to have the most benefit from a personalised HMI. All users tend to show more acceptance for the personalised HMI compared to the baseline HMI, but usually not to a significant extent. The acceptance has improved significantly in context of three statements in the FO group and regarding one statement in the cautious driver group (cf. Table 1.4). The self-confident drivers (FO group) rated a better appropriation of the HMI on a 10% significance level. In addition, for the self-confident drivers a personalised HMI helped to increase the situational awareness as well as the awareness what they should do next. Furthermore, a personalised HMI increased the trust into the system for the cautious drivers significantly.

Study 2: Evaluation of Transition Concepts for the Design of Phase 2

The second simulator study focused on the investigation of different designs of the steering support in transition phase 2. Therefore, the HMI was not altered between the drives and every participant experienced the same HMI, i.e. the one designed for the FO group (see Fig. 1.29). However, the steering support was varied and consisted of the following categories:

(1) Standard (S): gradual decrease of the steering support (parameter α_A) as implemented in study 1
(2) Manual (M): manual adjustment of the steering support by the driver through mounted clutch paddles on the steering wheel (see Fig. 1.41)
(3) Controller (C): automatic adjustment of the steering support executed by the controller based on the proposed concept of estimating the parameter α_H.

In order to compare the results to the first study the experimental drives were kept the same, i.e. in total four drives (see Fig. 1.42). The order of the drives was again randomised to eliminate learning effects. The only adjustment was in the second drive where participants experienced the RtI in a right-hand bended curve with a steering support of category C. As before, the second drive was fixed respectively the same for all participants and hence not part of the randomisation of the drives. In all drives, the automation started performing the lane change manoeuvre and adapted its support according to the corresponding category, when the participants had their

Table 1.4 Results of the interviews regarding the evaluation of the HMI concepts

Statement	Group	Self-confident		Balanced		Cautious	
	N	8		11		11	
	HMI	BASE	FO[a]	BASE	UO	BASE	SO
I found the system feedback (display and output) appropriate for the situation		4.1*	5.9*	5.7	5.3	4.5	5.0
I knew immediately what the system was doing after the feedback was displayed/given		4.1	4.4	3.5	4.4	4.4	4.0
With the help of the system feedback (display and output), I was immediately aware of the (dangerous) situation		4.5**	5.9**	4.6	5.3	4.5	4.5
I found information the system feedback contained (display and output) to be appropriate[b]		3.8	5.6	4.2	5.4	4.5	4.5
The system feedback (display and output) provided an appropriate solution[b]		2.4	6.0	2.4	5.2	3.3	4.7
The announcement (display and output) of the system was presented at the right moment		4.6**	5.4**	4.2	3.7	4.2	4.7
I found the action given in the system feedback (display and output) easy to execute		5.3	6.0	5.1	5.6	4.5	4.9
The system feedback (display and output) was absolutely certain about the suggested solution[b]		2.8	6.1	3.1	4.9	3.9	4.2
I knew immediately what I had to do after the system feedback was displayed/given		5.1**	6.3**	4.2	5.4	5.2	5.2
I found the system feedback (display and output) easy to understand		5.9	6.4	5.5	5.8	4.7	5.2
I wanted to overrule the system feedback (display and output[b]		1.6	1.1	1.7	1.6	1.7	1.6
It would have been difficult for me to exert my will against the system[b]		3.1	3.4	2.8	3.3	3.9**	4.1**

(continued)

Table 1.4 (continued)

Scaling						
Completely disagree	Disagree	Slightly disagree	Neutral	Slightly agree	Agree	Completely agree
1	2	3	4	5	6	7

Participants were divided according to three driver types
$^{**}p < .05$, $^{*}p < .1$, Wilcoxon signed-rank test
$^{a}N = 7$, ^{b}N varies between 4 and 10 if user was not able to rate statement (i.e. didn't read message)

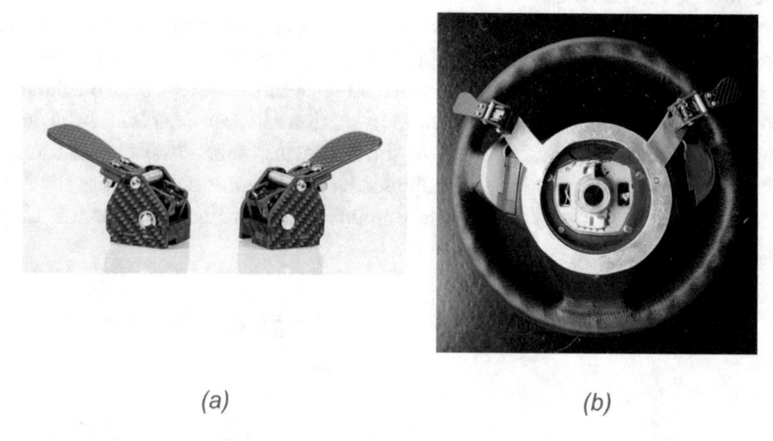

Fig. 1.41 **a** Clutch paddles used for the steering support of category M (image taken from Ascher Racing); **b** clutch paddles installed on the steering wheel of the static driving simulator [image adapted from (Kalb et al. 2020)]

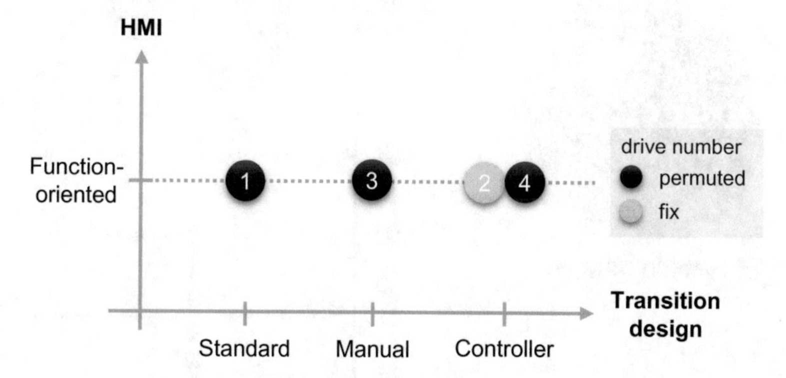

Fig. 1.42 Overview of the experimental drives and their respective HMI and transition design for study 2

hands on the steering wheel. We recorded valid datasets from 31 drivers in study 2. Figure 1.43 shows the gender and age distribution ($M = 24.7$ years, SD = 7.5).

Results of Study 2

In the following presentation of the results of study 2, we focus on the evaluation of the standard (linear decrease of transition parameter α_A) and the controller-based (adaption based on estimating α_H) transition concepts. Since most of the participants did not use the clutch paddles of category M, no interpretable and meaningful results can be deduced in this case.

One transition concept, which is part of the subsequent comparison, is to change the transition parameter α_A over time in a predefined way (linearly in our case), like for the cooperative concept in the previous study. For the other concept, the action-based model adaption, which is presented in section *Human–Machine Interface and Control Transition Concept*, is used. In the following, these two modes are called timebased and adaptive mode. The trajectories of all study participants for the test runs are illustrated in Fig. 1.44 for the timebased and in Fig. 1.45 for the adaptive mode.

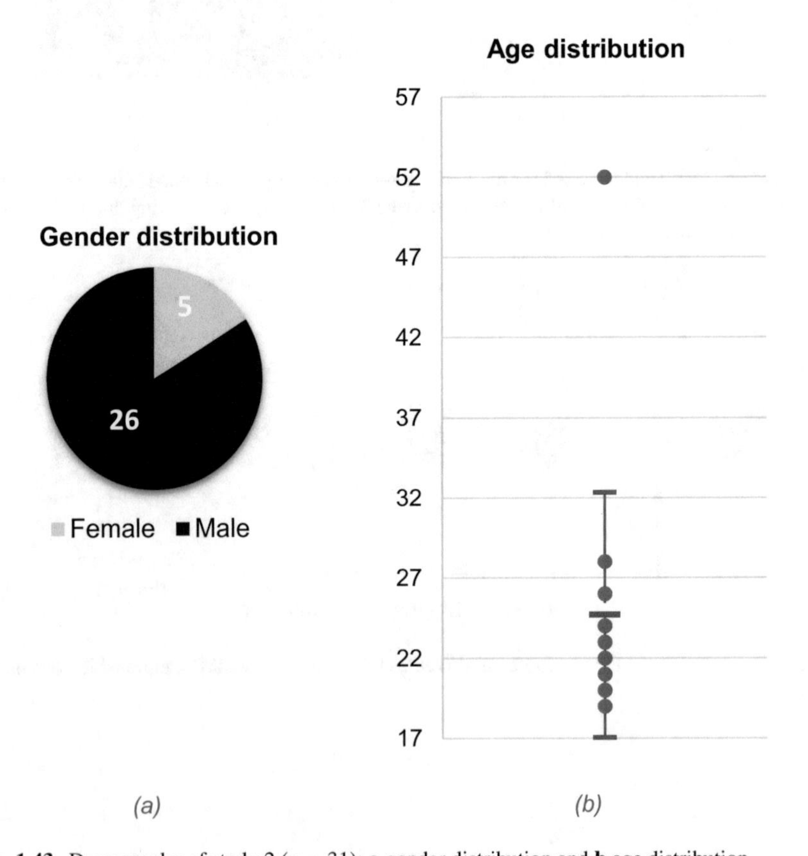

Fig. 1.43 Demography of study 2 ($n = 31$): **a** gender distribution and **b** age distribution

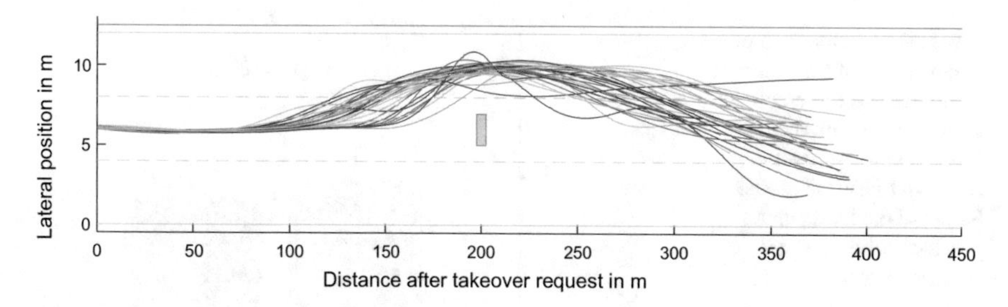

Fig. 1.44 Vehicle trajectories of all study participants achieved with the transition based on the timebased mode (linear decrease of α_A)

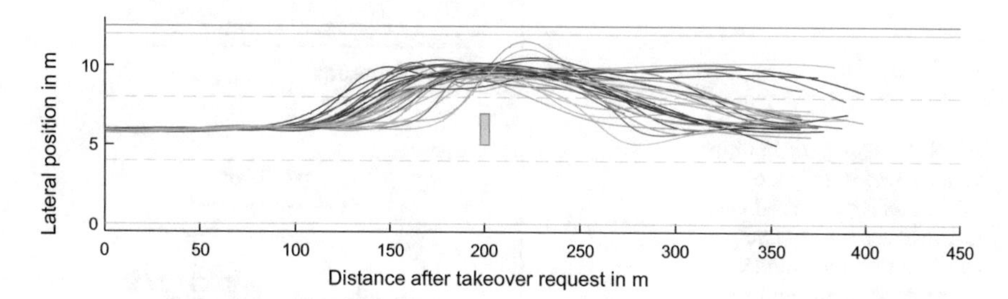

Fig. 1.45 Vehicle trajectories of all study participants achieved with the transition based on the adaptive mode (α_A results from the estimation of α_H and the relation of an ideal transition)

The *t*-tests to compare both methods regarding the same criteria as before show that there is no statistically significant improvement by using the adaptive method. This is shown in Figs. 1.46, 1.47, 1.48 and 1.49.

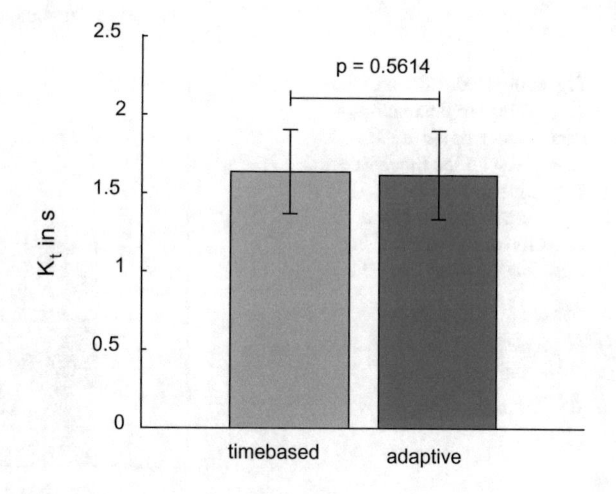

Fig. 1.46 Comparison of the mean values of the minimum time-to-collision K_t achieved with the timebased respectively adaptive transition mode. In addition, the standard deviation and the *p*-value of the *t*-test are illustrated

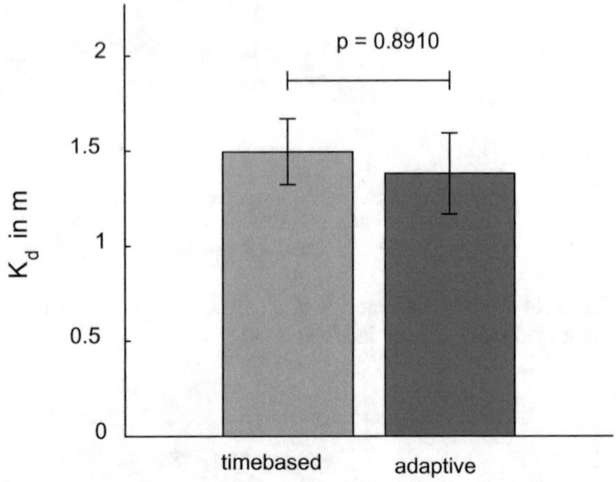

p = 0.8910
K_d in m
timebased adaptive

Fig. 1.47 Comparison of the mean values of the minimum lateral distance to an object K_{d} achieved with the timebased respectively adaptive transition mode. Standard deviation and *p*-value of the *t*-test are illustrated as well

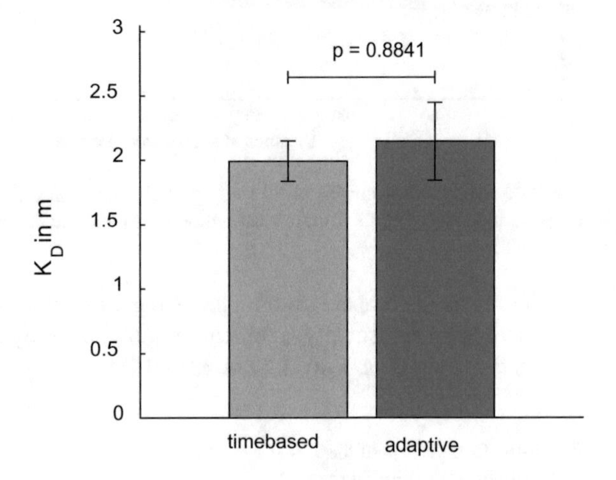

Fig. 1.48 Comparison of the mean values of the lane utilisation K_{D} achieved with the timebased respectively adaptive transition mode. Standard deviation and *p*-value of the *t*-test are illustrated as well

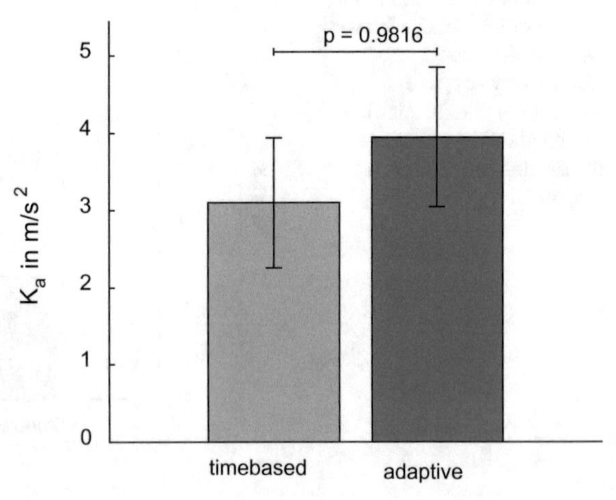

Fig. 1.49 Comparison of the mean values of the maximum lateral acceleration K_{a} achieved with the timebased respectively adaptive transition mode. Standard deviation and *p*-value of the *t*-test are illustrated as well

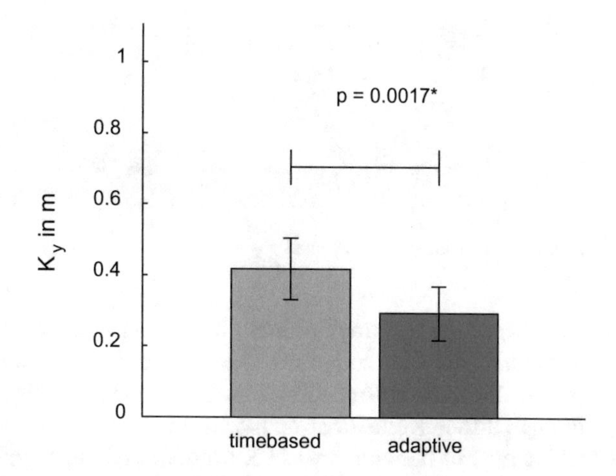

Fig. 1.50 Comparison of the mean values of the mean distance to the target trajectory K_y achieved with the timebased respectively adaptive transition mode. Standard deviation and *p*-value of the *t*-test are illustrated as well

However, the experimental results show that the adaptive transition method helps to reduce the distance to the predefined target trajectory (reference trajectory) of the automation. Because the automation only relinquishes control when the driver follows this goal, the driver is pushed towards this goal if he steers in a different direction. Therefore, this guiding behaviour of the adaptive method reduces the mean distance to the target trajectory K_y. This effect is statistically significant. Figure 1.50 depicts the respective values.

Overall, the results of the driving simulator studies clearly highlight that handing over a control task from an automation to the human abruptly, like in the current state of the art, is not sensible. Instead, the presented concept of a cooperative control transition based on a differential game formulation provides a way to make this transfer significantly safer, more comfortable and more intuitive for the human partner.

1.6.3 Driving Study—Perception of Use Cases in Personalised and Individualised Driving

The real-life study addressed if the driving behaviour of an automated car during the autonomous driving phase can be adapted to the needs and individual preferences of its passengers to increase acceptance and driving comfort. While safety-relevant situations had already be evaluated with the driving simulator studies, this real-world study is important to give present and future users the possibility to experience the PAKoS system in a real-life situation, as the vestibular and audio-visual perception in a simulator would not correspond to reality and thus to evaluate the very subjective measures of human preference and acceptance (Frey 2014, p. 13).

Due to the legal regulations in Germany, it was not possible to conduct the study in live traffic with realistic, not specially trained, users. Therefore, we decided to use a dedicated test track in Pferdsfeld, Germany. In addition to the safety and legal

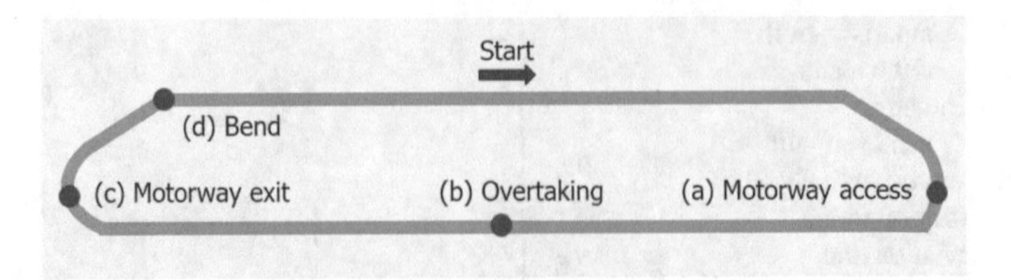

Fig. 1.51 Roadmap and use cases

advantages the test track offered, it also made it easier to design the scenarios in a standardised way compared to a less controllable situation on public roads. To provide the maximum experience within the test situation, users had to be able to run through different situations on the test track, including driving onto a motorway (a), taking over on a motorway at 120 km/h (b), taking a motorway exit (c) and taking a bend on a country road (d). See Fig. 1.51 for the test track. These situations have also been chosen as the most relevant use cases for automated driving and were selected in coordination with the project partners.

The implementation of the PAKoS system in the test vehicle was already explained in section *Implementation Aspects*. To ensure safety, a safety driver was present during the drives, so that he could intervene if a situation became dangerous. Moreover, a technician accompanied the drive to ensure the proper function of the additional sensor equipment and data recording, as did the interviewer, who was needed to observe and question the participant during the drive. In addition to the sensor system of the level-3 car two additional cameras were installed in the vehicle, one of which recorded the participant during the interview and the other the route driven from the windscreen to the outside.

1.6.3.1 Method

Subjective Evaluation—The SUXES Method

The study was conducted as a real-life test drive on a designated test track in Pferdsfeld in the form of an interviewer-supported usability study/acceptance study. Subjective evaluation techniques based on the SUXES User Experience Evaluation were used to evaluate the acceptance and perceived comfort of the driving style of the proposed human-centred highly automated car. The SUXES is a questionnaire developed to report user experience in challenging situations, i.e. with systems a user never used before. The user can agree or disagree to statements that are relevant for the system.

We used the SUXES procedure as orientation, testing not only the users' experiences, but also the users' expectations of such a system. In accordance with (Turunen et al. 2009), the rating is done by the participant before and after the usage of the PAKoS system to enable comparison of the expectations with the experiences.

An advantage of the SUXES method is an individual anchor that is not bound to a benchmark, which would not be available for a system like the proposed one. The users themselves define their own expectations, which makes it possible to test the acceptance. It shows whether the average expectations of a specific target group have been met or not.

A further reason to follow the approach of the SUXES is its quite flexible possibility of optional measurements. This allows the dimensions to be tailored directly to the product, and irrelevant items do not need to be considered by the participant. For example, an evaluation of a standardised scale (e.g. UEQ, Schrepp 2019) such as for "ease of use" in our setting would distort the result to the extent that the prototype is not the final system and does not correspond to the state of the product in all aspects (e.g. aesthetics etc.).

In the original SUXES, there is a twofold evaluation of the expectations (one value for desired level and one value for accepted level), but in accordance with (Keskinen 2015), we only used the value for the desired level (Keskinen 2015, p. 17) because we did not want to overload the user in conjunction with the length of the study. This also helped the user to focus more on the system instead of filling out the questionnaire too many times.

The following items (Table 1.5) were used for the usability evaluation of the user study. We formulated items that correspond to the dimensions relaxation (with aspects of Comfort and Discomfort) or tension (which refers to aspects of Safety and Energy) (Hajek 2017, p. 37).

The user did not rate only the experience of the system-defined highly automated drive (personalised drive) that was parametrised using the data of the manual ride, he or she was also able to individually adjust the driving behaviour to his or her own needs (individualised drive) using the APP. The user-adjustments helped to evaluate

Table 1.5 Usability evaluation items

Error-free error-prone	Safety/Energy
Natural unnatural	Comfort/Discomfort
Fast slow	Safety/Energy
Satisfying unsatisfying	Comfort/Discomfort
Pleasant unpleasant	Comfort/Discomfort
Good bad	Safety/Energy
Likeable repulsive	Comfort/Discomfort
Attractive unattractive	Comfort/Discomfort
Sympathetic unsympathetic	Comfort/Discomfort
Safe unsafe	Safety/Energy
Predictable unpredictable	Safety/Energy
In line with expectations Not in line with expectations	Safety/Energy
Assistive disabling	Safety/Energy

the automated driving behaviour of the car in order to increase the comfort in the automatic adjustment process.

Additional Activities: Secondary Task

Driver perception happens in interaction with various aspects of driving, above all of the driver him- or herself and his or her focus of perception. This raised the additional question of how participants will perceive the driving behaviour of the car when they focus on a different activity (secondary task). Literature shows little data on this question (compare Hajek 2017, p. 37ff), and existing data mainly focuses on the perception without additional activities.

Within our user study, we offer the user the possibility to turn away (the focus of perception) from traffic during the highly automated journey in the form of a standardised secondary task ("EndlessQuiz"). To make the drives comparable, this secondary activity was only made available after the first personalised drive. Thus, the participant first experienced all use cases of the personalised highly automated driving without the offer of a secondary task. In the following journey, the interviewer had the opportunity to activate the quiz on the tablet for the participant. The same applies to the subsequent individualised drives. The installed tablet was equipped with the Spiegel Institute's Mobile Wizard Tool. With this tool, the interviewer was able to activate the tablet remotely using a wireless Bluetooth connection. Once the tablet was activated, the participants were free to carry out the secondary activity. This helped to evaluate the automated driving behaviour while the user had the possibility of doing an additional task.

The whole procedure and the track scheme can be found in Fig. 1.52.

Objective Measurement Data

In addition to the subjective measurements, the test vehicle was capable of observing and collecting driving data (such as acceleration, vehicle speed, etc.) when driven manually by the user. This collected data were then used for the personalised and adaptive highly automated drive. The recorded data material also contained the changes made by the participants themselves in the individualised automated drive.

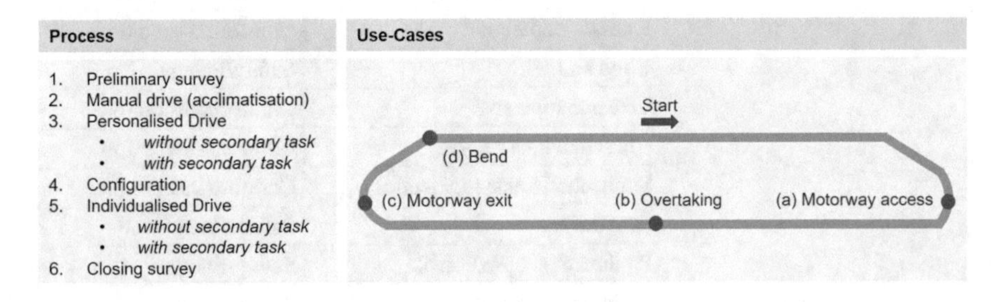

Fig. 1.52 Procedure and use cases on track

1.6.3.2 Sample

For the study sample, the recruiting process was supported by a standardised screener that allowed the recruiters to easily determine a participant's fit to a user group during the recruiting process.

The study sample ($N = 34$) is characterised by a good mix of ages ($M = 41$ years) and driving performance. Accordingly, both people who drive a lot (for private or business purposes) and people who do not drive very much took part in the study. In addition, both participants who use driver assistance systems little or not at all and/or seldom and those who use driver assistance systems a lot and/or frequently were included. 76% of the study participants are men, 24% are women. Details of the sample can be found in Table 1.6.

The participants characterised themselves as having a rather positive attitude towards autonomous driving. However, no participant described himself/herself as an expert in this field (regarding knowledge).

Set-up

As depicted in Fig. 1.52 the study consisted of three parts: manual driving, personalised driving and individualised driving. Each part of the study consisted of two conditions: without a secondary task and with a secondary task. On the test track, the participants experienced the following use cases in each of the three parts: a motorway access, an overtaking manoeuvre, a motorway exit and a bend on a country road. It was important to identify realistic and relevant use cases.

The participants were guided and accompanied by the interviewer during the interview.

A preliminary survey started the study with the collection of demographic and personal characteristics such as age, driving style, driving performance, level of knowledge on the subject of automated driving and attitudes and expectations towards automated driving. To get to know the track, the first step was a regular manual drive on the test track. During this manual drive, the driving style of the participants was recorded to form the personalisation. Afterwards, the participants started the personalised drive by activating the automation. After two rounds, the participants were provided with the option of doing the EndlessQuiz on the tablet. We did not force the users to execute this secondary task, but offered the possibility along with the information that it would be legal (in live traffic in future).

Table 1.6 Sample statistics

	Self-confident	Cautious	Balanced	Total
N	11	14	9	34
Avg. Age **[in years]**	42	33	54	41
% Men	91	71	67	76
% Women	9	29	33	24

At the end of the drive, a follow-up survey was conducted to record the user experience. The next part of the study focused on the topic of individualisation. For this purpose, the participants had the possibility to individually adjust the driving style of the automated vehicle. They were able to increase or decrease the speed. Once the participants had done the appropriate settings, the individualised automated drive took place. As with the personalised drive, the participants drove two rounds without any secondary activities and two rounds with the possibility of performing a secondary task. At the end of the individualised drive, a final recording of the users' experience was made.

The present study design is based on the necessity to keep a constant order of the study sequences. For example, the part of individualisation could only be done on the basis of previous personalisation.

1.6.4 Study Results—Perception of Use Cases in Individualised and Personalised Driving (Subjective) and the Effect of Secondary Tasks

All use cases are perceived as pleasant in both the personalised drive and the individualised drive. In general, a trend towards the perception of the use cases becoming more and more positive over time was observed (Table 1.7).

There was also a tendency with regard to secondary activities: in the personalised driving, secondary activities seem to have a greater influence on the perception of use cases than in the individualised driving. The following insights were gained in the respective use cases.

Motorway Access

Within the use case "motorway access", the performance of secondary activities during the personalised driving leads to a change in the perception of the situation: entering the motorway is perceived as more pleasant. Participants describe a distraction caused by the quiz, so that the attention previously devoted to controlling the course of the drive and the driving style is now devoted to the quiz, making the driving experience more pleasant. Within the individualised drive, the performance of secondary activities has no major influence on the perception of the motorway access.

However, the change from personalisation to individualisation has a greater effect. Participants perceive the motorway access as more pleasant in the individualised drive than in the personalised drive. A course of approach emerges on the level of user groups. In the personalised drive, the self-confident driver perceives the motorway approach as most pleasant, the cautious driver as most likely unpleasant. The evaluations of the motorway access converge by adding the secondary activities and the individualisation, so that the perceptions of the user groups in the individualised drive with secondary activities hardly differ.

Table 1.7 User-ratings in the UseCases (UCs)

	Self-confident				Balanced				Cautious			
	Personalised		Individualised		Personalised		Individualised		Personalised		Individualised	
UC	No Task	2nd Task	No Task	2nd Task	No Task	2nd Task	No Task	2nd Task	No Task	2nd Task	No Task	2nd Task
a	5.9	6.3	6.8	6.5	5.5	6.1	6.4	6.6	4.4	5.3	6.6	6.4
b	5.5	6.5	5.5	6.4	5.5	6	6.1	6.2	5.4	5.7	5.1	5.4
c	5.9	6.4	6.6	6.4	5.6	6.1	6.4	6.6	5.7	6.6	6.3	6.4
d	6.3	6.2	6.5	6.7	5.6	6.2	6.4	6.4	5.8	6	6.4	6.3

Scaling

Unplesant						Plesant
1	2	3	4	5	6	7

UCs (a) Motorway access. (b) Overtaking. (c) Motorway exit. (d) Bend

Motorway Exit

The ratings of the motorway exit are comparable to the ratings of the motorway access. Just like with the motorway access, the addition of a secondary activity in the personalised drive has an effect as well: The rating of the motorway exit becomes more positive.

In contrast to the motorway access, however, there is no major effect in changing from personalisation to individualisation. The evaluations of the use case are consistently positive; there is hardly any change through individualisation. No significant differences can be found on the level of the user groups as well. In addition, the evaluations of the user groups converge.

During the use case, participants emphasise the perception of a pleasant, smooth and fluent driving style of the system.

Bend

In the use case "bend", essentially the same effects can be seen as those identified for the motorway access (see Table 1.8). Within the personalised drive, an improvement in the assessment of the situation is discernible, and this development can also be observed in the transition from personalised to individualised driving. In contrast to the personalised drive, in which some participants describe the bend as unpleasant, rough and too sporty (without secondary activity $n = 8$; with secondary activity $n = 6$), the participants emphasise the appropriateness of the speed within the individualised drive. In addition, the driving style is described as pleasant, smooth and fluid.

In this use case, the evaluations of the user groups differentiate at the beginning (personalised drive without secondary activity), but then start to converge within the personalised drive with a secondary task and the individualised drive without a secondary task. Adding secondary activities in the individualised drive, however, seems to polarise: while the cautious and the balanced drivers perceive the situation with a secondary task as somewhat more unpleasant, the assessment of the self-confident drivers continues to get more positive.

Table 1.8 Average set-speeds in UCs

UC	Avg. set-set-speed		Adjustment of speed			
	Personalised (km/h)	Individualised (km/h)	All users (km/h)	Self-confident (km/h)	Balanced (km/h)	Cautious (km/h)
a	50	48	−2	−2	−3	−1
b	116	115	−1	+ 2	−2	−2
c	52	50	−2	−9	0	+ 4
d	76	76	0	+ 1	0	0

Overtaking

During the process of overtaking, a large influence of secondary activities and a small influence of individualisation on the driving sensation become evident. The addition of secondary activities increases the well-being of the participants during the use case, while individualisation causes almost no change. In short, the personalised drive without secondary activity and the individualised drive without secondary activity are rated similarly, and the personalised drive with secondary activity and the individualised drive with secondary activity are rated similarly (see Table 1.8).

Summarised for all scenarios the individualised driving is perceived as more harmless, controllable and uncritical compared to the personalised driving. Participants also state they feel more relaxed and confident during the drive.

1.6.5 Study Results—Perception of Use Cases in Individualised and Personalised Driving (Objective)/Driving Data

Basically, the adjustments made by all participants during the process of individualisation are rather small on average (e.g. for speed quantities ± 2 km/h). When looking at the set speed, however, a tendency can be seen: on sections of the road that can be classified as more critical (motorway exit, motorway), the speed tends to be slightly reduced. On sections of road that can be classified as largely uncritical (straight section), the speed is slightly increased or not changed at all.

When analysing the adjustment of speed at the level of user groups, larger individualisations can be observed. For example, the self-confident driver regulates the set speed by −9 km/h on the motorway on average, while the cautious driver increases speed by +4 km/h on the same section of road on average. But by making this adjustment, the groups are approximating their driving styles. Thus the confident driver reduces his speed on the motorway from 58 to 49 km/h, while the cautious driver increases the speed from 48 km/h to almost 52 km/h (on average). For details, see Table 1.8.

In the subjective assessment, participants generally ascribe a very high importance to individualisation options.

Secondary Task

The majority of participants carried out the offered secondary activity in the different use cases, if available.[3] Participants who did not take part in the secondary activity were mainly motivated by the desire to control and monitor the driving behaviour of

[3]The only exception to this is overtaking, which, for technical reasons, requires the participant to initiate it by setting the direction indicator, so it was not possible for the participants to consistently deal with the secondary activity.

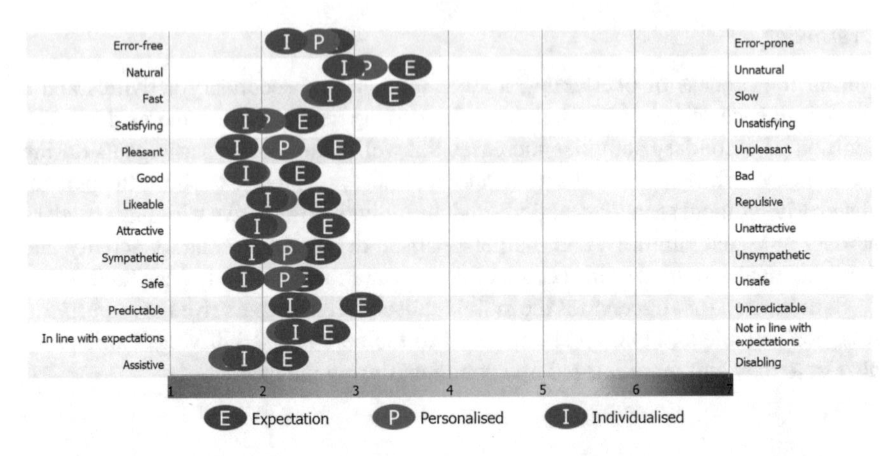

Fig. 1.53 Expectations and actual user experience

the automated vehicle. This desire prevailed both in the personalised ($n = 9$) and in the individualised driving ($n = 7$).[4]

Expectations and Actual User Experience, Development of Trust in the System and Intention to Use

Using opposing pairs of words in a semantic differential, the expectations of the experience of automated driving were queried at the beginning of the study. Overall, all expectations of the experience were exceeded by the actual experience (see Fig. 1.53). The experience of personalisation is even surpassed by the experience of individualisation. In the conducted *t*-tests, the actual experiences in the individualised drive showed significant better ratings in all but two items (error-free and in line with expectations). In the personalised drive, users rated 7 out of 13 items significantly better than they would have expected (Table 1.9 for details). The increase of positive ratings from the personalised drive compared to the individualised drive became still significant in three items: users rated in significantly more pleasant, more sympathic and more safe in the individualised drive compared to the personalised drive. The greatest differences between expected experience and actual experience can be seen in the topics "predictable–unpredictable", "attractive–unattractive" and "good–bad". In short, participants perceive the automated drive more predictable, better and more attractive than they had expected.

Trust in the system and the intention to use it also undergo a change. Both the trust in the system and the intention to use it increase from a basic value (first round personalised) in the personalised automated drive to a higher value in the individualised drive. With regard to the differentiation according to user groups, the cautious driver initially exhibits the lowest level of system trust and the lowest

[4]In case the participant had not performed a secondary activity in at least one of the four use cases, a request for an explanation regarding the renunciation of a secondary activity was made after the drive.

Table 1.9 Validation of expectations without having used the system and experienced ratings

Dimension (scaling)		User's expectation	User experience	
(1)	(7)	Before drive	Personalised	Individualised
Error-free	Error-prone	2.8	2.6	2.3
Natural	Unnatural	3.6	3.1	2.9**
Fast	Slow	3.4	2.6**	2.7**
Satisfying	Unsatisfying	2.4	2.0	1.8**
Pleasant	Unpleasant	2.8	2.2**	1.7**
Good	Bad	2.4	1.9**	1.8**
Likable	Repulsive	5.4	5.9**	5.9**
Attractive	Unattractive	2.7	2.0**	1.9**
Sympathetic	Unsympathetic	2.6	2.3	1.9**
Safe	Unsafe	2.4	2.2	1.8**
Predictable	Unpredictable	4.9	5.6**	5.7**
In line with expectations	Not in line with expectations	2.7	2.3	2.4
Assistive	Disabling	5.7	6.4**	6.2**

**$p < .05$

intention to use the system. However, both values increase due to the actual experience with the system in the personalised and individualised drives. Of all three user groups, the cautious driver seems to benefit most from individualisation in terms of system trust and intention to use.

In order to feel even more comfortable, participants predominantly wish for a softer, more natural, anticipatory driving behaviour. Overall, the majority of participants emphasise the positive overall impression of the system.

1.7 Summary and Discussion of the Studies

In general, the driving style in highly automated driving mode met the expectations of the participants. Nobody felt unpleasant about the way the car drove. It can be ascertained that possibilities of individualisation are generally used by the participants. When looking at the specific adjustments and modifications that the participants have made, an interesting picture emerges: on average, the participants do not force any major changes to the system. When looking at the data on the level of the user groups, they do differ in their driving behaviour in the personalised drive and in their adjustment behaviour during the individualisation phase. However, we have also seen that the different user groups converge in their driving behaviour through individualisation measures. Some use cases have also shown that the perceptions initially differ to a large extent (e.g. motorway access), but that they also converge. The balanced driving style seems to be most desirable for all user groups. This raises the question of whether individualisation options make sense at all, assuming that all participants prefer a similar driving style of the automated vehicle. It is certain that study participants rate the (subjective) relevance of individualisation options as very high. It is conceivable that the individualisation process helps them to get used to automation. Individualisation promotes interaction with the system and could support the building of trust accordingly. Individualisation can be viewed as a kind of creative process through which the customer can subconsciously build a bond. The ability to influence the final product and express oneself individually may lead to a personal bond with the product. In the end, the result of this activity reflects the personal taste of the user. Another point to be considered is the fact that the possibility to adjust the speed seems to have a different effect depending on the use case performed.

In the course of the individualisation, there was a possibility to adjust the parameter "speed". Due to the desire to avoid overstraining the participants, further driving parameters, such as lateral acceleration, were not adjustable.

To test the safety-relevant transition phases, two simulator studies have been conducted. As mentioned earlier, several perceptive deceptions can also have an influence on the evaluation in a simulator. Nevertheless, the driving behaviour of participants is quite comparable to a real-life situation. The results of both simulator studies reveal the advantage of a cooperative transition phase. Users will handle the situation better and more precisely compared to a standard transition. The advantage of the simulator studies was to test against a baseline, so we have clear differences in the users' behaviour and reaction.

Regarding the HMI concepts, the fact that the higher acceptance ratings of users from the balanced and cautious driver group did not become significant may result in the situation itself. It was a highly dangerous and surprising situation, and users usually tried to take over control of the car first, instead of concentrating on the HMI. In accordance with the characteristics of these groups, they do not want to concentrate on the HMI.

For the next iterations, it should be taken into account that individualisation in the real-life study enhanced the comfort and the user experience. We assume that the possibility of control over the system also helps to understand how and why it works. In terms of HMI and design, the challenge would be to integrate all possibilities into a complete HMI concept to ensure a good user experience.

Overall, the proposed-centred highly automated car that was designed in the PAKoS project exceeded the users' expectations when it came to the comfort they experienced, and it introduced a safer transition system compared to the state of the art. The adaptive transition helped users to handle critical situations and users with a desire for technical information profit from a personalised HMI.

1.8 Conclusion

In this contribution, the idea of a human-centred highly automated car has been presented. More precise, a concept of how to design a human-centred control transition and how to generally adapt the driving behaviour of highly automated cars to the passengers was introduced. For this, a module, which can understand the driver by determining his status and activities, was developed in a first step. Its technical realisation relies on a camera-based driver monitoring system and state-of-the-art neural network approaches. By dividing the actual control transition between the automation and the human into two phases, it is possible to support the driver during the task of taking over the control of the vehicle. Suitable and multimodal human–machine interfaces for informing the driver and establishing his physical readiness have been discussed to design the first phase. A novel control algorithm that directly takes the capabilities and the mental readiness of the driver into account coordinates the cooperative transition process between the automation and the human. In order to consider individual varieties among different drivers, the vehicle automation algorithm includes an adaption procedure. This individualisation and personalisation idea was also extended to the purely automated driving and implemented on a real test vehicle. In order to integrate such driver respectively user profiles into the digital architecture of the car, we propose a mobile phone application along with a data protection concept based on Face ID.

The results of various driving studies (simulator as well as real test vehicle) show a clear benefit of the driver oriented automation system concept compared to the state of the art. In context of the control transition phase, the cooperative transition concept increases the takeover performance and thus the driving safety compared to state-of-the-art concepts with statistical significance. In addition, the adaption procedure leads to better subjective ratings by the participants. For the regular highly automated driving scenario, the studies highlight that the personalised and individualised runs outmatch the expectations of the participants. Hence, adapting the highly automated driving function to the passengers and their activities acts as an important aspect towards a higher user acceptance.

In our opinion, an individualisation and personalisation of the control transition and the automated driving in cars of SAE level 3 or higher will not only be crucial but also be necessary. Considering the benefits that can already be achieved by means of the limited effort of the described project compared to the overall efforts made regarding the design, programming and safeguard of highly automated cars the results should be even more impressive.

References

Abbink DA, Mulder M (2010) Neuromuscular analysis as a guideline in designing shared control. Adv Haptics (IntechOpen)

Abbink DA, Mulder M, Boer ER (2012) Haptic shared control: smoothly shifting control authority? In: Cognition, technology and work, vol 14, no 1, S. 19–28

Abouelnaga Y, Eraqi HE, Moustafa MN (2018) Real-time distracted driver posture classification. Machine learning for intelligent transportation systems. In: Workshop in the conference on neural information processing systems (NeuroIPS)

Banks VA et al (2018) Is partially automated driving a bad idea? Observations from an on-road study. Appl Ergonomics 68

Brandenburg and Skottke (2014) Switching from manual to automated driving and reverse: are drivers behaving more risky after highly automated driving. ITSC

Carreira J, Zisserman A (2017) Quo vadis, action recognition? A new model and the kinetics dataset. In: Proceedings of the IEEE conference on computer vision and pattern recognition

Cao Z et al (2018) OpenPose: realtime multi-person 2D pose estimation using part affinity fields. ArXiv:1812.08008 [Cs]

Damböck D et al (2012) Übernahmezeiten Beim Hochautomatisierten Fahren. Tagung Fahrerassistenz. München 15

Deng J, Dong W, Socher R, Li L-J, Li K, Fei-Fei L (2009) ImageNet: a large-scale hierarchical image database

Endsley and Kris (1995) The out-of-the-loop performance problem and level of control in automation. Human Factors

Engwerda J (2005) LQ dynamic optimization and differential games. Wiley

Eriksson et al (2017) Rolling out the red (and green) carpet: supporting driver decision making in automation-to-manual transitions

Feldhütter A, Ruhl A, Feierle A, Bengler K (2019) The effect of fatigue on take-over performance in urgent situations in conditionally automated driving in 2019. In: IEEE intelligent transportation systems conference (ITSC), pp 1889–1894

Flad M, Otten J, Schwab S, Hohmann S (2014) Steering driver assistance system: a systematic cooperative shared control design approach. In: IEEE international conference on systems, man and cybernetics (SMC), S. 3585–3592

Flad M (2016) Kooperative Regelungskonzepte auf Basis der Spieltheorie und deren Anwendung auf Fahrerassistenzsysteme. KIT Scientific Publishing, Karlsruhe

Frey A (2014) Fahrsimulatorstudie im Rahmen des Verbundprojekts URBAN: Statischer und dynamischer Fahrsimulator im Vergleich: Wahrnehmung von Abstand und Geschwindigkeit. (Masterarbeit an der Heinrich-Heine-Universität Düsseldorf 2014), Bundesanstalt für Straßenwesen, Bergisch Gladbach

Gebert P et al (2019) End-to-end prediction of driver intention using 3D convolutional neural networks. In: 2019 IEEE intelligent vehicles symposium (IV). IEEE

Gold C et al (2013) 'Take over!' how long does it take to get the driver back into the loop? In: Proceedings of the human factors and ergonomics society annual meeting, vol 57, no 1

Hajek H (2017) Längsdynamik und Antriebsakustik von elektrifizierten Straßenfahrzeugen–Beschreibung und Gestaltung des emotionalen Erlebens. https://mediatum.ub.tum.de/doc/1343083/1343083.pdf, 30 Jan 2020

Hochreiter S, Schmidhuber J (1997) Long short-term memory. Neural Comput 9(8)

Jain A, Koppula HS, Raghavan B, Soh S, Saxena A (2015) Car that knows before you do: anticipating maneuvers via learning temporal driving models. In: Proceedings of the IEEE international conference on computer vision, pp 3182–3190

Kalb et al (2018a) What drivers make of directional maneuver information in a take-over scenario. In: IEEE international conference on systems, man, and cybernetics (SMC), pp 3859–3864

Kalb et al (2018b) Multimodal priming of drivers for a cooperative take-over. In: 21st international conference on intelligent transportation systems (ITSC), pp 1029–1034

Kalb et al (2020) Manual adaption of steering support in a take-over scenario—a technical evaluation. In: Tagungsband Uni-DAS 13. Workshop Fahrerassistenzsystem und automatisiertes Fahren, Walting im Altmühltal (2020)

Karakaya et al (2018) Cooperative approach to overcome automation effects during the transition phase of conditional automated vehicles. In: Tagungsband Uni-DAS 12. Workshop Fahrerassistenzsystem und automatisiertes Fahren, Walting im Altmühltal

Keskinen T (2015) Evaluating the user experience of interactive systems in challenging circumstances. Dissertations in Interactive Technology, no 22, Tampere

Körber (2018) Individual differences in human-automation interaction. Dissertation, Technical University of Munich, Munich

Lefèvre S et al (2015) Driver models for personalised driving assistance. Veh Syst Dyn 53(12)1705–1720

Lloyd D et al (2014) Reported road casualties Great Britain: 2014 annual report

Lorenz et al (2014) Designing take over scenarios for automated driving. How does augmented reality support the driver to get back into the loop? Proc Human Factors Ergonomics Soc Annu Meet 58(1):1681–1685

Ludwig J, Gote C, Flad M, Hohmann S (2017) Cooperative dynamic vehicle control allocation using time-variant differential games. In: IEEE international conference on systems, man and cybernetics, SMC, Banff, Alberta, Canada, pp 117–122

Ludwig J, Haas A, Flad M, Hohmann S (2018a) A comparison of concepts for control transitions from automation to human. In: IEEE international conference on systems, man and cybernetics, SMS, Miyazaki, pp 3201–3206

Ludwig J, Martin M, Horne M, Flad M, Voit M, Stiefelhagen R, Hohmann S (2018b) Driver observation and shared vehicle control: supporting the driver on the way back into the control loop. Automatisierungstechnik 66(2):146–159

Martin M et al (2018) Body pose and context information for driver secondary task detection. Intel Veh Symp (IV)

Martin M, Roitberg A et al (2019) Drive & act: a multi-modal dataset for fine-grained driver behavior recognition in autonomous vehicles. In: Proceedings of the IEEE international conference on computer vision

Merat et al (2014) Transition to manual: driver behaviour when resuming control from a highly automated vehicle. Transp Res Part F: Traffic Psychol Behav

Misra I et al (2016) Cross-stitch networks for multi-task learning. In: Proceedings of the IEEE conference on computer vision and pattern recognition (2016)

Na X, Cole DJ (2013) Linear quadratic game and non-cooperative predictive methods for potential application to modelling driver–AFS interactive steering control. In: Vehicle system dynamics 51(2013), no 2, S. 165–198

Ohn-Bar E, Martin S, Tawari A, Trivedi MM (2014) Head, eye, and hand patterns for driver activity recognition. In: 2014 22nd international conference on pattern recognition, pp 660–665. IEEE

Petermann-Stock I et al (2013) Wie Lange Braucht Der Fahrer? Eine Analyse Zu Übernahmezeiten Aus Verschiedenen Nebentätigkeiten Während Einer Hochautomatisierten Staufahrt. *6. Tagung Fahrerassistenzsysteme. Der Weg Zum Automatischen Fahren*

Petermeijer et al (2017a) Take-over again: investigating multimodal and directional TORs to get the driver back into the loop. Appl Ergonomics 62:204–215

Petermeijer et al (2017b) Comparing spatially static and dynamic vibrotactile take-over requests in the driver seat. Accid Anal Prev 99(Pt A):218–227

Qiu Z, Ting Y, Tao M (2017) Learning spatio-temporal representation with pseudo-3d residual networks. In: Proceedings of the IEEE international conference on computer vision

Radlmayr J et al (2014) How traffic situations and non-driving-related tasks affect the take-over quality in highly automated driving. In: Proceedings of the human factors and ergonomics society annual meeting, vol 58

Radlmayr J et al (2019) Take-overs in level 3 automated driving—proposal of the take-over performance score (TOPS). In: Proceedings of the 20th congress of the international ergonomics association (IEA 2018), pp 436–446

Roitberg A et al (2018) Informed democracy: voting-based novelty detection for action recognition. BMVC

Roitberg A et al (2019) Analysis of deep fusion strategies for multi-modal gesture recognition. In: Proceedings of the IEEE conference on computer vision and pattern recognition workshops

SAE J3016 (2018) Taxonomy and definitions for terms related to driving automation systems for on-road motor vehicles

Schrepp M (2019) User experience questionnaire handbook. https://www.ueq-online.org/Material/Handbook.pdf, 30 Jan 2020

Simonyan K, Zisserman A (2014) Two-stream convolutional networks for action recognition in videos. In: Advances in neural information processing systems, pp 568–576

Singh S (2015) Critical reasons for crashes investigated in the national motor vehicle crash causation survey. No. DOT HS 812:115

Streit (2018) Ergonomic evaluation of concepts for automatic action recommendations in take-over situations during conditionally automated driving. Master's Thesis, Technical University of Munich

Tamaddoni SH, Ahmadian M, Taheri S (2011) Optimal vehicle stability control design based on preview game theory concept. In: American control conference (ACC), S. 5249–5254

Todorov E, Jordan MI (2002) Optimal feedback control as a theory of motor coordination. In: Nature neuroscience, vol 5, no 11, S. 1226–1235

Tran D et al (2015) Learning spatiotemporal features with 3D convolutional networks. In: Proceedings of the IEEE international conference on computer vision

TUM (2020) https://www.lfe.mw.tum.de/forschung/labore/statischer-fahrsimulator/

Turunen et al (2009) SUXES—user experience evaluation method for spoken and multimodal interaction. University of Tampere, Department of Computer Sciences

Wang H and Wang L (2017) Modeling temporal dynamics and spatial configurations of actions using two-stream recurrent neural networks. E Conference on Computer Vision and Pa Ern Recognition (CVPR)

Wolf E et al (2018) Estimating mental load in passive and active tasks from pupil and gaze changes using bayesian surprise. In: Proceedings of the workshop on modelling cognitive processes from multimodal data

Yan C, Coenen F, Zhang B (2016) Driving posture recognition by convolutional neural networks. IET Comput Vision 10(2):103–114

Chapter 2
KomfoPilot—Comfortable Automated Driving

Matthias Beggiato, Franziska Hartwich, Patrick Roßner, André Dettmann, Stephan Enhuber, Timo Pech, Dagmar Gesmann-Nuissl, Klaus Mößner, Angelika C. Bullinger, and Josef Krems

Abstract Automated driving is expected to bring several benefits such as improved traffic safety, reduced congestions and emissions, social inclusion, enhanced accessibility and higher driving comfort. A central human–machine interaction issue addresses the question of how automated vehicles should drive to ensure comfort and a positive driving experience. The project KomfoPilot aimed at investigating, assessing and enhancing comfort during automated driving by two driving simulator studies and a test track study. Sensors such as wearable devices, eye tracking, face tracking and motion tracking allowed for the integration of driver state data with information about the vehicle and the surroundings. Various driving styles as well as display solutions were evaluated for reducing discomfort. In addition, privacy issues were continuously monitored for all aspects over the project lifetime. Section 2.1 gives an overview on the background and aims of the project, definitions of central concepts and an overall summary of key results. Methodological details on the experimental design, participants, assessment of discomfort, sensors and questionnaires of the three studies are presented in Sect. 2.2. A first research objective was to find factors that affect comfort on a rather general level, such as driving situations and driving style parameters (e.g. speed, longitudinal/lateral distance, driving style familiarity). These results are presented in Sects. 2.3 and 2.4. A second objective was the development of algorithmic approaches for real-time discomfort detection based on

M. Beggiato (✉) · F. Hartwich · J. Krems
Cognitive and Engineering Psychology, Chemnitz University of Technology, 09107 Chemnitz, Germany
e-mail: matthias.beggiato@psychologie.tu-chemnitz.de

P. Roßner · A. Dettmann · A. C. Bullinger
Ergonomics and Innovation, Chemnitz University of Technology, 09107 Chemnitz, Germany

S. Enhuber · T. Pech · K. Mößner
Communications Engineering, Chemnitz University of Technology, 09107 Chemnitz, Germany

D. Gesmann-Nuissl
Private Law and Intellectual Property Rights, Chemnitz University of Technology, 09107 Chemnitz, Germany

G. Meixner (ed.), *Smart Automotive Mobility*, Human–Computer Interaction Series,
https://doi.org/10.1007/978-3-030-45131-8_2

sensor data. Results on physiological discomfort indicators are presented in Sect. 2.5, whereas Sect. 2.6 gives an overview of the work on algorithms. Privacy and liability aspects are discussed in Sect. 2.7.

2.1 Background, Aims and Summary of Results

2.1.1 Background

Automated driving is expected to bring several benefits such as reduction of traffic accidents, increase of traffic efficiency and urban accessibility, reduced environmental pollution, economic competitiveness and social inclusion by new mobility options, e.g. in the context of demographic changes (Smith et al. 2018; Meyer and Biker 2018). However, to exploit the potential of automated driving, several challenges need to be addressed in different areas and on local, European and sometimes international level (ERTRAC 2019): Next to the development and deployment of vehicle technology, new mobility services, shared economy, connected digital and physical infrastructure including big data and artificial intelligence need to be established at the area of systems and services. The overarching area related to users and society includes socio-economic assessment and sustainability, policy and regulation needs, safety validation, user awareness, ethics, driver training, human factors issues as well as user and societal acceptance.

Within the human factors and acceptance topic, comfort is considered as one of the main drivers for higher levels of automation (ERTRAC 2019). Ensuring a positive driving experience is a prerequisite for the purchase, usage and acceptance of automated vehicles as well as the fulfilment of promises regarding new opportunities during driving such as relaxation, work and entertainment (Bellem et al. 2018; Riener et al. 2016). Comfort aspects in vehicles such as noise, temperature, vibrations or comfort of seats have a long research tradition (Qatu 2012; Silva 2002). However, additional psychological determinants are discussed in automated driving conditions, such as trust in the system, familiarity of driving manoeuvres, apparent safety, feelings of control and information about system actions and states (Beggiato et al. 2015; Bellem et al. 2016; Constantin et al. 2014; Elbanhawi et al. 2015; Festner 2019). Although there is no common comfort definition in the scientific community, existing definitions share some central ideas: comfort (1) is considered a subjective construct that therefore differs between individuals; (2) is affected by physiological, physical and psychological determinants; and (3) emerges from an interaction with the environment (Looze et al. 2003; Hartwich et al. 2018). The International Organisation for Standardisation defines comfort as a "complex subjective entity depending upon the effective summation of all the physical factors present in the induced environment, as well as upon individual sensitivity to those factors and their summation, and such psychological factors as expectation." (ISO 5805 1997, p. 9). Thus, in the KomfoPilot project comfort is understood as a subjective, pleasant state of relaxation resulting from confidence in safe vehicle operation (Constantin et al. 2014), "which is achieved by the removal or absence of uneasiness and distress" (Bellem et al. 2016, p. 45). This rather broad definition of comfort (with the opposite part of discomfort) shows overlaps and similarities to related concepts of stress, mental workload, alertness, motion sickness, fear or anger. Within the KomfoPilot project, the conceptual approach started with a "bottom up" strategy by collecting empirical results on reported driving situations that cause any type of unpleasant experience. Discomfort as the opposite part of comfort is the attempt to integrate all these unpleasant experiences into one concept. Mental workload and stress show probably the closest relation to discomfort; however, both concepts mainly aim on a perceived discrepancy between (task) demands and available resources/coping strategies (Backs and Boucsein 2000; Charles and Nixon 2019). This could hold true for some uncomfortable situations, but, e.g., driving style-related issues provoking motion sickness or "traditional" comfort issues such as seat comfort would hardly fit into these concepts.

2.1.2 Project Aims

KomfoPilot was an extended follow-up project from the pre-project DriveMe (BMBF, grant no. 16SV7119, Hartwich et al. 2016). The main aim of KomfoPilot was to investigate comfort in automated driving. An interdisciplinary team consisting of the four professorships—Cognitive and Engineering Psychology, Ergonomics and Innovation, Communication Engineering and Private Law and Intellectual Property Rights at Chemnitz University of Technology cooperated with the company FusionSystems GmbH (specialised in sensor data fusion) and the associate industry partners BMW and Delphi/Aptiv. The underlying project idea was the metaphor of a driver-vehicle-team, where both partners know each other's limitations, strengths and current states, and are able to adapt their behaviour accordingly (Klein et al. 2004). The overall project aim based on this idea of human-automation-teaming was divided into two research areas:

(1) One objective was to find aspects that affect comfort on a more general and aggregated level, located on the behavioural (personality, preferences), strategical (trip planning, route choice) and tactical (manoeuvres) driving level (Michon 1985; Cacciabue 2007). This includes situations and driving manoeuvres that are perceived as uncomfortable (in manual driving), personal characteristics and preferences, relevance of information about the automation, driving style familiarity as well as the importance of single driving parameters for comfort such as speed, longitudinal/lateral distance, acceleration and jerk. Corresponding project results are reported in Sects. 2.3 and 2.4. These findings provide important insights into higher-level contributing factors to comfort/discomfort. However, according to the comfort definition as a subjective construct, individuals are supposed to perceive a different extent of comfort/discomfort in, for example, the same driving situation.

(2) Thus, a second complementary objective was the development of algorithmic approaches for real-time discomfort detection, working at the operational driving level in milliseconds time horizon. Detected discomfort could finally be used to adapt the automated driving style and the presentation of information. The investigation of real-time discomfort predictors was achieved by using a feedback mechanism for gradually and continuously reporting discomfort (handset control, see Sect. 2.2.2.2) as well as sensors for driver monitoring such as eye tracking, face tracking, motion tracking, a seat pressure mat as well as a smartband for physiological parameters. Detailed results can be found in Sects. 2.5 and 2.6.

Privacy and liability aspects have been considered throughout the whole project and are presented in Sect. 2.7. It should be noted that these two main project objectives are just complementary perspectives at different levels and not independent of each other. For example, situations with a high probability of provoking discomfort (strategical/tactical level) need to be identified for creating experimental situations that potentially elicit discomfort. Identified parameters/algorithms at the operational level could in turn help to discover new uncomfortable situations in automated driving conditions that were not considered beforehand.

2.1.3 Summary of Results

This section summarises the main findings of the two driving simulator studies and the test track study of KomfoPilot as well as key results from relevant pre-projects such as DriveMe. Each research topic includes references to the corresponding sections and publications for further details.

2.1.3.1 Secondary Data Analysis on Uncomfortable Situations in Manual Driving and Potential Driver Indicators of Discomfort

Secondary data analysis was the first project work package and included literature analyses as well as re-analyses of existing datasets from four previous projects at TU Chemnitz with a new specific focus on discomfort indicators. The previous projects were DriveMe (Hartwich et al. 2016; Beggiato et al. 2017; Scherer et al. 2015), UR:BAN (Beggiato et al. 2017; Beggiato and Krems 2013b; Bocklisch et al. 2017; Leonhardt et al. 2017), DriveFree (Beggiato and Krems 2013a) and a naturalistic driving study on demanding driving situations (Simon 2017). The main findings are summarised below. Further details on the literature analysis can be found in Beggiato et al. 2019.

- Infrastructure-related discomfort factors: situations with high complexity such as intersections, roundabouts, construction zones, enter and exit situations at highways.
- Discomfort-inducing behaviour of other road users: unpredictable/unclear behaviour of others, manoeuvres of larger vehicles such as trucks, short distances, behaviour of vulnerable road user such as pedestrians, motorcycles and bicycles.
- Uncomfortable own driving manoeuvres: turning, entering and exiting roads, overtaking (including avoiding obstacles), distance keeping in complex situations such as high traffic density and/or complex intersections.
- External factors that could increase discomfort: darkness, adverse weather conditions, icy roads, obstacles on the road as well as bad road conditions such as potholes.
- Discomfort-inducing situations as co-driver: short overtaking on rural roads, abrupt braking, short distances to the vehicle in front, high speed in curves and high speed on the highway. Major cause for discomfort as co-driver relates to missing control and intervention possibilities, coupled with unpredictability of subsequent actions.
- Main potential indicators of driver discomfort can be grouped into (1) psychophysiological measures such as heart rate, heart rate variability, skin conductance level, skin temperature and pupil diameter, (2) glance behaviour, e.g. control glances to monitor automation behaviour in a certain situation, (3) facial expressions such as pressed lips, (4) body posture and movements such as pushing backwards and (5) person characteristics such as anxiety, sensation seeking, risk tolerance, need for control, age and driving experience.
- Regarding personality, higher comfort in automated driving was reported by individuals with higher extraversion scores (active, optimistic, sociable), higher mileage and less need for control (Beggiato et al. 2017). On the other hand, less comfort was reported when showing higher scores on neuroticism (anxiety, nervousness, uncertainty).

2.1.3.2 Users' Driving Experience, Attitudes and Design Preferences on a General Level

The main findings on users' perceptions and feedback are summarised below. Details are reported in Sect. 2.3 and in Hartwich et al. 2015, 2016, 2018, 2019.

- Perceived discomfort during driving strongly depended on the driving situation, peaking in complex situations such as intersections.
- Of different predefined automated driving styles, drivers preferred a defensive driving style, which provoked the least amount of discomfort, but the highest perceived safety in the driving simulator and on the test track.
- Driving style did not affect the enjoyment of automated driving, both in the driving simulator and on the test track.

- Driving enjoyment was generally low for younger drivers, likely based on boredom due to giving up the driving task, but considerably higher for older drivers.
- Even though drivers gave references to their own driving styles when evaluating automated driving styles, they did not prefer their own driving style in comparison to predefined driving styles. This effect might have had methodological reasons, since the own driving styles represented replays of the participants' manual drives along the test track and therefore might have been confounded by the quality of the individual manual driving performances.
- Driving style differences regarding discomfort and perceived safety did not translate into user attitudes towards automated driving: Irrespective of the driving environment (driving simulator vs. test track), neither trust in automation nor the acceptance of automated driving varied depending on the previously experienced driving style. Reasons might be individually different driving style preferences based on other user characteristics than age, for example personality traits or a priori trust in automation.
- In contrast to the automated driving style, the display-based presentation of system information during driving had consistently positive effects on users' driving experience and their attitudes towards automated driving.
- Supplementing the automated driving system by such an HMI resulted in lower discomfort, higher perceived safety and higher driving enjoyment, which also translated into higher trust in automation and a higher acceptance of automated driving.

2.1.3.3 Discomfort Related to Driving Situations and Driving Style

The main findings at the level of driving situations and driving parameters are summarised below. Details are reported in Sect. 2.4 and in Roßner and Bullinger 2018, 2019a, b, c.

- Results show that the older age group tended to feel more comfortable with the defensive driving style.
- As comfort-critical driving style parameters (mainly characteristics of the dynamic driving style) could be identified: (1) high acceleration rates, (2) high deceleration rates and late brake initiation and (3) low longitudinal distance to the vehicle driving ahead.
- Within the traffic sign regulated intersection, discomfort was mainly related to the appearance of crossing traffic. Both age groups felt uncomfortable because of the unclear further behaviour of the crossing traffic as well as the unclear action of the own automated vehicle (hereafter referred as ego vehicle). The slight braking during the defensive driving style and the resulting lower speed hardly influenced discomfort.
- A clear analysis of the obstacle situation is difficult due to its increased complexity: different driving style components occurred at different times or overlapped at the same time. Both dynamic and defensive acceleration caused discomfort. Within

the older age group, discomfort was related to the appearance of the obstacle and the unclear further actions of the ego vehicle.

- At the traffic light regulated intersection, the older age group felt significantly more uncomfortable than the younger group. With the dynamic driving style, the late reaction as well as the resulting high braking rate caused a large part of the reported discomfort.
- During the highway-entering situation, both driving styles evoked similar discomfort. A combination of the fast cornering (dynamic driving style) and the lower acceleration rate (defensive driving style) could reduce the discomfort experienced by younger participants.
- The discomfort-inducing element while passing the construction site (the construction site vehicle) was approached too timidly in defensive driving style and too aggressively in the dynamic driving style.
- In most situations, a driving behaviour in between or a combination of defensive and dynamic driving style should lead to less discomfort in the younger age group.
- The differences within the study sample cannot only be explained by age differences, but require further studies on influencing factors, e.g. mood, fatigue or experience with highly automated driving. In addition, a stronger discomfort feedback frequency of the younger age group has been identified. Further studies should clarify whether this is a random effect of the experiment or a factor to be considered in long term.

2.1.3.4 Physiological Indicators of Discomfort

The main findings on physiological correlates of discomfort in all three studies are summarised below. Details are reported in Sect. 2.5 and in Beggiato et al. 2019, 2018a, b, 2020.

- Overall, physiological reactions could be observed for scenarios with specific events that provoke moderate to high discomfort. Longer lasting and slowly evolving situations with lower reported discomfort only showed weak or missing changes in physiological parameters. No divergent results could be found for age groups and gender.
- Common findings in uncomfortable situations were an increase of pupil diameter, a decrease of eye blink frequency and a decrease of heart rate and heart rate variability (may related to the phenomenon "preparation for action").
- Body motion measured by motion tracking as well as seat pressure mat showed the expected push-back movement during the close approach to a truck driving ahead. However, this body movement is specific for such approach situations and cannot be observed for other uncomfortable scenarios such as overtaking an obstacle or merging into a highway.
- The use of the consumer smartband MS Band 2 was an explicit project goal in KomfoPilot to estimate the potential and challenges of such cheap and easy-to-use

devices in an applied context. Overall, there is potential for contributing to discomfort detection; however, sophisticated data filtering, checking and standardisation procedures need to be applied to maximise the signal-to-noise ratio.

- Skin conductance level only showed a remarkable steeper increase in uncomfortable situations when measured at the inner side of the wrist. If the smartband is turned by 180° and thus skin conductance level is measured at the outer side of the wrist, only weak effects are present. However, the same issue occurs for heart rate (sensor at the opposite side of skin conductance level sensor): If heart rate is measured at the outer side of the wrist, no situation-specific heart rate decrease is observable (Driving Simulator Study 2).
- First attempts in facial expression analysis for the close approach situation showed a rise of inner brows as well as upper lids, indicating surprise. Lips were stretched and pressed as sign for tension. In line with the results from the eye tracking glasses, eyes were kept open and eye blinks were reduced in the approach situation.
- Specific situation-related physiological effects could not be identified in the test track study. However, this was mainly related to the short driving area at the closed test track which did not allow to have longer rest periods between driving manoeuvres. Longer tests on the road are required to validate the findings.
- All physiological effects were identified by aggregating data over participants in specific situations. Individual discomfort detection requires data fusion of all relevant driver parameters including vehicle kinematic and situation information. Even though the presented results including overall relevance, direction, and strength of physiological effects as well as timings and variability serve as valuable basis for developing individualised discomfort detection algorithms.

2.1.3.5 Algorithm Development for Discomfort Detection

The main procedures, approaches and findings on algorithm development are summarised below. Details are reported in Sect. 2.6.

- Based on an artificial neural network approach, a suitable structure for a multi-layer perceptron (MLP) was derived.
- MLPs were able to classify discomfort from physiological data. Based on the analyses described in Sect. 2.5, the input feature vector included heart rate, pupil diameter and interblink interval time to distinguish between no discomfort = class 0 versus discomfort = class 1.
- Input raw values needed to be standardised by z-score calculation for assessing individual changes. A ten-second running window was chosen for real-time recognition.
- The MLP approach and structure were validated with regard to the indicated discomfort of the participants in different automated driving situations acquired in both driving simulator studies (standardised truck approaches, intersection and traffic light approach). Estimation results reproduced statistical tendencies matching the participants' feedback in questionnaires as well as the shape of handset control values.

- The output of the developed MLP networks can be interpreted as relative rating with the ability to estimate given situations with regard to peaks of discomfort. In addition, two arbitrary situations could be compared to estimate which one provoked more discomfort to the driver.
- The proposed approach is a step towards a quantitative operationalisation of discomfort with relation to automated driving contextual situations and driving styles. Further research should focus on deeper reflection of individual variations in discomfort perception by acquiring more data from the identical person in specific situations.

2.1.3.6 Legal Aspects of Camera-Supported Driving Comfort Recording

The main issues, processes and solutions concerning legal and privacy aspects of KomfoPilot are summarised below. Details are reported in Sect. 2.7.

- For real-time comfort measurement during automated driving, a great deal of information about the driver and his immediate environment is collected, evaluated, processed and documented. Part of this information is personal data, so that the data protection law and its observance is of great importance.
- In this respect, it was necessary to examine where conflicts with data protection regulations could arise in the context of data collection and the specification of requirements and how these conflicts could be resolved in a targeted manner (e.g. by means of consent, anonymisation, pseudonymisation).
- The aim was to ensure that the rights of data subjects to information, correction and deletion provided in data protection law were already incorporated into the requirements specification for sensor technology. This also applies to all other requirements of data protection law, e.g. with regard to data avoidance and economy, purpose limitation and transparency, as well as the various concretisations of the right to informational self-determination which have been made by case law, such as the right to not know, the prohibition of creating personality profiles, the prohibition of data retention and the prohibition of all-round monitoring.
- A comprehensive data protection evaluation and the resulting recommendations for action ensured that the applicable data protection regulations were consistently taken into account in data acquisition and the specification of sensor technology requirements in accordance with the principles "Privacy by Design" and "Privacy by Default".
- The evaluation of the test results also included an analysis and quantification of liability risks. After all, a high level of legal certainty is regularly demanded, particularly with regard to the introduction of such innovations. For this reason, the (product) liability risks with regard to the implementation of such a measure in automated vehicles were described and quantified in advance. In particular, the scope of the road safety obligations—first and foremost the duty to instruct—was addressed. In this context, it was clarified, among other things, to what extent

specific attention had to be paid to groups of potential users or how the group of users could be enabled to use the systems safely in road traffic.

2.2 Methodology of the KomfoPilot Project

2.2.1 Study Overview

The research aims of KomfoPilot were examined within the framework of three user studies. Two of these studies were carried out in a driving simulator, where automated driving could be implemented in a safe manner for a wide variety of driving situations and manoeuvres. The first driving simulator study was aimed at the evaluation of different automated driving styles. In the second driving simulator study, effects of an in-vehicle HMI informing about the automated driving behaviour were investigated. The test track study was conducted in order to validate the results of the first driving simulator study in a more realistic driving environment. Aim was the evaluation of automated driving styles for a selected set of driving situations that could be implemented safely in an automated vehicle on a test track. Beyond these user evaluations, data of all studies were used to set up the sensor database for the identification of discomfort indicators and the development of discomfort detection algorithms.

Since KomfoPilot additionally took into account different user groups, the driving simulator studies were conducted with drivers of two age groups: A group of *younger drivers*, defined as persons aged between 20 and 45 years, and a group of *older drivers*, which included persons aged 60 years or older. When possible, all user studies were based on similar methods in order to allow for cross-study comparisons. Methods that were applied to all studies are described in Sect. 2.2.2. Specifics of each study are presented in the subsequent Sects. 2.2.3–2.2.5.

2.2.2 Cross-Study Assessment Methods

Dependent variables were specified and assessed in a comparable manner in all user studies. Strategies for the enhancement of driving comfort in automated driving were evaluated in terms of their effects on driving experience and user attitudes towards automated driving. These variables were assessed at trip level using the questionnaires described in Sect. 2.2.2.1. In addition, a method for a continuous assessment of discomfort during driving, which is presented in Sect. 2.2.2.2, was applied to all studies. This approach served as a basis for analyses of driving experience at trip level and at driving situation level, and as ground truth for the development of discomfort detection algorithms. These algorithms were based on sensor data, which were recorded in all studies using a comparable sensor set-up (outlined in Sect. 2.2.2.3).

Table 2.1 Item wording used for the assessment of driving experience

Aspect of driving experience	Wording of agreement scale item
Perceived safety	During the drive, I was sure that the vehicle was able to handle traffic situations safely at any time
Driving enjoyment	Overall, the drive felt highly enjoyable to me

2.2.2.1 Questionnaires on Driving Experience and User Attitudes

In all user studies, we applied German-language questionnaires to evaluate aspects of driving experience (perceived safety, driving enjoyment) and user attitudes towards automated driving (acceptance, trust) after each drive. Driving experience was assessed using single-item measurement: Participants rated their *perceived safety* and *driving enjoyment* regarding a drive on a continuous agreement scale ranging from 0 (totally disagree) to 100 (totally agree) each. The wording of both items is presented in Table 2.1.

User attitudes towards automated driving were assessed using standardised questionnaires. To evaluate the *acceptance of automated driving* in terms of attitudes, we applied the procedure presented by Van Der Laan et al. (1997). This questionnaire allows evaluating users' satisfaction with a system as well as its perceived usefulness with a total of nine 5-point rating-scale items. For the assessment of *trust in automation*, we applied the unidimensional scale by Jian et al. (2000). It consists of twelve 7-point-agreement scale items, which can be averaged into one score indicating trust in an automated system.

2.2.2.2 Continuous Assessment of Discomfort During Driving

Representing the central aspect of driving experience addressed by KomfoPilot, perceived *discomfort* was evaluated on a deeper level based on continuous, nonverbal self-reporting via manual input device during each automated drive (for Details, see Hartwich et al. 2018). Therefore, all driving environments were equipped with a professional handset control (ACD pro 10, see Fig. 2.1a). Participants were instructed

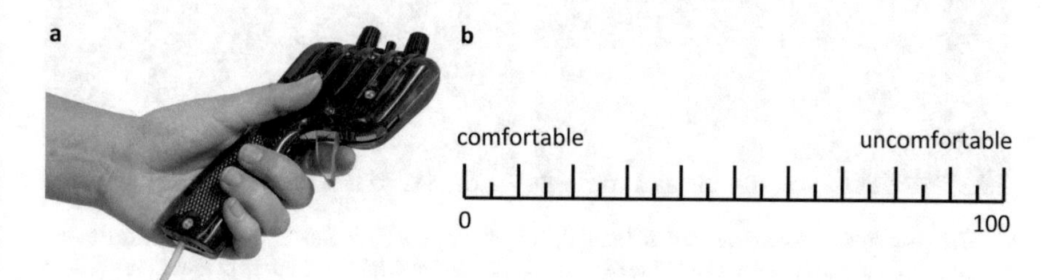

Fig. 2.1 Handset control (**a**) and corresponding response scale (**b**) used for the continuous assessment of perceived discomfort during automated driving

to press the lever of the handset control gradually in accordance with the extent of their currently perceived discomfort throughout an entire drive. Stronger pressing of the lever corresponded to stronger discomfort. The handset control signal was recorded and transformed into a continuous scale of 0 (lever not pressed = comfortable) to 100 (lever fully pressed = uncomfortable) (see Fig. 2.1b).

2.2.2.3 Sensor Set-Up for Real-Time Detection of Discomfort During Driving

In order to develop algorithms for the real-time detection of user discomfort during automated driving, it was necessary to identify discomfort indicators that are measurable and processible in real-time while driving. Potential indicators included characteristics of the driving environment (e.g. distance to other road users), the automated vehicle (e.g. current speed) and the user (e.g. heart rate). Therefore, the test vehicles and participants of the user studies were equipped with a variety of sensors for the continuous assessment of potential discomfort indicators.

In the driving simulator studies, information on the driving environment and the vehicle was provided by the simulation software SILAB 5.1. In the test track study, such information was derived from a data logging system integrated in the test vehicle. In addition, different sensors were set up to capture user characteristics: *Physiological parameters* (heart rate, heart rate variability, skin conductance level) were assessed using the smart band Microsoft Band 2 (Fig. 2.2a). The smart band

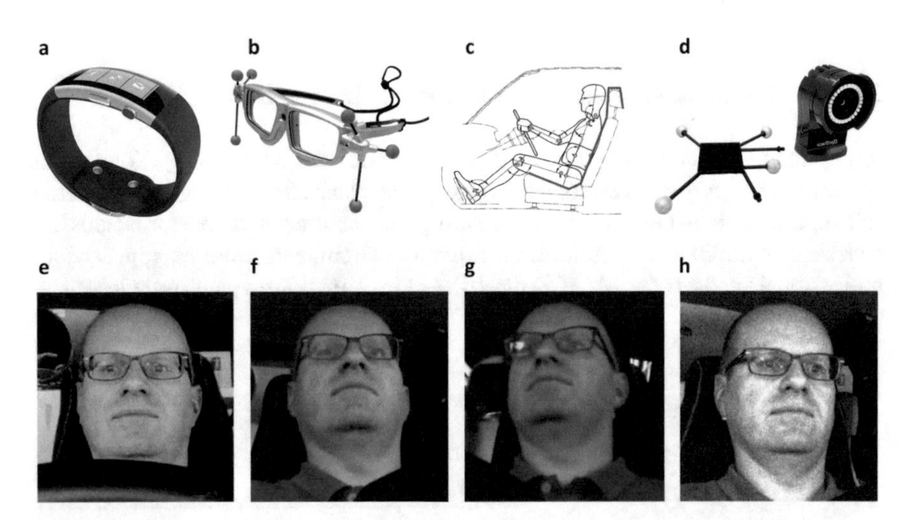

Fig. 2.2 Sensors (**a** Microsoft Smart Band 2, **b** SMI eye tracking glasses 2, **c** seat pressure mat, **d** rigid body and infrared camera of the OptiTrack motion tracker) and camera perspectives (**e** Intel RealSense SR300, **f** GoPro Hero 5 right side, **g** GoPro Hero 5 left side, **h** AVT Mako G-234B) used for the assessment of discomfort indicators. Written informed consent was obtained from the individual for the publication of this image

Fig. 2.3 Exemplary visualisation of synchronised video data (left side and upper right side) and sensor data (lower right side: handset control data and driving data). Written informed consent was obtained from the individual for the publication of this image

also recorded accelerometer and gyroscope data, which were used to identify and correct for hand movements. For the assessment of *eye tracking data* (pupil diameter, eye blinks), participants not wearing eyeglasses during driving were equipped with the SMI Eye Tracking Glasses 2 (Fig. 2.2b). A seat pressure mat (Fig. 2.2c) was attached to the driver's seat of the test vehicles in order to capture *body motion.* The mat, which was developed by the project partner FusionSystems GmbH, included eight pressure sensors at different positions (Fig. 2.34). Body motion was additionally assessed in driving simulator study 1 using the marker-based motion tracking system from OptiTrack. The system consisted of four Flex 13 infrared cameras installed in the vehicle cabin and four rigid bodies attached to the participants' left and right hand, right shoulder and head (Figs. 2.2d and 2.3). The rigid bodies served as markers for tracking the position and orientation of these parts of the participants' bodies. *Facial expressions* were analysed using video recordings by four cameras (one Intel RealSense SR300, two GoPros Hero 5, one AVT Mako G-234B), which were installed on the vehicle dashboard and directed at the participants' faces from different angles (Fig. 2.2e–h).

2.2.2.4 Data Recording and Synchronisation

Data captured by each sensor were recorded by independent data loggers, with an individual recording frequency each. Driving data of the driving simulator, which included handset control data, were logged by the SILAB software with a frequency of 60 Hz. In the test track vehicle, the integrated data logging system stored driving data with a frequency of 30 Hz. MS Band 2 data were sent to a self-developed logging application via Bluetooth connection. The application was programmed based on the Software Development Kit coming with the MS Band 2 and was used to log physiological data with a frequency of 10 Hz. Recording frequencies were 60 Hz for the SMI ETG 2 eye tracking data, 10 Hz for the data captured by the seat pressure mat and 120 Hz for the OptiTrack motion tracking data. System time for all data logging devices was continuously synchronised with a software tool based on the network time protocol (Meinberg NTP Software).

Participants' facial expressions were recorded in colour by the Intel RealSense SR300 camera with a resolution of 1280 × 720 pixel and 30 frames per second (fps) as well as by both GoPro Hero 5 video cameras with a resolution of 1920 × 1080 pixel with 50 fps. The AVT Mako G-234B camera provided greyscale recordings of facial expressions with a resolution of 640 × 480 pixel and 30 fps.

All sensor data were inspected, synchronised and processed in a data storage and analysis framework based on the relational open-source database management system PostgreSQL (for details, see Beggiato 2015). For data synchronisation, values of all sensor data streams were added to the corresponding timestamps of the driving simulator data (for the driving simulator studies) or the vehicle logging system (for the test track study), respectively. Video files were stored on a separate fileserver and referenced in the database. Synchronised video data and sensor data are exemplarily visualised in Fig. 2.3.

2.2.3 Driving Simulator Study 1—Driving Style Evaluation

In the first driving simulator study, 40 participants of two age groups experienced several differently parameterised automated drives in a driving simulator. The data gathered in this study were used to develop algorithms for the real-time detection of discomfort during driving and to identify factors influencing the automated driving experience. Special emphasise was put on the role of the automated driving style, including the question whether users prefer to experience an automated driving style that corresponds to their individual manual driving styles.

2.2.3.1 Research Objectives

In detail, we conducted the study to pursue the following research objectives:

- Identify driving situations and manoeuvres that are perceived as uncomfortable in automated driving. Corresponding results are presented in Sect. 2.4.
- Examine the effects of different automated driving styles on users' driving experience and evaluation of automated driving. While Sect. 2.3 gives an overview of the results at trip level, Sect. 2.4 contains corresponding results at driving situation level.

- Identify parameters of the user, the vehicle and the driving environment that are suitable to indicate discomfort during automated driving in real time. Statistical analyses regarding this objective are reported in Sect. 2.5. The implementation of these analyses within the development of discomfort detection algorithms is described in Sect. 2.6.

2.2.3.2 Study Design

The study consisted of two distinct driving sessions with approximately two months delay in between. Each session included four different drives in the simulator, which were executed either manually by the participants or automated (see Fig. 2.4). In both sessions, drivers' age was included as a between-subjects factor by dividing the sample in two age groups (younger vs. older drivers). In Session 2, three different driving styles (defensive vs. dynamic vs. each driver's own) were presented to all participants. All additional drives were executed for training purposes (manual training drive), to record the participants' manual driving style as basis for the comparison of automated driving styles (manual test drive) and to gather sensor data for the development of discomfort detection algorithms (automated standardisation drives, automated test drives).

2.2.3.3 Participants

Initially, 46 participants attended the study. However, study conduct was discontinued for six persons due to minor simulator sickness symptoms reported at the beginning of Session 1. Consequently, the final sample consisted of 40 participants (15 females, 25 males), of which 21 belonged to the younger driver group (25–39 years) and 19 belonged to the older driver group (65–84 years). Demographics of both age groups are presented in Table 2.2. All participants held a valid driver's licence, but had no prior experience with highly automated driving. They gave written informed consent prior to study conduct and received a monetary reward for their participation.

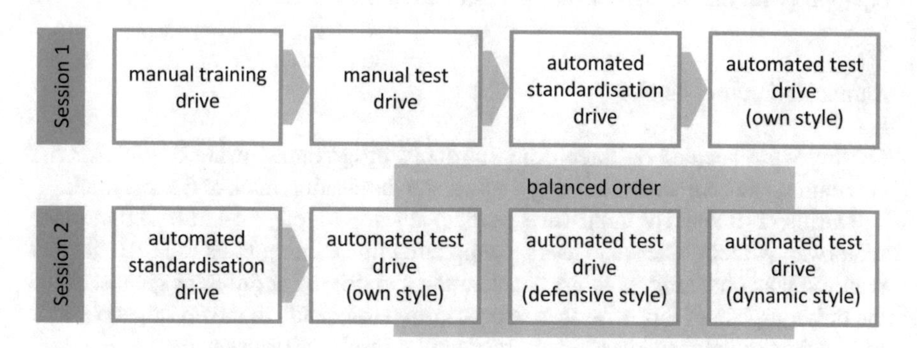

Fig. 2.4 Sequence of drives experienced by each participant of driving simulator study 1

Table 2.2 Demographics of the younger and older driver group of driving simulator study 1

Characteristic	Younger drivers ($N = 21$)	Older drivers ($N = 19$)
Gender	9 females, 12 males	6 females, 13 males
Age in years	$M = 29.9$, $SD = 4.4$	$M = 71.6$, $SD = 6.0$
Number of years obtaining a driver's licence	$M = 12.5$, $SD = 4.6$	$M = 48.6$, $SD = 14.1$

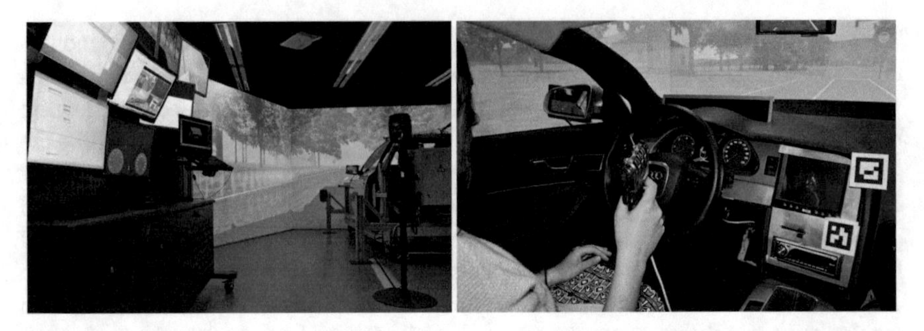

Fig. 2.5 Exterior (left) and interior (right, including handset control) of the driving simulator

2.2.3.4 Driving Environment and Apparatus

Driving Simulator

All drives took place in a fixed-base driving simulator with a fully equipped interior and a 180° horizontal field of view, which included a rear-view mirror and two side mirrors (see Fig. 2.5). For all drives, participants sat in the driver's seat. Driving was performed either manually by themselves or automated. For automated driving, driving data of pre-recorded drives were replayed in the simulator, pretending that all aspects of the driving task were executed by an automated vehicle. Pedals and steering wheel were inoperative during these drives. For this study, the driving simulator was equipped with the full sensor set-up described in Sect. 2.2.2.3.

Simulated Driving Tracks

Driving was executed on three different tracks programmed using the SILAB 5.1 simulation environment: a training track, a standardisation track and a test track.

During the manual training drive, participants drove along a 4-km-long three-lane motorway without other road users. Completing this training track took 10 min and required lane changing, braking, accelerating and driving at different speeds. Since the only purpose of this drive was to accustom participants to the driving simulator and avoid simulator sickness, no corresponding results are presented.

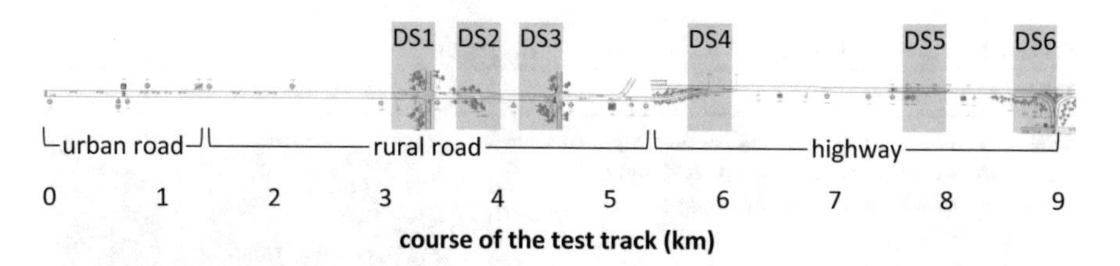

Fig. 2.6 Overview of the test track including six discomfort situations (DS; DS1 = non-signalised intersection, DS2 = parking truck blocking the lane, DS3 = red traffic light, DS4 = entering highway, DS5 = construction zone with slow vehicle ahead, DS6 = exiting highway)

The two automated standardisation drives, which took 3 min each, were conducted to accustom participants to the handset control (see Sect. 2.2.2.2) and to gather sensor data in a highly standardised driving situation. The standardisation track consisted of a two-lane rural road and included three identical, potentially critical and discomfort-inducing driving situations. In these situations, the own vehicle drove automated at a speed of 100 km/h and approached a truck driving ahead at a constant speed of 80 km/h (see Fig. 2.3). Automated breaking was initialised comparably late at a distance of 9 m between the own vehicle and the truck ahead. Before the distance was increased by reducing the speed of the own vehicle, a minimum distance of 4.2 m was reached, which corresponded to a minimum time to contact (TTC) of 1.1 s.

All test drives were performed manually or automated along the 9-km-long test track. This track consisted of three sections (see Fig. 2.6): urban road (speed limit: 50 km/h), rural road (speed limit: 100 km/h) and highway (speed limit varying between 80 km/h in a construction zone and no speed limit as common on German Autobahn). All sections included a variety of driving situations, among them six potentially discomfort-inducing situations, which are marked in Fig. 2.6 and described in Table 2.3. Along the whole track, turning manoeuvres were reduced to a minimum in order to avoid simulator sickness.

Automated Driving Styles

For automated driving along the test track, three different driving styles were implemented: a "defensive" style, a "dynamic" style and each participant's own style. The defensive driving style was characterised by cautious driving with a strong emphasis on traffic safety, implemented by lower speed, smoother acceleration and earlier deceleration. In contrast, the dynamic driving style was characterised by higher speed and sharper acceleration and deceleration, taking more advantage of the capabilities of an error-free automated driving system. The own driving style represented an exact replay of each participant's manual driving style, which was recorded during the manual test driving in Session 1. While the defensive and dynamic driving style were identically for all participants, the own driving style differed between individuals correspondingly to their manual driving styles. Speed profiles of the three driving styles are shown in Fig. 2.7.

Table 2.3 Discomfort situations (DS) included in the test track

DS	Description	Screenshot of simulation
DS1	Non-signalised intersection with right of way for the own vehicle on a rural road, vehicle approaching fast from the right	
DS2	Parking truck blocking the own lane on a rural road, provoking an overtaking manoeuvre on the opposite lane	
DS3	Approaching a red traffic light on a rural road	
DS4	Entering a highway with traffic running on it	
DS5	Approaching a slow vehicle ahead in a construction zone on a highway	
DS6	Exiting a highway by following a narrow band	

2.2.3.5 Procedure

The experimental procedure was divided into two sessions with approximately two months delay in between (see Fig. 2.4). On average, it took 90 min to complete one session. Participants received instructions and gave written informed consent at the beginning of each session. Upon completion of Session 2, they received a monetary compensation for their attendance.

Session 1 started with a questionnaire on demographic variables and a 10-min manual training drive in order to accustom participants to the driving simulator and avoid the occurrence of simulator sickness. Subsequently, participants were equipped with the sensor set-up for the assessment of discomfort indicators (see Sect. 2.2.2.3)

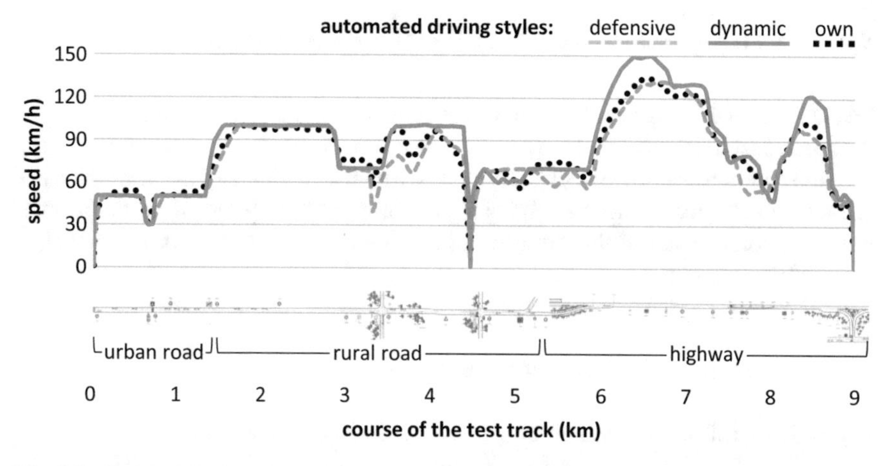

Fig. 2.7 Speed of the three presented automated driving styles over the course of the test track (own driving style represents averaged values of all 40 participants)

and performed a manual drive along the test track, which was used to record the participants' individual driving styles. Instructions for this drive demanded to drive in accordance with one's usual driving habits and to obey traffic laws. After manual driving, participants trained the usage of the handset control (see Sect. 2.2.2.2) during a 3-min automated standardisation drive. They were instructed to press the lever of the handset control in accordance with the extent of their perceived discomfort throughout the entire drive. A display in the instrument cluster provided visual feedback of the currently entered value on the corresponding response scale. Upon handset control training, participants experienced their first automated drive along the test track based on their own driving style while using the handset control to indicate discomfort (this time without visual feedback provided in the instrument cluster).

At the beginning of Session 2, participants were equipped with the sensor set-up again. They rehearsed the usage of the handset control during another automated standardisation drive with visual feedback of the entered handset control values provided in the dashboard. Subsequently, they experienced three automated drives along the test track based on different driving styles (defensive, dynamic, own; see Sect. "Automated Driving Styles") in a balanced order to avoid sequence effects. During all drives, discomfort was indicated using the handset control (without visual feedback provided in the instrument cluster). Each drive was followed by a questionnaire on driving experience and user attitudes towards highly automated driving.

In case one of the participants reported symptoms of simulator sickness, the procedure was immediately cancelled and a proportionate compensation was paid.

2.2.4 *Driving Simulator Study 2—HMI Evaluation*

While the first driving simulator study put emphasis on the automated driving style, the second driving simulator study was focussed on the effects of in-vehicle information presentation on users' driving experience. Therefore, a human–machine interface (HMI) was integrated into the driving simulator and presented to 41 participants of two age groups. Data of this study were also used for the further development of discomfort detection algorithms.

2.2.4.1 Research Objectives and Questions

In detail, the following research objectives were pursued in this study:

- Examine the effects of presenting system information via HMI during automated driving on users' driving experience and evaluation of automated driving. Corresponding results are reported in Sect. 2.3.
- Gather additional data for the identification of parameters that are suitable to indicate discomfort during automated driving in real time. See Sect. 2.5 for statistical analyses and Sect. 2.6 for the implementation of these analyses within the development of discomfort detection algorithms.

2.2.4.2 Study Design

All participants experienced three automated drives in fixed order (see Fig. 2.8). HMI-effects were evaluated by comparing two identically parameterised drives along the test track. In the second of these drives, automated driving was complemented by a display that informed participants about the driving behaviour of the automated system. The preceded automated standardisation drive was conducted for training purposes and to gather sensor data for the development of discomfort detection algorithms. Driver's age was included as a between-subjects factor by executing all drives with drivers of two age groups (younger vs. older drivers).

2.2.4.3 Participants

The study was conducted with a sample of 41 participants (17 females, 24 males), who had no prior experience of highly automated driving, but held a valid driver's

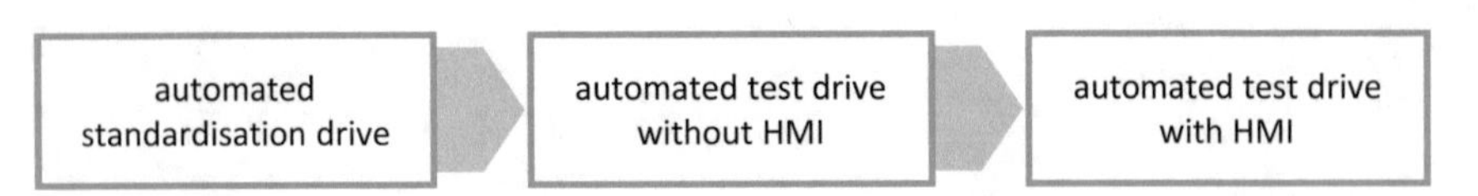

Fig. 2.8 Sequence of drives experienced by each participant of driving simulator study 2

Table 2.4 Demographics of the younger and older driver group of driving simulator study 2

Characteristic	Younger drivers ($N = 20$)	Older drivers ($N = 21$)
Gender	10 females, 10 males	7 females, 14 males
Age in years	$M = 28.1$, $SD = 0.7$	$M = 68.3$, $SD = 1.0$
Number of years obtaining a driver's licence	$M = 9.7$, $SD = 1.1$	$M = 45.6$, $SD = 2.0$

licence. The sample was divided into 20 younger drivers (24–38 years old) and 21 older drivers (61–77 years old), whose demographics are presented in Table 2.4. All participants gave written informed consent at the beginning of the study. Upon study completion, they were monetarily rewarded for their attendance. In this study, no drop-outs due to simulator sickness occurred.

2.2.4.4 Driving Environment and Apparatus

Driving Simulator and Simulated Driving Tracks

The second driving simulator study took place in the same driving simulator the first driving simulator study was conducted in (see Sect. "Driving Simulator"). Tracks for the automated standardisation drive and both automated test drives (with and without HMI) were also adopted from the first study (see Sect. "Simulated Driving Tracks"). Again, participants sat in the driver's seat for all drives. In this study, manual driving was not required, since all the drives were carried out automated. For both test drives, the dynamic driving style was implemented (see Sect. "Automated Driving Styles"). Except for the motion tracking system, the complete sensor set-up described in Sect. 2.2.2.3 was integrated in the driving simulator for this study.

Human–Machine Interface (HMI)

The HMI consisted of a display which was added to the instrument cluster of the driving simulator in the second automated test drive. Aim of the display was to increase system transparency and thereby help users to understand and predict the behaviour of the automated vehicle. The HMI was activated when approaching each of the six discomfort situations appearing along the test track (see Sect. "Automated Driving Styles") to inform users about the status of the automated system (referred to as AP = autopilot), traffic elements detected by the vehicle, the planned driving manoeuvre for this situation and the time remaining to the start of this manoeuvre (as countdown). One exemplary display for each discomfort situation is presented in Fig. 2.9.

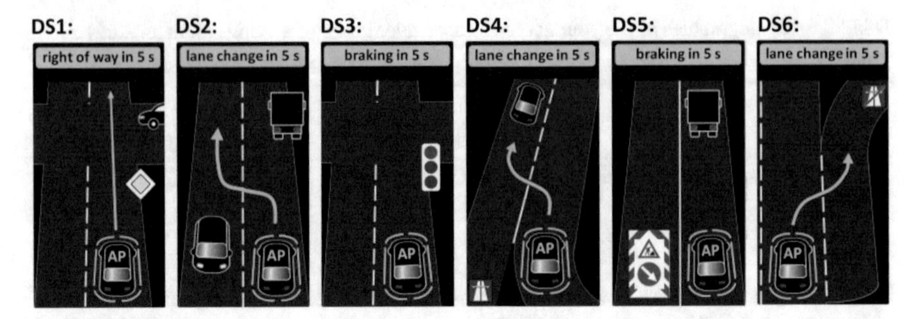

Fig. 2.9 HMI displayed in the instrument cluster during the six discomfort situations (DS) of the text track (DS1 = non-signalised intersection, DS2 = parking truck blocking the lane, DS3 = red traffic light, DS4 = entering highway, DS5 = construction zone with slow vehicle ahead, DS6 = exiting highway) (AP = auto pilot)

2.2.4.5 Procedure

At the beginning of the study, all participants received an introduction, gave written informed consent and answered a questionnaire on demographic variables. Participants were then equipped with the eye tracker and the smart band (see Sect. 2.2.2.3) and introduced to the handset control (see Sect. 2.2.2.2). The usage of this tool was trained in the following automated standardisation drive (see Sect. "Simulated Driving Tracks"), during which participants were instructed to press the lever of the handset control in accordance with the extent of their perceived discomfort. A display in the instrument cluster provided visual feedback of the currently entered value on the corresponding response scale. This handset control training was followed by two automated drives along the identical test track (see Sect. "Simulated Driving Tracks"). During both drives, participants gave continuous feedback on their level of discomfort by using the handset control (this time without visual feedback provided in the instrument cluster). Before the second of these drives, an HMI informing about the automated driving behaviour in complex situation (see Sect. "Human–Machine Interface (HMI)") was explained to the participants. It was subsequently presented in the instrument cluster during driving. Both automated test drives were followed by questionnaires on driving experience and user attitudes towards highly automated driving. On average, it took participants 90 min to complete this study. They received monetary compensation for their attendance at the end of the procedure.

2.2.5 Test Track Study—Driving Style Evaluation

Aim of the test track study was to examine the transferability of selected results from the first driving simulator study to real-world conditions. The similarity lies in the selected driving manoeuvres, which had already been tested in the driving simulator (see Sect. "Simulated Driving Tracks"). In addition, the method of using two differently parameterised driving styles as well as the assessment of discomfort during driving based on the handset control (see Sect. 2.2.2.2) were adopted.

2.2.5.1 Research Objectives

The main research objective was to validate the findings from the first driving simulator study in a real vehicle. Results from this study indicated that the parametrisation of an automated driving style evokes different subjective ratings regarding discomfort (see Sect. 2.3.2). This was especially apparent for longitudinal manoeuvres in which the automated vehicle accelerated or decelerated (see Sect. 2.4). Typical situations of this type in the driving simulator study were approaching a red traffic light or a sudden change in speed at a construction zone.

As these results were obtained using a fixed-base driving simulator with missing inertia forces, the research questions arose as to what extent the findings from the driving simulator can be transferred into the real world. Literature indicates that especially for longitudinal manoeuvres, results should not differ between the experimental environments (Jentsch 2014).

2.2.5.2 Study Design

The study was designed as a three-stage experiment in order to investigate the subjective perception of discomfort in highly automated driving on a standardised test track. During the experiment, the participants experienced three automated drives: (1) a baseline ride to experience the system, (2) a "dynamic" and (3) a "defensive" ride. The participants were randomly assigned to a condition defining the order of the presented drives. Including the baseline, this resulted in the following pattern of presented driving styles: Participant n: dynamic-defensive-dynamic, Participant n + 1: defensive-dynamic-defensive, etc.

2.2.5.3 Participants

The sample included 30 subjects (12 females, 18 males). The age ranged between 21 and 69 years ($M = 41.1$, $SD = 15.4$), without a separation of the sample into two distinct age groups. All participants had a valid driver's licence and covered an annual mileage of 14,950 km ($SD = 12{,}161$ km). The length of time a driver's licence was held was $M = 22.3$ years ($SD = 13.6$). After study completion, the participants received a monetary compensation for their attendance.

Fig. 2.10 Test track layout (left) and test vehicle (right) of the test track study

2.2.5.4 Driving Environment and Apparatus

Test Track and Driving Manoeuvres

The test track (Fig. 2.10 left) was located on a parking lot with the approximate size of 220 m by 70 m. The length of the course was ~650 m, and the speed limit was 30 km/h. The course contained two manoeuvres adopted from the simulator study:

(1) Traffic light approach with a crossing pedestrian: A light barrier 50 m in front of the traffic lights triggered a change of the traffic lights from green to red. A pedestrian started walking as soon as the traffic light switched to yellow and crossed the road while the vehicle continued to approach.
(2) Speed reduction when approaching a construction site: A speed limit sign at the beginning of the construction site demanded a speed reduction from 30 km/h to 10 km/h. The width of the road was reduced by one metre.

Test Vehicle and Automated Driving Styles

The test vehicle was a partially automated BMW i3 (see Fig. 2.10, right). The lateral control was implemented by the "Wizard of Oz" method, with a technical supervisor steering the vehicle in curves by using an adjusting wheel between the passenger seat and the door. The longitudinal guidance worked automated. Two driving styles, which differed regarding the parameters for acceleration and deceleration, were implemented:

(1) *Dynamic driving style* (Fig. 2.11): When approaching the traffic light, the dynamic profile initiated a deceleration 20 m before the traffic light with an average deceleration of about 0.7 m/s^2. After the traffic light had switched back to green, the vehicle accelerated over a course of approximately 25 m with an average acceleration of 0.28 m/s^2.
(2) *Defensive driving style* (Fig. 2.12): In this profile, the braking process was initiated earlier, already 40 m before the traffic light, with an average deceleration of 0.2 m/s^2. Shortly before the traffic light, braking intensity was increased to

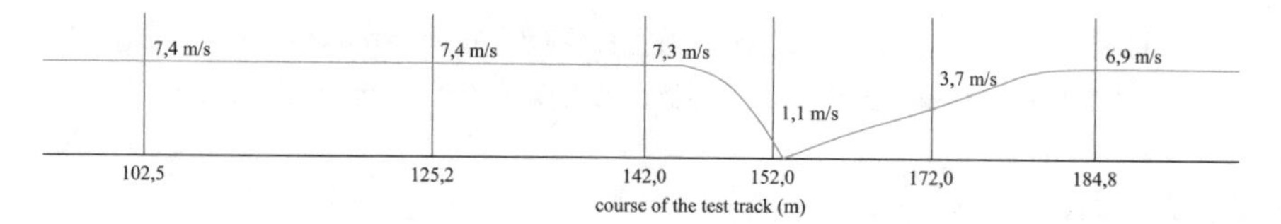

Fig. 2.11 Parameterisation of the dynamic driving style when approaching a red traffic light

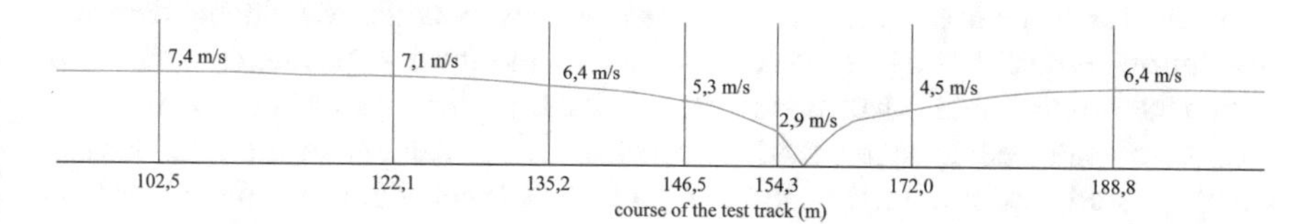

Fig. 2.12 Parameterisation of the defensive driving style when approaching a red traffic light

0.7 m/s^2. After the red traffic light phase, starting up took place with an average acceleration of 0.25 m/s^2.

Both driving styles used the same dynamics for the speed reduction manoeuvre in front of the construction site.

2.2.5.5 Procedure

Before starting the experiment, the procedure was explained to the participants. It was emphasised that this is a study on highly automated driving by stating that "The vehicle takes over all driving tasks by itself and the driver is driven and does not have to steer by himself" (excerpt from the instructions). Additionally, each subject was informed about the data processing, the voluntary of participation, the duration and the monetary compensation, which was handed over at the end of the procedure. Participants were asked to sign an informed consent in order to agree to the collection of personal data and the further processing of anonymously assigned data. Subsequently, they answered the first questionnaire assessing demographics.

After the automated baseline drive without handset control, the subjects answered the first questionnaires on their driving experience and their attitudes towards automated driving (see Sect. 2.2.2.1. for a detailed description of applied questionnaires). This procedure was repeated for both automated drives (dynamic and defensive driving style). During these two drives, participants continuously indicated discomfort via handset control (see Sect. 2.2.2.2). At the end of the study, all participants received a monetary compensation and were informed that a technical supervisor in the vehicle carried out the lateral control.

2.3 Enhancing Comfort in Automated Driving—User Evaluation of Different Strategies on a General Level

2.3.1 Overview

This section is focussed on characteristics of automated vehicles that can be adjusted in order to enhance users' driving experience, e.g. by avoiding or reducing detected discomfort during driving. Therefore, results of the three user studies conducted within the KomfoPilot Project are reported. In these studies, two enhancement strategies were evaluated from a users' perspective: while subject of the first driving simulator study (Sect. 2.3.2) and the test track study (Sect. 2.3.4) was the automated driving style, information presentation via HMI was topic of the second driving simulator study (Sect. 2.3.3). For all three studies, effects of the respective enhancement strategy on users' driving experience (discomfort, perceived safety) and evaluation of automated driving (acceptance, trust) are summarised.

2.3.2 Driving Simulator Study 1—Driving Style Evaluation

In the first driving simulator study, we examined user preferences regarding the driving style of automated vehicles. Therefore, 40 participants experienced three different automated driving styles (defensive, dynamic, each participants' own driving style) along an identical test track in a fixed-base driving simulator. Own driving styles were included in order to investigate whether users prefer to experience an automated driving style that corresponds to their individual manual driving styles (for details, see Hartwich et al. 2018). For methodological details of the study, please consult Sect. 2.2.3. Data presented in this section are based on the three automated test drives executed in Session 2 of the study.

2.3.2.1 Data Analysis

Effects of driving style on driving experience (discomfort, perceived safety, driving enjoyment) and user attitudes (acceptance, trust in automation) were analysed using mixed design ANOVA. Discomfort was estimated by summing up the value of all handset control responses given by each participant during each drive. Perceived safety and driving enjoyment were derived from the single items applied after each drive. Trust in automation was calculated in accordance with Jian et al. (2000) by averaging all items of the questionnaire. Internal consistency reliability was high, with Cronbach's α ranging between 0.91 and 0.92. In contrast to the two subscales of acceptance postulated by Laan et al. (1997), an analysis of the questionnaire's factorial structure did not confirm these two factors for the sample of this study. Accordingly, all nine items were averaged to one score indicating the participants' overall acceptance of automated driving. This dimension exhibited a high internal consistency reliability, with Cronbach's α values ranging between 0.93 and 0.95. Necessary parametric assumptions (normal distribution, sphericity, homogeneity of variances) were inspected for all dependent variables. In case an assumption could not be verified, additional nonparametric tests confirmed ANOVA results.

2.3.2.2 Results

Effects of Automated Driving Style on Driving Experience

Figure 2.13 displays the discomfort indicated by the handset control during each drive, summed up for all drivers of each age group.

The progression of these data over the course of the test track illustrates that perceived discomfort strongly depended on the driving situation, peaking in complex situations such as intersections. A detailed analysis at driving situation level is presented in Sect. 2.4. At trip level, additional effects of the automated driving style became apparent, $F(2, 54) = 3.63$, $p = 0.033$, $\eta_p^2 = 0.12$. According to pairwise comparisons, the dynamic driving style was perceived as significantly more uncomfortable than the defensive driving style ($p = 0.028$). However, a significant interaction between driving style and age group, $F(2, 54) = 3.69$, $p = 0.031$, $\eta_p^2 = 0.12$,

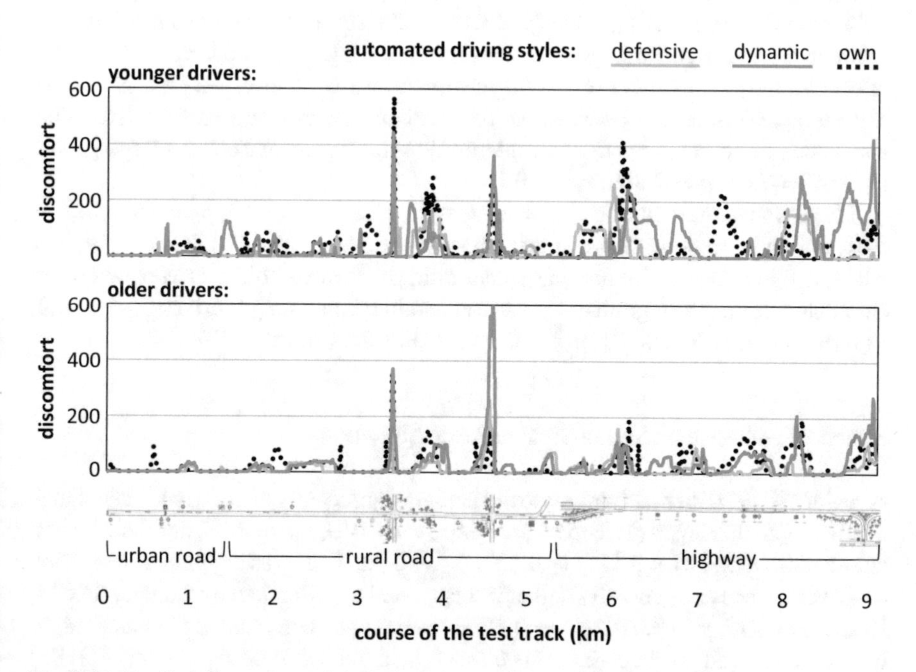

Fig. 2.13 Totalled discomfort values indicated by younger and older drivers via handset control during three automated drives based on different driving styles

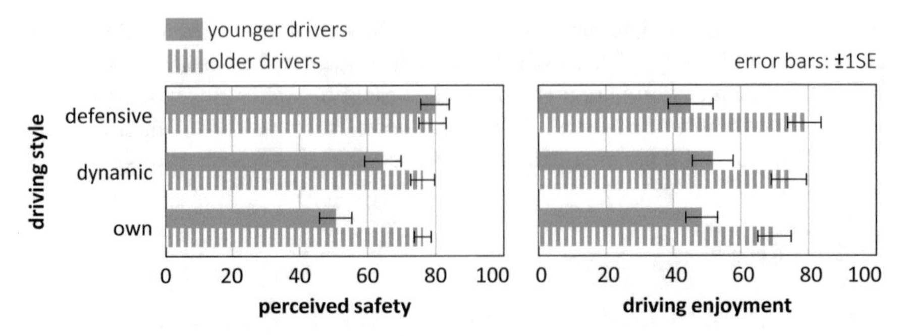

Fig. 2.14 Perceived safety and driving enjoyment rated after automated driving with different driving styles, separated by age group

clarifies that this effect was more pronounced for older than for younger drivers. While the dynamic driving style provoked the highest discomfort for older drivers, younger drivers perceived their own driving style as the most uncomfortable one. Both age groups indicated the defensive driving style to be the most comfortable one. There was no overall effect of driver's age on discomfort, $F(1, 27) = 0.46$, $p = 0.505$, $\eta_p^2 = 0.02$.

Perceived safety was also significantly affected by driving style, $F(2, 68) = 9.38$, $p < 0.001$, $\eta_p^2 = 0.22$, with the defensive driving style being perceived as significantly safer than the own driving style ($p < 0.001$). Based on the significant interaction between driving style and age group, $F(2, 68) = 6.23$, $p = 0.003$, $\eta_p^2 = 0.16$, this effect can only be assumed for younger drivers, while older drivers did not report considerable differences between the three driving styles (see Fig. 2.14 left). This also needs to be considered when interpreting the significant effect of age group, $F(1, 34) = 7.65$, $p = 0.009$, $\eta_p^2 = 0.18$.

There was neither an effect of driving style, $F(2, 72) = 0.53$, $p = 0.591$, $\eta_p^2 = 0.01$, nor an interaction between driving style and age group, $F(2, 72) = 1.54$, $p = 0.221$, $\eta_p^2 = 0.04$, on driving enjoyment ratings. However, older drivers perceived automated driving as significantly more enjoyable than younger drivers, in general, $F(1, 36) = 15.64$, $p < 0.001$, $\eta_p^2 = 0.30$ (see Fig. 2.14 right).

Effects of Automated Driving Style on User Attitudes

As evident in Fig. 2.15, driving style did neither significantly affect users' acceptance of automated driving, $F(1.56, 51.48) = 3.03$, $p = 0.069$, $\eta_p^2 = 0.08$, nor their trust in automation, $F(2, 68) = 2.11$, $p = 0.129$, $\eta_p^2 = 0.06$. Trust was significantly affected by drivers' age, with generally higher ratings stated by older than by younger drivers, $F(1, 34) = 6.15$, $p = 0.018$, $\eta_p^2 = 0.15$. Even though a comparable tendency might be visible regarding acceptance, this effect turned out not to be significant, $F(1, 33) = 2.27$, $p = 0.141$, $\eta_p^2 = 0.06$. There were also no significant interactions between

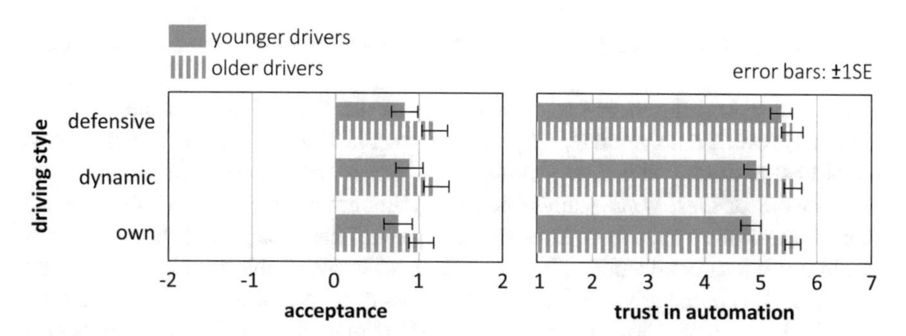

Fig. 2.15 User acceptance and trust in automation rated after automated driving with different driving styles, separated by age group

driving style and age group regarding acceptance, $F(1.56, 51.48) = 0.20, p = 0.766,$ $\eta_p{}^2 = 0.01$, or trust, $F(2, 68) = 2.39, p = 0.100, \eta_p{}^2 = 0.07$.

2.3.2.3 Discussion

Summing up these results, first-time users' driving experience during automated driving in the simulator was affected by driver's age as well as by automated driving style. However, results were not entirely consistent. Both age groups preferred the defensive driving style, which provoked the least amount of discomfort, but the highest perceived safety. In contrast, no differences between the three presented driving styles were stated with regard to driving enjoyment. This variable was only affected by age group, with generally low ratings made by younger drivers and considerably higher ratings by older drivers. This is consistent with previous studies indicating that, in contrast to older drivers, younger drivers do not assess automated driving without secondary activities as joyful, likely based on boredom due to giving up the driving task (Hartwich et al. 2018). This lack of enjoyment does not appear to be compensable by means of the automated driving style. The approach to enhance driving experience by implementing automated driving styles that correspond to each user's manual driving style was not supported by the results of this study. The own driving style was not preferred regarding any of the examined variables and even provoked the most discomfort for younger drivers. However, this effect might be attributable to the design of the study. In contrast to the defensive and the dynamic driving styles, which were pre-recorded and parameterised identical for all drivers, the own driving styles represented replays of the participants' manual drives along the test track and therefore might have been confounded by the quality of the individual manual driving performances. Comparable studies without these differences between the presented driving styles support the approach of designing automated driving styles under consideration of users' manual driving styles at least for younger drivers (Hartwich et al. 2018). Overall, a defensive driving style, characterised by lower speed, smoother acceleration and earlier deceleration, appeared to be most suitable to provide a pleasant driving experience for potential future users of automated vehicles.

Driving style differences regarding driving experience did not translate into user attitudes towards automated driving. Neither trust in automation nor the acceptance of automated driving varied depending on the previously experienced driving style. One reason for these results might be the driving environment. Given the absence of physical motion cues as well as the participants' knowledge that system failures would not have actual safety effects, differences between the driving styles might have been less perceptible in the driving simulator than in a more realistic setting. For this reason, real-world data were collected in the subsequent test track study (see Sect. 2.3.4 for corresponding results). Beyond this methodological limitation, these results imply the existence of individually different driving style preferences based on other user characteristics than age. As potentially relevant characteristics, personality traits (Beggiato et al. 2017) and a priori trust in automation (Beggiato et al. 2015) have already been identified. Consequently, automated driving style preferences might not be addressable by a one-for-all approach for the majority of potential users. Instead, driving experience could by enhanced by the individual adaptability of automated driving styles, for instance based on manual user input or the automated detection of discomfort during driving (see Sect. 2.6).

2.3.3 Driving Simulator Study 2—HMI Evaluation

In the second driving simulator study, we examined whether driving experience and user attitudes towards automated driving can be enhanced by the display-based presentation of system information during driving. Therefore, a new sample of 41 participants experienced two identical automated drives based on a dynamic driving style along an identical test track. During one of these drives, system information was presented in an HMI integrated in the instrument cluster of the driving simulator in order to increase system transparency. In this section, user evaluations of both drives are compared. For methodological details of the study, please consult Sect. 2.2.4.

2.3.3.1 Data Analysis

Effects of an HMI presenting system information during driving on users' driving experience (discomfort, perceived safety, driving enjoyment) and attitudes (acceptance, trust in automation) were analysed using mixed-design ANOVA. Data preparation followed the procedure of the first driving simulator study (see Sect. 2.3.2.1). Internal consistency reliability was high for the questionnaire assessing trust in automation (Jian et al. 2000), Cronbach's $\alpha = 0.91$–0.92, and the overall score of the acceptance questionnaire (Laan et al. 1997), Cronbach's $\alpha = 0.86$–0.91. In case parametric assumptions (normal distribution, homogeneity of variances) could not be verified, additional nonparametric tests confirmed ANOVA results.

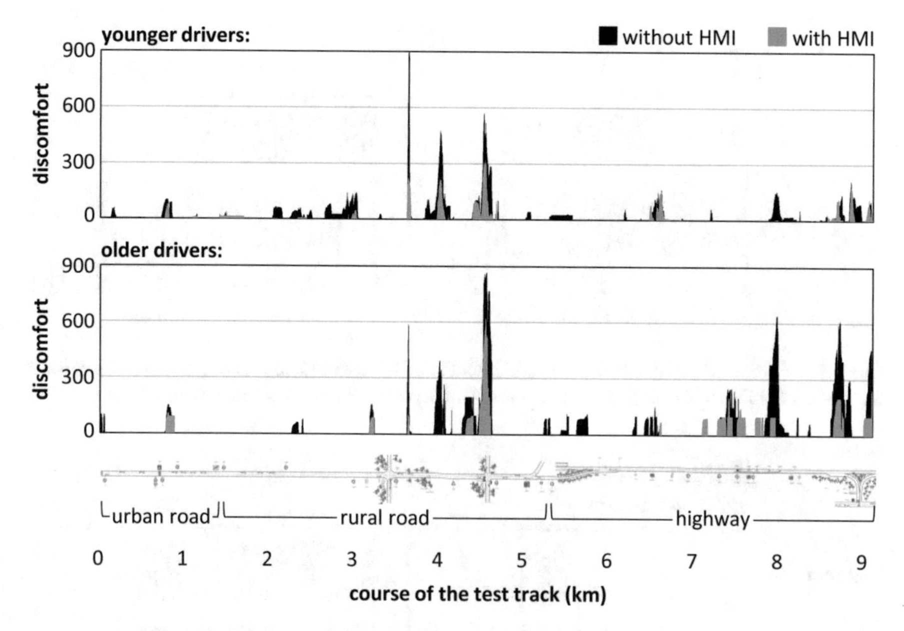

Fig. 2.16 Totalled discomfort values indicated by younger and older drivers via handset control during two automated drives with vs. without HMI

2.3.3.2 Results

Effects of HMI on Driving Experience

Figure 2.16 shows the discomfort indicated via handset control during both drives. The progression of these data over the course of the test track affirms the strong situational effects on driving experience, which were already identified in the first driving simulator study (see Sect. "Effects of Automated Driving Style on Driving Experience"). Participants reported significantly less discomfort during the drive with HMI compared to the drive without HMI, $F(1, 34) = 50.36$, $p < 0.001$, $\eta_p^2 = 0.60$. This effect can be assumed for younger and older drivers given the absence of an age effect, $F(1, 34) = 0.00$, $p = 1.000$, $\eta_p^2 = 0.00$, and the absence of an interaction between HMI and age group, $F(1, 34) = 0.42$, $p = 0.523$, $\eta_p^2 = 0.01$.

In parallel with the results regarding discomfort, participants perceived the drive with HMI as significantly safer, $F(1, 9) = 18.05$, $p < 0.001$, $\eta_p^2 = 0.32$, and significantly more enjoyable, $F(1, 39) = 5.18$, $p = 0.028$, $\eta_p^2 = 0.12$, than the drive without HMI (see Fig. 2.17). Given the absence of an interaction between HMI and age group for both variables (perceived safety: $F(1, 39) = 0.09$, $p = 0.796$, $\eta_p^2 = 0.00$, driving enjoyment: $F(1, 39) = 1.22$, $p = 0.276$, $\eta_p^2 = 0.03$), these HMI-effects can be assumed for younger and older drivers. Both age groups did not differ significantly in their assessment of the perceived safety, $F(1, 39) = 0.61$, $p = 0.441$,

Fig. 2.17 Perceived safety and driving enjoyment rated after automated driving with versus without an in-vehicle HMI informing about the automated driving system, separated by age group

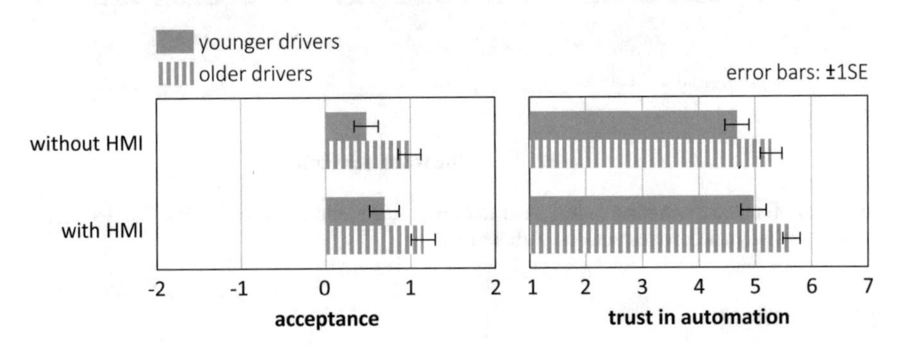

Fig. 2.18 User acceptance and trust in automation rated after automated driving with versus without an in-vehicle HMI informing about the automated driving system, separated by age group

$\eta_p{}^2 = 0.02$, and the driving enjoyment of both drives, $F(1, 39) = 2.12$, $p = 0.153$, $\eta_p{}^2 = 0.05$, even though driving enjoyment ratings of older drivers appeared to be higher than those of younger drivers on a descriptive level.

Effects of HMI on User Attitudes

Irrespective of the HMI condition, older drivers reported a significantly higher system acceptance, $F(1, 37) = 8.24$, $p = 0.007$, $\eta_p{}^2 = 0.18$, and trust in automation, $F(1, 39) = 28.15$, $p < 0.001$, $\eta_p{}^2 = 0.42$, than younger drivers. In addition, both variables were significantly affected by HMI condition, with higher acceptance, $F(1, 37) = 8.24$, $p = 0.007$, $\eta_p{}^2 = 0.18$, and higher trust, $F(1, 39) = 28.15$, $p < 0.001$, $\eta_p{}^2 = 0.42$, reported after the drive with HMI in comparison to the drive without HMI. These effects can be assumed for all drivers, since neither acceptance $F(1, 37) = 0.17$, $p = 0.681$, $\eta_p{}^2 = 0.01$, nor trust in automation, $F(1, 37) = 0.17$, $p = 0.681$, $\eta_p{}^2 = 0.01$, were significantly affected by an interaction between age group and HMI condition. Mean values of all conditions are presented in Fig. 2.18.

2.3.3.3 Discussion

In contrast to the automated driving style, the display-based presentation of system information during driving had consistently positive effects on users' driving experience and attitudes towards automated driving. Supplementing the automated driving system by such an HMI resulted in lower discomfort, higher perceived safety and higher driving enjoyment. This enhancement of driving experience translated into higher trust in automation and a higher acceptance of automated driving. All of these results applied to drivers of both age group, indicating that a display-based increase of system transparency might be an enhancement strategy of higher general application than driving style modifications.

Given the fixed order of the two automated test drives, sequence effects confounding the presented HMI effects cannot be excluded entirely. The fixed order was chosen to avoid overstraining (especially older) participants and instead giving them the opportunity to gradually accustom themselves to the different elements of the study (1. usage of handset control during the standardisation drive, 2. automated driving in default mode during the first test drive, 3. HMI during the second test drive). Previous studies state that only the first one of several consecutive automated drives along an identical track resulted in a significant increase of users' trust and acceptance, which reach a plateau after this first system experience (Hartwich et al. 2019). This knowledge indicates that the differences between the first and second system experience presented in this section exceed the effects of repeated drives along the same track and therefore are attributable to the HMI. Nevertheless, these effects will be verified in follow-up studies based on a randomly ordered presentation of different HMI conditions.

2.3.4 Test Track Study—Driving Style Evaluation

In this study, we examined the transferability of selected results of the first driving simulator study on user preferences regarding automated driving styles to real-world conditions. Therefore, we compared the assessment of two driving styles (dynamic versus defensive) made by 30 participants after experiencing automated driving in selected driving situations on a test track (see Sect. 2.2.5. for exact methods).

2.3.4.1 Data Analysis

All presented data were gathered during and after the two automated test drives based on the dynamic versus defensive driving style. For each metre driven, the mean value of the handset control data was calculated and then summed up over all test subjects. Data preparation of questionnaire data followed the procedure of the first driving simulator study (see Sect. 2.3.2.1). If not stated otherwise, all *t*-tests conducted for statistical analyses met the necessary requirements.

2.3.4.2 Results

Effects of Automated Driving Style on Driving Experience

At track level, the data analysis revealed that the driving style had no overall effect on the discomfort indicated during driving via handset control, $t(22) = 1.33$, $p = 0.961$, ${\eta_p}^2 = 0.11$, which is illustrated in Fig. 2.19.

Regarding perceived safety, a significant difference between driving styles was found, $t(1, 29) = 3.08$, $p = 0.005$, indicating that overall, the defensive driving style was perceived as safer than the dynamic driving style. Driving enjoyment was not affected in statistically significant manner by the driving style (Fig. 2.20).

Next to the above analysis of handset control data for the entire test track, discomfort was also analysed separately for the two central driving situations of the track: (1) approaching a red traffic light and (2) speed reduction in front of a construction zone. In the traffic light situation, a statistical significant difference between the two presented driving styles was found, $t(22) = 2.37, p = 0.031$. As presented in Fig. 2.21 (left), the dynamic driving style provoked more discomfort when approaching the

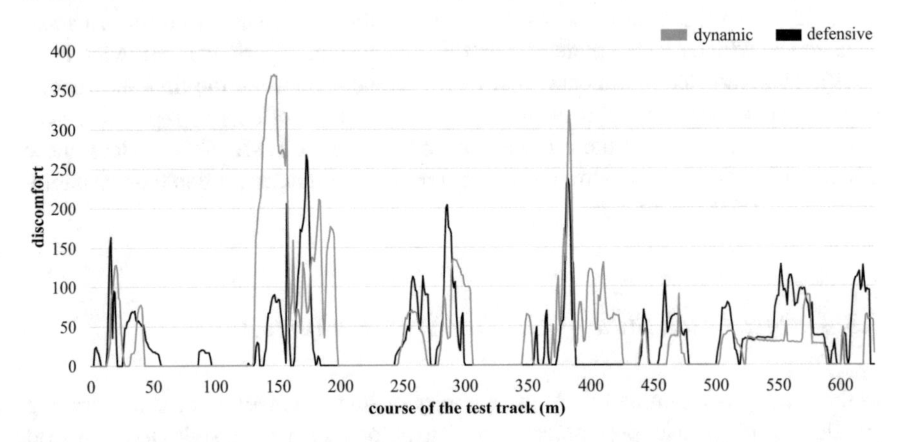

Fig. 2.19 Totalled discomfort values indicated by drivers via handset control during two automated drives based on different driving styles

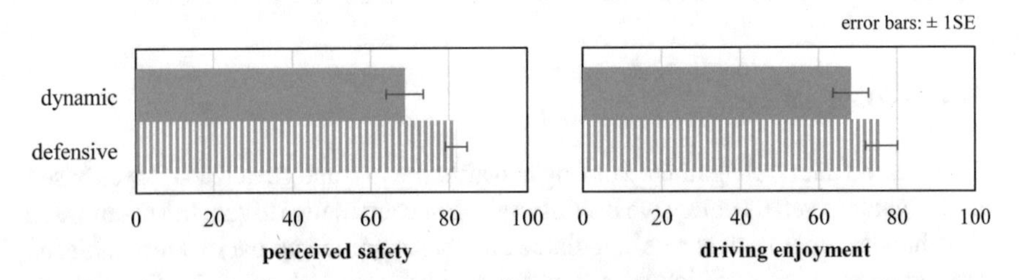

Fig. 2.20 Perceived safety and driving enjoyment rated after automated driving based on a dynamic vs. defensive driving style

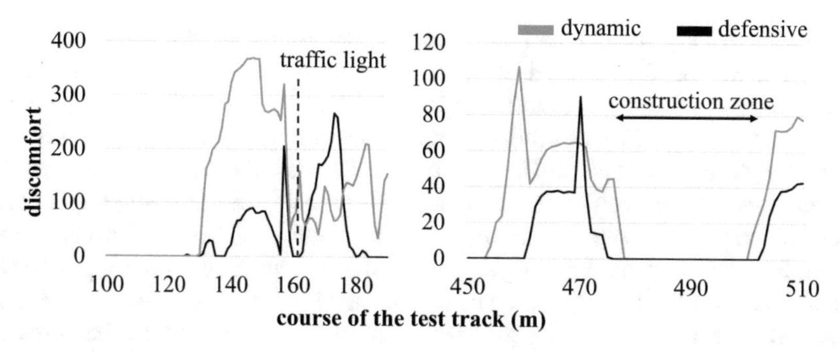

Fig. 2.21 Totalled discomfort values indicated via handset control during two driving situations (left: traffic light approach, right: construction zone) based on different driving styles

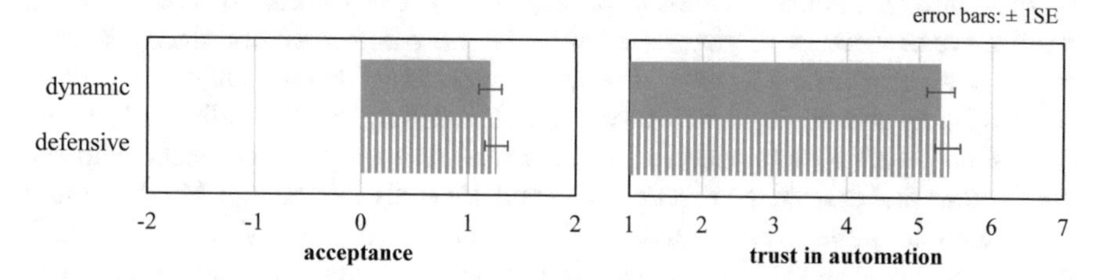

Fig. 2.22 User acceptance and trust in automation rated after automated driving based on a dynamic versus defensive driving style

traffic light than the defensive driving style. Even though a comparable effect is visible for the construction zone situation when looking at the handset values over the course of this situation (Fig. 2.21 right), these differences turned out to be not statistically significant overall, $t(22) = -1.19, p = 0.250$.

Effects of Automated Driving Style on User Attitudes

For participants' acceptance of automated driving as well as trust in automation, no statistically significant differences between the two driving styles could be identified (see Fig. 2.22).

2.3.4.3 Discussion

From a methodological point of view, the use of a real vehicle on the test track afforded the participants a more realistic assessment of automated driving than the driving simulator. Overall, results obtained in this environment verified the results of the first driving simulator study for selected driving situations. The analysis of the test track data showed that a defensive driving style felt generally safer than a dynamic driving style. When examining the handset control data for the whole test track, we found that the driving style had no effect on the overall perceived discomfort, which might be attributable to methodological constraints. Since the space on the test track was limited, driving was only possible at low speeds (10–30 km/h). As Bellem et al. (2016) already demonstrated, the sensation is different at higher speeds. Especially the feeling of safety is probably higher at lower speeds and thus influences the perceived discomfort (Elbanhawi et al. 2015). However, when examining the two central driving situations in detail, an effect of driving style was found for the traffic light approach and in tendency for the construction zone. In both situations, the dynamic driving style provoked more discomfort than the defensive driving style. Like in the corresponding driving simulator study (Sect. 2.3.2), driving style differences regarding driving experience did not translate into user attitudes towards automated driving.

In summary, a holistic statement on whether users of automated vehicles prefer a defensive or dynamic driving style cannot be made unrestrictedly, since test track results were inconsistent as well. However, it was shown that the subjectively experienced safety and driving comfort can be influenced by varying individual driving parameters, with a preference for a defensive driving style. Consistent with the conclusions drawn from the first driving simulator study (see Sect. 2.3.2.3), the sum of these results implies the existence of other factors than driving style affecting users' driving experience and attitudes towards automated driving. Such factors appear to include specifics of the driving situation (e.g. speed range, presence of pedestrians), the vehicle (e.g. presence of an HMI) and the user (e.g. personality traits, a priori trust in automation).

2.4 Discomfort Related to Driving Situations and Driving Style

2.4.1 Data Preparation

This section aims at identifying driving situations and manoeuvre characteristics that are perceived as uncomfortable in automated driving. Therefore, the analyses are focused on users' evaluation of the dynamic and the defensive driving style in the potentially uncomfortable driving situations DS1 to DS5 (Sect. "Simulated Driving Tracks"), which were gathered during the second session of the first driving simulator study. In this section, data of 40 participants experiencing both automated driving styles along an identical test track in a fixed-base driving simulator are presented. For any further methodological details, please consult Sect. 2.2.2.

Effects of driving style on discomfort ratings reported via handset control were analysed using mixed design ANOVA. Discomfort was estimated by summing up the value of all handset control responses given by each participant during each drive.

The discomfort maximum is 1900 (100 as maximum per participant × 19 participant) for the older group and 2100 (100 × 21) for the younger drivers. For example, a data point of 400 can arise from four participants feeling complete uncomfortable or eight participants experiencing mid discomfort. Necessary parametric assumptions (normal distribution, sphericity, homogeneity of variances) were inspected for all dependent variables. In case an assumption could not be verified, additional nonparametric tests confirmed ANOVA results.

2.4.2 *Results*

2.4.2.1 Non-Signalised Intersection

In this situation (DS1), the participants' automated vehicle (hereafter referred to as ego vehicle) had the right of way at the intersection. At 40 m ahead of the intersection's centre, a vehicle approaching from the right was visible. This corresponds to a Time to Contact of about 2 s, in case crossing traffic would not decelerate. Starting from that point, the defensive driving style reduced velocity with a target speed of 40 km/h at the centre of the intersection, while the dynamic driving style passed the intersection with a constant speed of 70 km/h. The crossing traffic stopped at the stopping line when the ego vehicle was still 26.5 m ahead of the intersection's centre. Figure 2.23 shows both situations from a driver's perspective.

Totalled discomfort values indicated by younger and older driver via handset control during the traffic sign regulated intersection are shown in Fig. 2.24. Zero point on the abscissa defines the intersection centre. Both age groups showed a similar distribution of discomfort. There was a strong increase of discomfort about 40–50 m ahead of the intersection's centre, when the crossing traffic got perceptible. Discomfort ratings reached their maximum about 20–25 m ahead of the intersection's centre, when the crossing traffic reached the stopping line. At the end of the driving situation, the acceleration manoeuvre to 100 km/h caused discomfort within the younger age group. A total of 18 subjects felt uncomfortable during the whole

Fig. 2.23 Driver's perspective at the traffic sign regulated intersection (40 m ahead of the intersection's centre on the left, 26.5 m on the right)

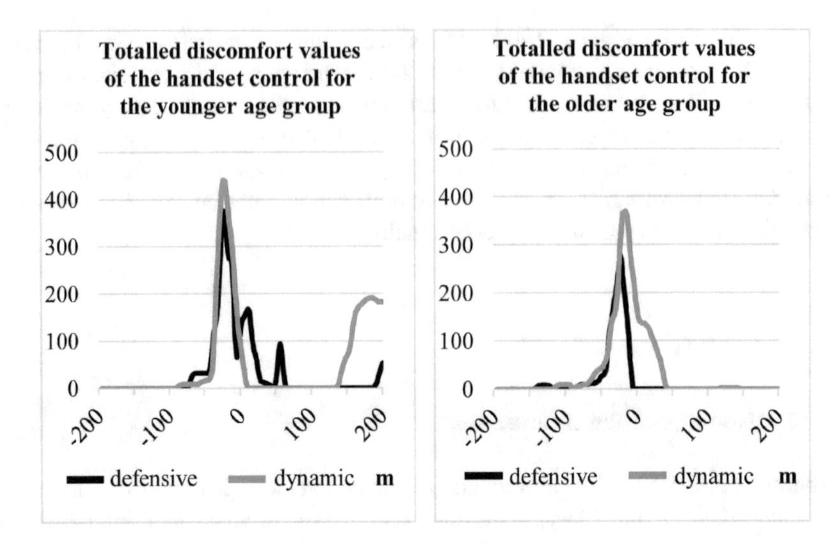

Fig. 2.24 Totalled discomfort values indicated by younger and older drivers via handset control during the traffic sign regulated intersection

situation, whereof only 5 were assigned to the older and 13 to the younger age group. No statistically significant differences for driving style and age groups could be found.

2.4.2.2 Parking Truck Blocking the Lane

In the obstacle situation (DS2), the ego vehicle passed a truck that partially blocked the ego vehicle's lane. At a longitudinal distance of about 85 m to the truck, a lane change manoeuvre was initiated in both driving styles. At a longitudinal distance of about 20 m to the truck (Fig. 2.25), the lane change manoeuvre was completed. During the dynamic driving style, the ego vehicle passed the truck with a constant

Fig. 2.25 Driver's perspective in the obstacle situation (left when initiating, right after completing the lane change manoeuvre to the left)

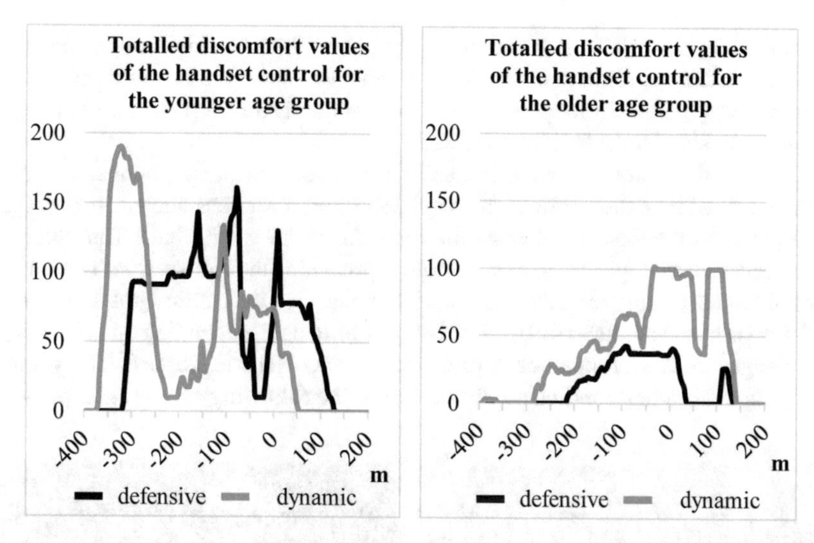

Fig. 2.26 Totalled discomfort values indicated by younger and older drivers via handset control during the obstacle situation

speed of 100 km/h, while it went by with 65 km/h during the defensive drive. When passing the truck, lateral distance to the truck was 0.4 m greater during the dynamic drive than during the defensive drive. Immediately after passing the truck, a lane change manoeuvre to the right was initiated and completed about 100 m later for both driving styles.

Totalled discomfort values indicated by younger and older drivers via handset control during the obstacle situation are presented in Fig. 2.26. Zero point on the abscissa defines the end of the truck. No clear tendency could be identified within the younger age group, whereas the older participants felt more uncomfortable with the dynamic driving style. Within the older age group, discomfort ratings steadily increased for both driving styles, starting at a longitudinal distance of about 200–300 m to the obstacle. When passing the truck, discomfort ratings reached their maximum and returned to 0 about 150 m after the truck got passed. A total of 15 participants felt uncomfortable during the whole situation, whereof only 3 were assigned to the older and 12 to the younger age group. No statistical significant differences for driving style and age groups could be found.

2.4.2.3 Approach to Red Traffic Light

When the ego vehicle approached the intersection regulated by traffic light (DS3), a traffic light sign was passed about 370 m ahead of the traffic light. During the defensive drive, the ego vehicle reacted by rolling out at this point. While further approaching the intersection, traffic lights changed from green to red. At this point, the braking manoeuvre was intensified during the defensive drive in order to brake from the remaining 80 km/h to stop within a distance of 170 m. During the dynamic drive, the ego vehicle also initiated a braking manoeuvre when the traffic light changed from red to green. At this point, there were about 100 m left to decelerate from 100 km/h to stop (see left side of Fig. 2.27).

Totalled discomfort values indicated via handset control by younger and older participants during the traffic light regulated intersection are shown in Fig. 2.28. Zero point on the abscissa defines the location of the traffic light. The older age group experienced higher discomfort while approaching the intersection. The handset control values rapidly increased about 150 m ahead of the traffic light and reached their maximum at about 30–80 m ahead of the traffic light, depending on driving style and age group, and dropped back to 0 when the ego vehicle stopped. The younger age group additionally indicated discomfort in the following acceleration phase. A

Fig. 2.27 Driver's perspective at the traffic light regulated intersection (left at initiation, right at end of braking manoeuvre in dynamic driving style)

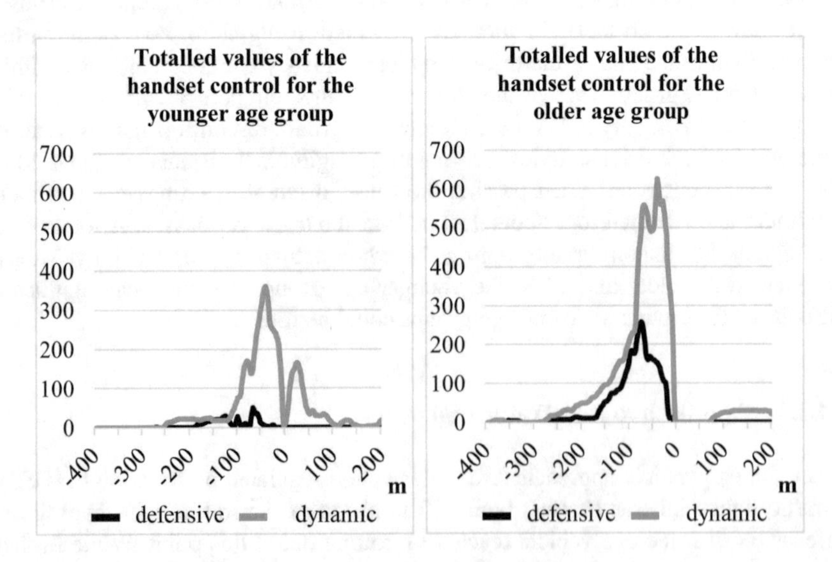

Fig. 2.28 Totalled discomfort values indicated by younger and older drivers via handset control during the traffic light regulated intersection

total of 20 participants felt uncomfortable during the whole situation, whereof 10 were assigned to the older and 10 to the younger age group. Statistical significant differences could be found for driving style, $F(1, 38) = 15.70, p < 0.001, \eta_p{}^2 = 0.29$, but not for age groups.

2.4.2.4 Entering Highway

Approaching the highway-entering situation (DS4) was parameterised differently in the two driving styles. During the dynamic drive, the ego vehicle entered the highway with about 70 km/h, drove at the same speed through the slight right-hand bend, quickly changed to the right lane of the highway and accelerated to about 125 km/h within the next 300 m (end point diagram = 300 m, Fig. 2.29). During the defensive drive, the ego vehicle also approached the highway with about 70 km/h, passed through the slight right-hand bend with about 60 km/h, quickly changed to the right lane of the highway and accelerated to about 100 km/h within the next 300 m.

Totalled discomfort values indicated via handset control by younger and older participants during highway entering are shown in Fig. 2.30. Zero point on the abscissa defines the location of the beginning of the entrance lane. For the younger age group, the defensive and the dynamic driving style were evaluated similarly, in total. The occurrence of discomfort, however, differed considerably. While the braking behaviour before the slight right turn and the resulting low speed when entering the highway were rated negatively in the defensive driving style, the strong acceleration manoeuvre on the highway was perceived as uncomfortable in the dynamic driving style. The older age group evaluated the situation similarly in both driving styles. A total of 14 subjects felt uncomfortable during the whole situation, whereof only 3 were assigned to the older and 11 to the younger age group. No statistically significant differences for driving style and age groups could be identified.

Fig. 2.29 Driver's perspective while entering the highway, screenshots at zero point on the abscissa (left defensive, right dynamic driving style)

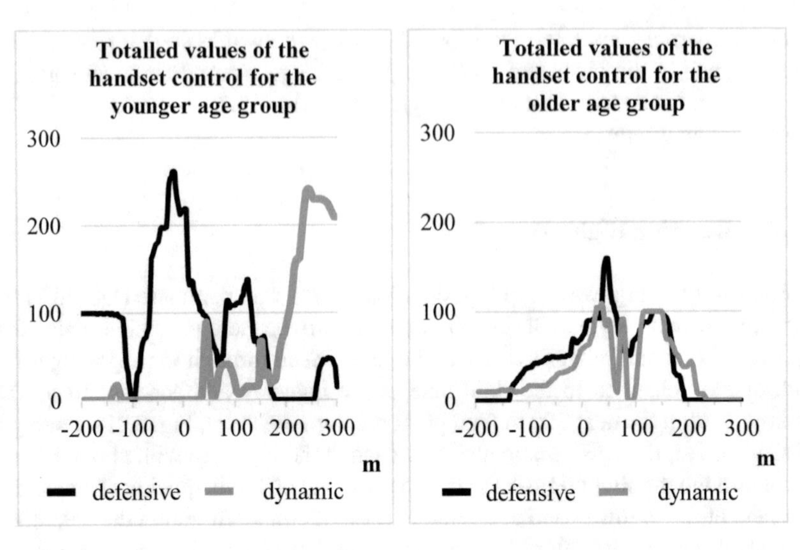

Fig. 2.30 Totalled discomfort values indicated by younger and older drivers via handset control during entering the highway

2.4.2.5 Construction Zone

When entering the construction zone (DS5) with the dynamic driving style, the ego vehicle slowed down from 130 km/h to 80 km/h with very strong deceleration rates. Within the construction site, it followed a construction site vehicle that decelerated from 80 to 30 km/h and turned left into the construction site (Fig. 2.31). The ego vehicle maintained the minimum safety distance (about 25 m at 50 km/h) and quickly accelerated out of the construction site at the end of the situation. During the defensive driving style, the ego vehicle performed a smooth braking manoeuvre while approaching the construction site. Within the construction site, it maintained a five times safety distance (about 125 m at 50 km/h). At the end of the situation, the ego vehicle smoothly accelerated out of the construction site.

Fig. 2.31 Driver's perspective while following the construction site vehicle (left defensive, right dynamic driving style)

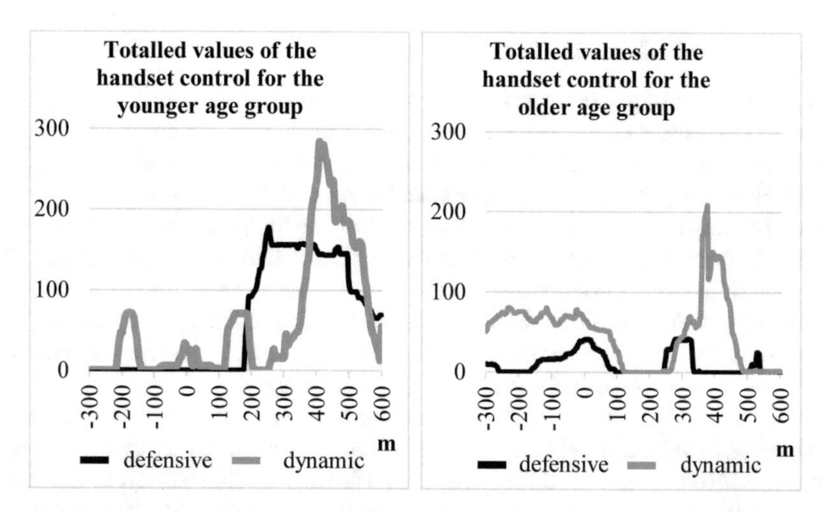

Fig. 2.32 Totalled discomfort values indicated by younger and older drivers via handset control during the construction zone

Totalled discomfort values indicated via handset control by younger and older participants during the construction zone are presented in Fig. 2.32. Zero point on the abscissa defines the start of the construction site. A different distribution of discomfort between the age groups could be observed. While the younger age group felt uncomfortable during both driving styles, the older group perceived more discomfort during the dynamic driving style. A total of 19 subjects felt uncomfortable during the whole situation, whereof 7 were assigned to the older and 12 to the younger age group. No significant differences for driving style and age group could be found.

2.4.3 Discussion

This section aimed at identifying driving situations and manoeuvre characteristics that are perceived as uncomfortable in automated driving. The focus was on the comparison between a dynamic and a defensive driving style in order to derive globally preferred driving style characteristics as well as on the investigation of age-specific effects in driving style preferences.

Results showed that the older age group tended to feel more comfortable with the defensive driving style, as also demonstrated in Sect. 2.3. When examining the handset control values in more detail, global "bad" driving style elements (that are mainly characteristics of the dynamic driving style) could be identified (in conformity with Griesche et al. 2016):

- High acceleration rates
- High deceleration rates and late brake initiation

- Low longitudinal distance to the front vehicle.

Further age group-specific driving style preferences as well as different options to avoid discomfort could be identified at situational level. Within the traffic sign regulated intersection, discomfort was mainly related to the appearance of crossing traffic. Both age groups felt uncomfortable because of the unclear future behaviour of the crossing traffic (stopping in time at the stopping line, crossing the intersection, turning into the main road) as well as the unclear action of the ego vehicle. The slight braking during the defensive driving style and the resulting lower speed hardly influenced the discomfort feeling. Possible discomfort reducing recommendations in this situation are the integration of an HMI (see Sect. 2.3.3) showing that crossing traffic has been detected and passing the intersection is safe.

A clear analysis of the obstacle situation was difficult due to its increased complexity, because different driving style components occurred at different times or overlapped at the same time. Both dynamic (too fast and too aggressive) and defensive (too smooth and slow) acceleration caused discomfort. Within the older age group, discomfort was related to the appearance of the obstacle and the unclear further actions of the ego vehicle. Again, an integration of information has the potential to reduce discomfort (see Sect. 2.3.3).

At the traffic light regulated intersection, the older age group felt more uncomfortable than the younger group. With the dynamic driving style, the late reaction as well as the resulting high braking rate caused a large part of the discomfort. Discomfort-reducing recommendations are an adaptation of the driving style (earlier reaction, lower deceleration rate), but also the presentation of information via HMI (traffic light detected, braking process initiated; see Sect. 2.3.3).

During the highway-entering situation, both driving styles evoked similar discomfort. A combination of the fast cornering of the dynamic driving style and the lower acceleration rate of the defensive driving style could reduce the discomfort experienced by younger subjects. There are similar recommendations for the construction site. The discomfort-inducing element (the construction site vehicle) was approached too timidly in defensive driving style and too aggressively in the dynamic driving style. A behaviour between these two extremes should lead to less discomfort in the younger age group.

In summary, globally preferred as well as situation and group specific driving style characteristics could be identified. The differences within the sample cannot only be explained by age differences, but require further studies on influencing factors, e.g. mood, fatigue or experience with highly automated driving. In addition, a stronger feedback frequency of the younger age group has been identified. Again, further studies should clarify whether this is a random effect of the experiment or a factor to be considered in long term.

2.5 Physiological Indicators of Discomfort

2.5.1 *Physiology and Discomfort*

Inferring mental states from physiological indicators has long research tradition. Secondary data analyses at the start of the KomfoPilot Project aimed at collecting potential physiological discomfort indicators from literature and previous own studies. A special focus was on finding least intrusive measures to increase chances for getting the project results applied into vehicles. Thus, results from complex and necessarily lab-based studies such as brain imaging or EEG were collected, but not considered further on. In line with this approach, the use of a commercially available smartband (MS Band 2) was an explicit project goal to estimate the potential of such widespread devices in everyday situations like driving. Results of this initial work package revealed heart rate (HR), heart rate variability (HRV), skin conductance level (SCL), skin temperature, pupil diameter, eye blinks, glance behaviour, facial expressions and body posture/movements as promising physiological measures for discomfort.

The cardiovascular parameters HR and HRV are often used in driving studies as indicators for stress, mental effort, task demands and workload (overview of studies in Backs and Boucsein 2000; Brookhuis and Waard 2011; Schmidt et al. 2016). Common findings report an increase in HR and decrease in HRV with higher workload, stress and invested effort. Several of the discomfort-inducing situations investigated in KomfoPilot could be seen as similar to stress situations with an uncertainty about the system capabilities to successfully complete the driving task. Thus, an increase in HR and decrease in HRV are expected during uncomfortable situations.

Pupil diameter has been studied intensively as indicator for cognitive workload, stress, affective processing and attention (Andreassi 2000; Cowley et al. 2016). Common findings report an increase in pupil diameter with mental workload, emotionality of stimuli, enhanced information-processing demands and task difficulty (Andreassi 2000; Backs and Boucsein, 2000; Cowley et al. 2016). Thus, pupil diameter is expected to increase during uncomfortable situations.

Changes in eye blink rate are considered an indicator for mental and visual workload, task demands, mood states and fatigue (Andreassi 2000; Cowley et al. 2016; Marquart et al. 2015). In complex situations requiring visual attention, blink rate has been found to decrease for fighter pilots as well as for car driver (Backs and Boucsein 2000). Thus, a reduction of eye blinks is expected for uncomfortable situations that are visually monitored.

Changes in electrodermal activity are reported in relation to higher arousal, emotional load, alertness, stress, workload, mental effort and task difficulty (Dawson et al. 2017). Two signal components are usually analysed: skin conductance level (SCL) represents the slowly varying tonic signal, whereas skin conductance response (SCR) refers to the rapidly changing phasic component, superimposed on the SCL (Cowley et al. 2016). For discomfort, an increase in SCL is expected due expected higher arousal and alertness.

Facial expressions could be an additional indicator of potential discomfort during automated driving. Analysis of individual face muscle or muscle groups (Action Units) can be performed by manual coding (FACS, Ekman et al. 2002) as well as nowadays by automatic analysis using computer-vision algorithms (Ko 2018).

Body motion during driving has mainly been investigated with regard to hand movements for estimating driver distraction (Tran and Trivedi 2009), trapezius muscle tension as an indicator for stress (Morris et al. 2017) or head movements for predicting driver intentions (Pech et al. 2014). The whole 3D-driver posture is considered useful for extracting information about distraction, affective states and intentions (Tran and Trivedi 2010). However, posture dynamics are strongly related to specific situations and should therefore be integrated with context information (Tran and Trivedi 2010). During close approach situations with the danger of a potential rear-end collision, a push-back movement is expected that should be reflected in seat pressure mat and motion tracking data.

2.5.2 The Smartband MS Band 2 as Sensor for Physiology

Smartbands/fitness trackers gain increasing popularity (Wade 2017) and provide a cheap and easy-to-use mobile assessment of physiological parameters in everyday live situations. When connected with vehicles, smartbands could provide valuable user state information (such as discomfort) to improve human–machine interaction. Thus, the use of a commercially available smartband (MS Band 2, Fig. 2.2a) was an explicit goal in KomfoPilot to estimate the potential and problems of such a sensor. As smartbands are no medical grade/laboratory devices, main issues are related to data quality, reliability, stability and accessibility. Regarding accessibility, the MS Band 2 was provided with a Software Development Kit that allowed for accessing sensor data and programming a dedicated logging application in KomfoPilot. Concerning data quality issues, the MS Band 2 has already been validated in several studies with laboratory devices, with sports devices using a chest strap as well as with other smartbands. In addition, the MS Band 2 has already been used in applied research for stress detection, assessing mental workload in different environments, activity recognition in home settings, and for regulating and predicting personal thermal comfort in buildings (a comprehensive and detailed overview of all studies can be found in Beggiato et al. 2019). Overall, data reliability of consumer smartbands is of course lower in comparison to medical grade/laboratory devices. A common finding is that wrist devices provide rather accurate results in stationary settings with low physical movement; however, accuracy deteriorates when participants are mobile. One main problem is the way (e.g. how tightly closed) people wear the devices, with resulting differences in skin contact. This issue of (temporarily) loosing skin contact is reported as one potential reason for poor sensor data quality. A second issue, especially related to the MS Band 2, is the "optimal" measurement position at the inner side of the wrist for measuring HR and SCL. The optical photoplethysmography sensor to detect HR from peaks in the blood flow under the skin is located under the display of the MS Band 2. Two electrodes to measure skin resistance (inverted for SCL) are located on the opposite side of the display. Thus, only one of these sensor can be placed at the (better) inner side of the wrist for optimal measures. Microsoft recommends to place the display at the inner side for better HR measurements. To test for these differences in KomfoPilot, the MS Band 2 has been applied with the display at the inner side in driving simulator study 1 (better HR measures) and turned by 180° in driving simulator study 2 (better SCL measures).

Overall, the main question for using smartbands in applied settings is not if they are equally good as laboratory devices (they won't), but if they are good enough for a specific research question or application (with all the benefits of mobility, comfort, wide daily usage, cheapness and easy-to-use assessment). Balancing these aspects and considering that driving/being driven in automated mode is an almost stationary setting with little physical movement, data reliability of the MS Band 2 was considered sufficient to be used in the studies.

2.5.3 Processing of Physiological Data

HR, HRV and SCL: The parameters HR, HRV and SCL were assessed by the smartband MS Band 2 in all three KomfoPilot studies. Raw HR values in beats per minute were directly recorded from the Band. A second data channel from the MS Band 2 containing the interbeat interval times (IBI) in seconds was used to calculate HRV. The values of HR and IBI are not exact reciprocal in the case of the MS Band 2, but IBI is recommended for HRV calculations (Cropley et al. 2017). The time-domain metric RMSSD (root mean square successive difference) was calculated, as this metric is recommended for measuring high-frequency HRV and when time intervals to compare are not equally long (Berntson et al. 2017). Nonlinear HRV measures and frequency domain analyses were not applied due to the relatively short time periods investigated. Two electrodes on the opposite side of the MS Band 2 display measured skin resistance level in kilo ohm. To obtain SCL in micro Siemens, these resistance values were inverted and multiplied by 1000. All SCL values were highly sensitive to changes in the arm/hand position such as placing a hand on the knees. Thus, based on the MS Band 2 accelerometer and gyroscope data, SCL values during high movement episodes were excluded (missing data).

Pupil diameter and eye blinks: The SMI Eye Tracking Glasses 2 (SMI ETG 2; Fig. 2.2b) were used to record eye tracking data including pupil diameter, fixations, saccades, and blinks. Raw pupil diameter values (mm) for the left and right eye were averaged to obtain a single diameter from both eyes. To correct for diameter fluctuations (especially close to blinks), a moving average over ± 300 ms was calculated. Eye blink rate was computed in two different ways: first, blinks per second were calculated for intervals before, during, and after reported discomfort (Fig. 2.33d). However, this blink rate does not provide information about timings of increase/decrease as well as significance levels over time. Thus, a second continuous blink-metric was calculated by transforming distinct eye blinks into a running

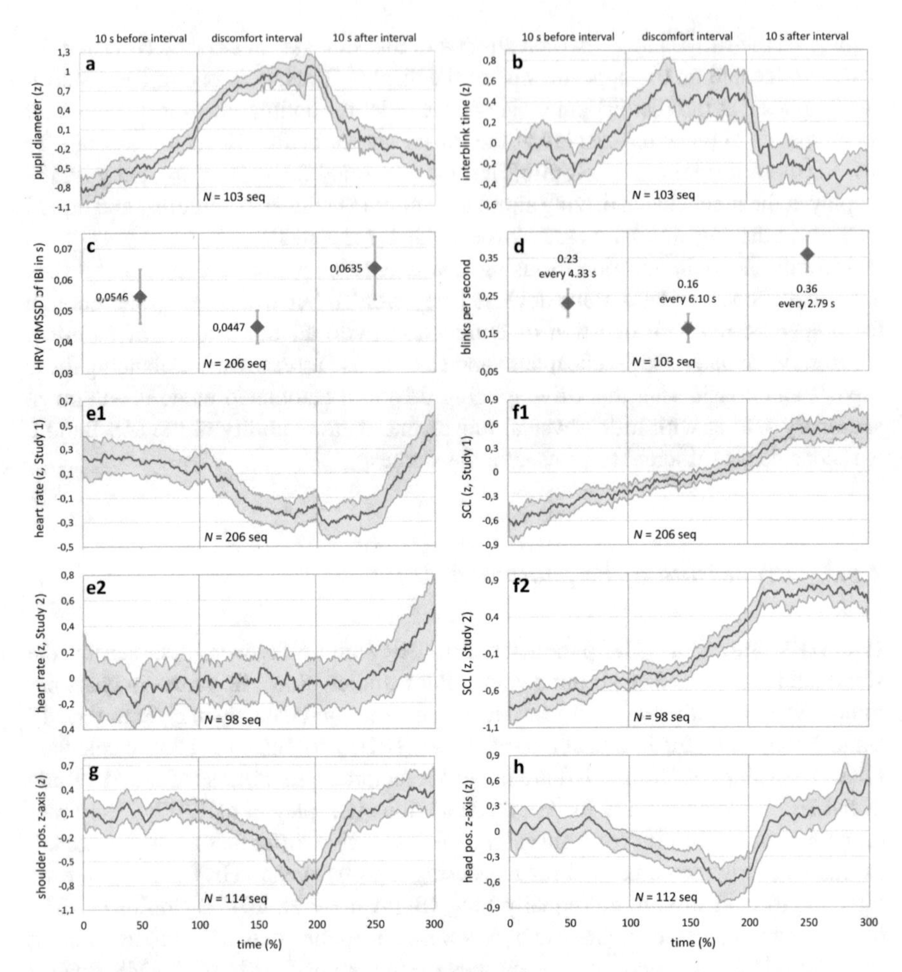

Fig. 2.33 Changes in physiological parameters during the standardised truck approach situations. All charts show mean values (bold blue line) and 95% pointwise confidence intervals (light red area) of physiological parameters before, during and after discomfort intervals in all truck approach situations of both driving simulator studies

"interblink interval time" (Johnston et al. 2013; Fig. 2.33b). A milliseconds-timer was set to zero at every moment a new blink was detected by the SMI ETG 2 and increased until the subsequent eye blink started. Blink duration itself entered the running time. Overall, eye tracking could not be applied in all cases: participants already wearing eyeglasses could not wear the SMI ETG 2, some trips in the driving simulator studies were recorded without the glasses because of testing camera-based, facial-feature recognition algorithms and the glasses were not applied in the test track study due to potential sunlight distortions. All result charts contain the total number of sequences that entered the analyses.

Body movements: Body movements were simultaneously captured by two sensor systems. The first device (only used in Driving Simulator Study 1) was a marker-based motion tracking system from OptiTrack (Fig. 2.2d). A total of four body regions were tracked (left and right hand, right shoulder and head; see Fig. 2.3) including position and orientation in six degrees of freedom with a precision of up to 0.1 mm. As the absolute body part positions in the 3D-space differed for each individual subject and each drive, only differences were computed for each sequence starting with zero at the beginning. The second sensor system for body movements was a seat pressure mat developed by the project partner FusionSystems GmbH (Fig. 2.2c). The mat was applied in all three studies, could easily be placed on top of the seat and reported data from eight pressure sensors located at different positions (see Fig. 2.34).

Facial expression analysis: A video-based software tool for facial expression analysis was acquired in the last project year (Visage facial-feature detection and face analysis software, version 8.4, visagetechnologies.com). The tool is delivered as Software Development Kit (SDK) and provides head- and eye tracking, estimates of age, gender and seven basic emotions as well as values for 23 facial Action Units (AU). AU represents movements of an individual face muscle or muscle group (Ekman et al. 2002). Due to limited time for integrating the SDK in an own logging application, only preliminary results were available at the time of writing (Beggiato et al. 2020). The 23 AU were analysed using the Intel RealSense SR300 camera in the truck approach situation of Driving Simulator Study 2 (Fig. 2.2e). Only video frames with a face tracking quality of 30% or higher entered the analysis. A moving average over $\pm$ 2 s was calculated for each AU raw score to correct for high frequency fluctuations.

2.5.3.1 Common *x*-axis

To show the development of all assessed parameters over all situations, trips and persons, a common *x*-axis had to be established. Three different approaches for a unified *x*-axis were applied: (1) elapsed time, (2) metres driven and (3) per cent of time. The procedures and specific pro's and con's of each approach are discussed subsequently.

The time-based *x*-axis allows for displaying "true" evolvement of sensor values as they were measured over time. However, if driven speed differs (e.g. different driving styles), data cannot be aggregated on the same time scale, even if it is the same route. Results from the test track study (Fig. 2.36) are displayed in the true time domain for the dynamic driving style.

Metres driven on the other hand allow for aggregating data over the same track, even if driven with different speed/driving styles. Drawbacks are reduced information about the exact timings of rise and fall of sensor values as well as problems when having longer phases of complete standstill (e.g. in the test track study at the traffic light). Data from the test drive of the two driving simulator studies are shown in metres domain to allow for judging changes in sensor values during the different driving situations.

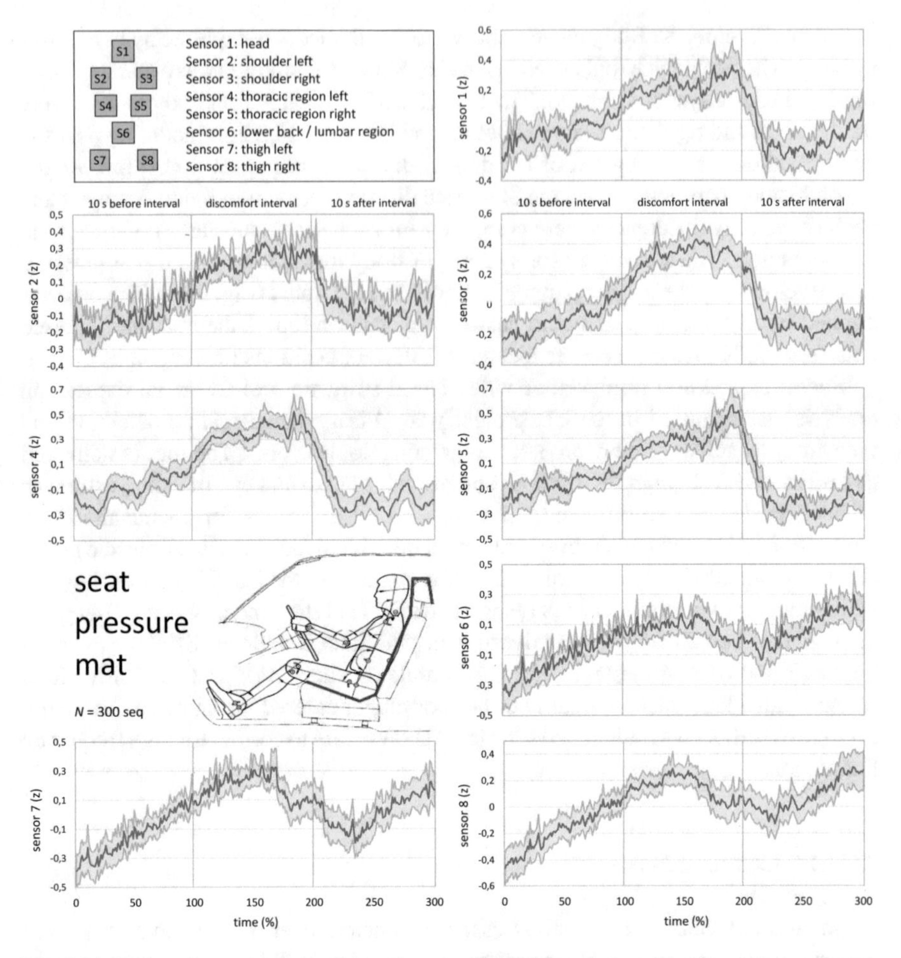

Fig. 2.34 Body movements during the truck approach situations captured by the seat pressure mat. All charts show mean values (bold blue line) and 95% pointwise confidence intervals (light red area) of the 8 seat pressure mat sensors before, during and after discomfort intervals in all truck approach situations of both driving simulator studies. Sensor placement on the mat is depicted at the top left

A per cent scale is useful for aggregating and comparing data of different sequence length, in either time or metres driven. Percentages are applied for the close approach situation to the truck driving ahead in all driving simulator studies (standardisation drive, Fig. 2.33). In these situations, each time period of continuous pressing the handset control was selected as one discomfort sequence with different durations. Fixed time periods of 10 s before and after each handset sequence were combined to a 300% stretch in total. In the first and last 100%, each per cent corresponds to 0.1 s and the handset sequence from 100% to 200% was divided into per cent slices according to the specific duration. This procedure allows for strictly considering individual reported discomfort and at the same time combining each approach situation of all studies into on chart. However, a drawback of the percentage scale is reduced time-related information, as subjectively reported intervals in per cent represent the unit of measurement.

2.5.3.2 Z-Transformation and 95% Confidence Intervals

A crucial issue in processing physiological data is distinguishing the signal from noise (Gratton and Fabiani 2017). Almost all physiological parameters such as HR, SCL, body position or pupil diameter have a strong individual component. Thus, absolute values can hardly be averaged and compared between subjects, whereas relative changes within a person or a situation provide a better signal-to-noise ratio. However, these relative changes need to be transformed into a common scale to be compared across situations and individuals. A frequently used transformation is the z-transformation (Jennings and Allen 2017), which expresses all values as the distance to the mean in units of standard deviations (z-score). The procedure creates an overall mean of zero and a standard deviation of one. Z-transformation was applied in all studies for each sensor channel and each drive (test drives and test track study), respectively each sequence (standardisation drives), resulting in relative changes over time in units of standard deviations. These z-scores were averaged per sensor over participants/drives/sequences. The bold blue line in the middle of all result charts represents this mean. In addition, the 95% confidence interval (CI) of each of these means was calculated pointwise and plotted around the means. The CI provides an estimation about the statistical significance of changes: if the CI-band does not overlap between two points on the x-axis, these two means differ in a statistically significant manner at $p < 0.01$ (Field 2013).

2.5.4 Results

The presented analyses aim to provide information about the specific changes of each physiological parameter during reported discomfort in various automated driving situations. Analyses combine data from all participants/drives whenever possible and reasonable. The results are split into the three different driving environments/tracks: (1) the standardisation drive with the truck approach situations from both driving simulator studies, (2) the 9.1-km-long test drives of both driving simulator studies, driven in various driving styles and (3) the test track study consisting of the 650-m-long closed test track. For methodological details of the three studies, please consult Sect. 2.2.

2.5.4.1 Standardisation Drive with Truck Approach Situations

Figures 2.33 and 2.34 present the combined results on changes in physiological parameters of both driving simulator studies during the standardised truck approach situations. The data basis consists of 81 different participants, whereas the 41 participants in Study 2 experienced the approach drive once, the 40 people in Study 1 twice at two distinct driving sessions (details in Sect. 2.2). Each single drive consisted of three subsequent truck approach situations; thus, a total of $41 \times 3 + 40 \times 6 = 363$ approach sequences were recorded. In 57 situations, participants did not press the handset control at all; thus, a total of 306 sequences entered the analyses with a mean duration of $M = 7.8$ s ($SD = 5.2$ s). As not every sensor was active every time, each result chart contains the total number of sequences building the basis for aggregation. HR and SCL data are shown separately for Driving Simulator Study 1 (Fig. 2.33e1/f1) and Driving Simulator Study 2 (Fig. 2.33 e2/f2) due to the 180° turned MS Band 2.

Pupil diameter (Fig. 2.33a), combined from participants of both driving simulator studies, increased significantly during the discomfort interval and decreased steadily after reported discomfort. Eye blink rate was computed in two different ways: The continuous interblink interval time (Fig. 2.33b) showed a noticeable increase during discomfort intervals (i.e. less blinks) and a return to the prior level after the discomfort interval. In addition, static blinks per second were calculated for each whole interval before, during and after reported discomfort (Fig. 2.33d). Blink rate decreased in a statistically significant manner from 0.23 blinks per second before discomfort to 0.16 blinks per second during discomfort and increased afterwards to 0.36 blinks per second ($F(1.81, 184.7) = 41.73$, $p < 0.001$, ${\eta_p}^2 = 0.290$).

HR in Study 1 (Fig. 2.33 e1, HR sensor at the inner side of the wrist) showed a steady decrease at the beginning of the discomfort interval. The bottom plateau was kept until about 5 s after reported discomfort (250%). Afterwards, HR rapidly increased to approximately the prior level. HRV of Study 1 (Fig. 2.33c) showed the expected u-shaped pattern with a statistical significant decrease during reported discomfort (Chi^2 (2) $= 40.05$, $p < 0.001$; nonparametric Friedman's ANOVA). In contrast, when turning the MS Band 2 by 180° (Fig. 2.33 e2, HR sensor at the outer side of the wrist), the HR decrease disappeared. In addition, substantially larger CI-bands indicate higher data instability. As the HR decrease was a relative consistent finding in both standardisation drives as well as in the test drives of Study 1, the missing HR effect in Study 2 was related to the changed sensor position with high probability.

SCL showed a continuous linear increase in both studies (Fig. 2.33 f1/f2). During reported discomfort, the increase tends to be steeper than beforehand and after the discomfort interval. This trend is stronger in Study 2 (Fig. 2.33 f2, SCL-sensor at the inner side of the wrist) then in Study 1 when measuring SCL at the outer side of the wrist (Fig. 2.33 f1).

First analyses of facial expressions (Beggiato et al. 2020) in driving simulator Study 2 showed a rise of inner brows (AU1) as well as upper lids (AU5), indicating surprise. Lips were stretched (AU20) and pressed (AU24) as sign for tension. In line with the results from the eye tracking glasses, eyes were kept open and eye blinks were reduced (AU43).

Body movements captured by the motion tracking system in Driving Simulator Study 1 showed a push-back movement when coming close to the truck. This backward motion is represented by the u-shaped decrease of the shoulder (Fig. 2.33g) and head position (Fig. 2.33 h) on the z-axis during the discomfort interval. The push-back movement was also present in the data of the seat pressure mat, captured in both studies with a total number of 300 sequences (Fig. 2.34). The movement did not only affect the upper part of the body, but was represented in all eight pressure sensors from the thighs over the back to the head (pressure increase during discomfort). Thus, the situation provoked an increase in tension throughout the body.

2.5.4.2 Test Drives

Figure 2.35 presents the mean handset control values reporting discomfort as well as the z-transformed values of HR, pupil diameter, interblink interval time, SCL and the combined eight pressure mat sensors for the entire test drives of 9.1 km, along metres driven. The data basis included all trips of all participants in both driving simulator studies (4 trips of 40 participants in Study 1, 2 trips of 41 participants in Study 2 = 242 trips in total). All analyses have been carried out separately for the different driving styles/conditions (defensive, dynamic, own manual driving style and manoeuvre-display). The effects are more pronounced for the dynamic driving style compared to the other driving styles and less pronounced when driving with the HMI showing manoeuvre information (Sect. 2.3.3). However, the effect directions are always the same, thus, all data were aggregated in Fig. 2.35.

Based on the literature research and secondary data analysis, six situations were programmed into the driving simulation to potentially provoke discomfort. Driving Situations (DS) included an non-signalised intersection with a fast vehicle approaching from the right (DS1), overtaking a truck parked on the right roadside (DS2), an approach to a signalised section with a red traffic light (DS3), a transition from the highway ramp to the two-lane highway with free-flowing vehicles (DS4), a construction zone on the highway with an uncritical approach to a slower driving construction vehicle ahead (DS5) and the exit from the highway on a narrow exit bend (DS6). Detailed pictures and descriptions of all situations can be found in Beggiato et al. (2019) and in Sect. "Simulated Driving Tracks". Physiological data patterns revealed another unplanned situation as potentially uncomfortable (DS0), which consisted of approaching a zebra-crossing in urban area including a speed limit change from 50 to 30 km/h (but no pedestrian actually crossing the street).

HR decreased significantly in almost all uncomfortable situations, including the non-intended DS0. As in the standardised truck approach situation, the HR decrease was consistently delayed by some seconds with regard to the reported discomfort peak in the handset control data (e.g. DS1). Pupil diameter showed the expected increase in DS1 and DS3, a slight increase could be noted as well for DS2. However, fluctuations in pupil diameter did also occur between the situations, which were probably related

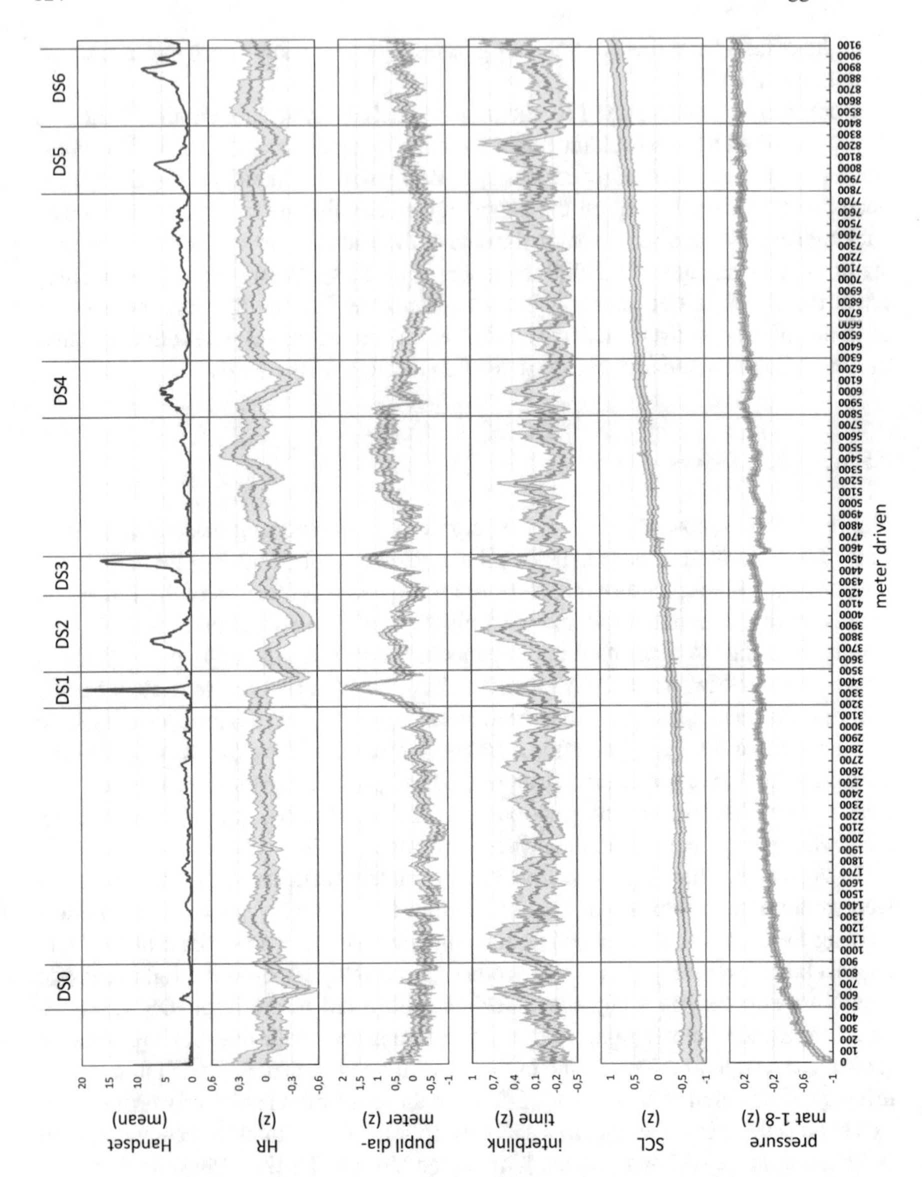

Fig. 2.35 Changes in physiological parameters and body movements during the test drives of both driving simulator studies. The top chart shows mean handset values. All other charts show mean values (bold blue line) and 95% pointwise confidence intervals (light red area) for z-scores of HR, pupil diameter, interblink interval time, SCL and the combined eight pressure mat sensors. (DS0 = zebra-crossing, DS1 = non-signalised intersection, DS2 = parking truck blocking the lane, DS3 = red traffic light, DS4 = entering highway, DS5 = construction zone with slow vehicle ahead, DS6 = exiting highway)

to changes in ambient light of the simulation. Interblink interval time increased in DS0 to DS3. However, CI-bands are quite broad, which suggest relatively high variability in the interblink data. SCL showed the already observed linear tendency to increase over time, which could be explained by the fact that participants got warm during the trips (also confirmed by the MS Band 2 skin temperature sensor, which showed a linear increasing trend without relation to specific situations). A small steeper increase in SCL could be observed in DS1 and DS3. It should be noted that 4 of the 6 trips were acquired with the SCL-sensor at the less sensitive outer side of the wrist. Body motion data, combined of all eight pressure mat sensors, did not show noticeable and consistent situation-specific effect. Very small tendencies in pressure increase could be noted for DS1 to DS4. It seems that the pressure sensors need some time at the beginning of the trip to become stabilised at a certain baseline pressure level.

2.5.4.3 Test Track Study

Figure 2.36 presents the mean handset control values, driven speed and steering wheel angle as well as the *z*-transformed values of HR, SCL and the combined eight pressure mat sensors for the test track drive in the dynamic driving style, along the time axis. As most physiological reactions were expected for the dynamic driving style, only these trips were analysed. The data basis was composed of all 37 trips with completely recorded data of all 30 participants. Due to technical problems with the logging system, data from eight dynamic trips were missing.

As intended, handset control data showed highest reported discomfort in the deceleration phase before the red traffic light (see as well Sect. 2.3.4). However, discomfort was reported as well in almost all subsequent bends/turns. Obviously, these automated bend drives were (unintentionally) uncomfortable as well. On the other hand, driving through the construction zone did not elicit as much discomfort as planned. Thus, the test track drive of 144 s included too many uncomfortable situations to really allow for resting in between them. This issue makes analyses and interpretation of situation-specific changes in physiological parameters difficult.

For HR, a decreasing trend could be observed in the approach phase to the red traffic light. However, CI-bands are rather broad, indicating high variability and a rather weak effect. Almost all increases and decreases of HR over time are located within the CI-bands, and thus, no particular strong situation-related HR effects could be identified. SCL showed again the linear increasing trend, indicating that participants got warm during driving. However, no situation-specific changes in SCL could be observed for the whole drive. Body motion measured by the aggregated data of all eight pressure mat sensors showed strong changes along the drive. However, these changes can all be related to longitudinal and lateral forces as a consequence of braking/acceleration/steering manoeuvres. It was not possible to extract situation-specific body movements independent of driving manoeuvre forces.

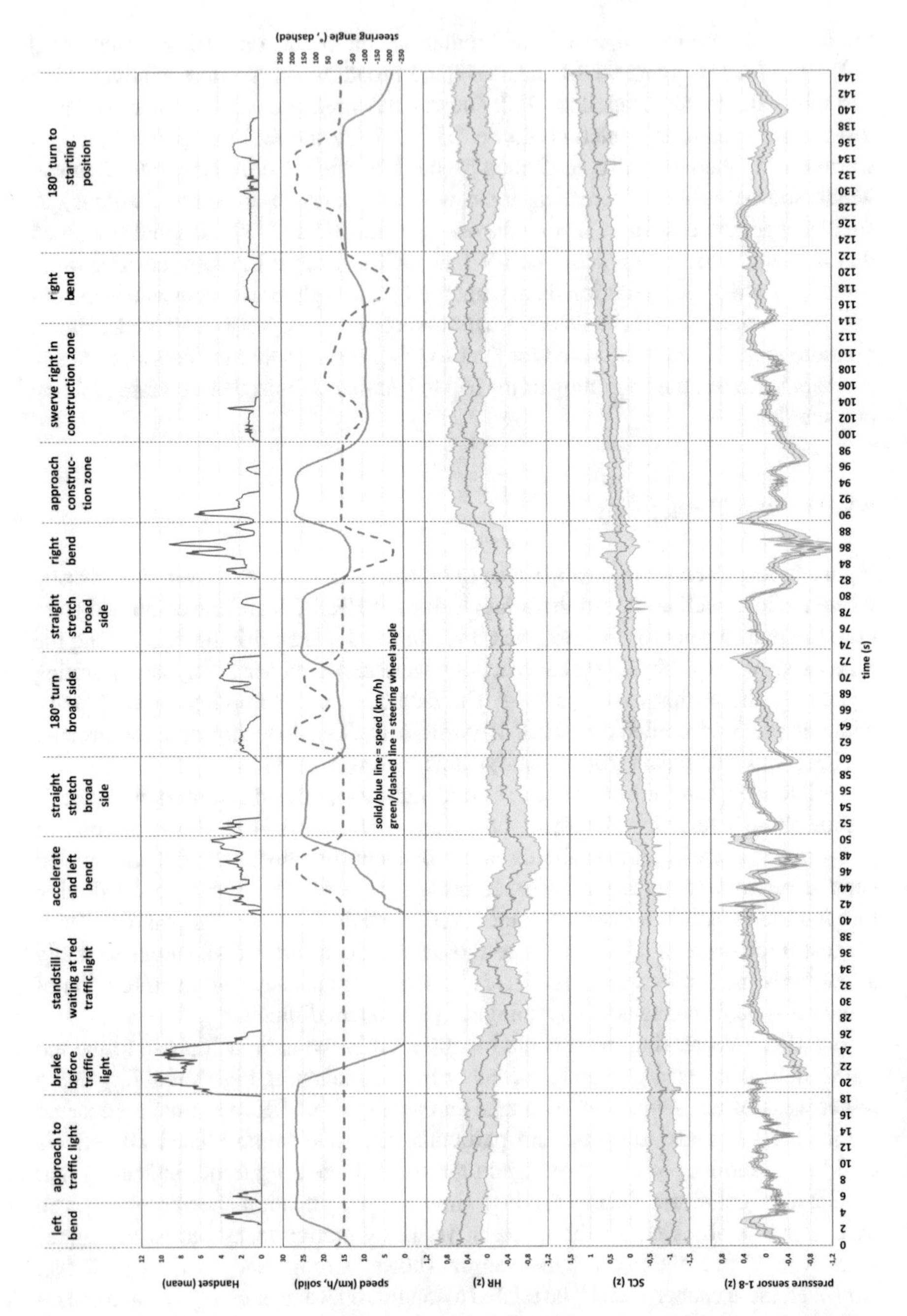

Fig. 2.36 Changes in physiological parameters and body movements during the test track trip, driven in the dynamic driving style. Mean values (bold blue line) and 95% pointwise confidence intervals (light red area) for *z*-scores of HR, SCL and the combined 8 pressure mat sensors

2.5.5 *Discussion and Conclusions*

The presented analyses aimed at providing information about discomfort-specific changes in physiological parameters during simulated as well as real automated driving situations. Based on the metaphor of a vehicle-user-team, automated vehicles could react to detected discomfort by changing information presentation and driving style parameters and thus, provide a better and personalised automated driving experience.

Overall, specific physiological reactions could be observed for situations with specific events that provoke moderate to high reported discomfort (e.g. close approach to vehicle driving ahead, intersection with a fast vehicle approaching from the right, approach to red traffic light, entering highway). Slowly evolving and longer lasting situations with lower reported discomfort only showed weak to missing effects in physiology (e.g. construction zone with uncritical approach to a slower driving vehicle ahead, exit from the highway on narrow exit bend). Separate analyses of age groups and gender did not reveal divergent results; however, further analyses on groups showing stronger/weaker physiological reactions are ongoing. A generally important learning of all KomfoPilot studies was that after each potentially uncomfortable situation, enough time of "normal" driving shall be provided to offer the chance of getting back to baseline conditions in all parameters. This was a main issue for physiological data analysis in the test track study, in which the constraints of the closed test track did not allow for a longer driving route. Thus, the two constructed uncomfortable situations (1) traffic light approach and (2) construction zone were too close to other (non-planned) uncomfortable situations (e.g. automated driving in narrow bends). Longer on-road tests are required to validate the findings of the driving simulator studies.

The project goal of using a commercially available smartband (MS Band 2) for assessing physiological measures demonstrated the potential of such cheap and easy-to-use devices in an applied context. Challenges are mainly related to rather little control on how people wear the device (i.e. how tight the band is worn and, in the specific case of MS Band 2, in which one of the two allowed positions). This issue is connected with the problem of having few information about the quality/uncertainty of each sensor value at a certain moment. However, due to the relatively high number of sensors on the band, an estimation of data quality could be inferred based on the knowledge gathered in these studies. Values of inertial band sensors give indications on the position of the band, absolute values of skin resistance provide information on how tight the band is closed and gyroscope sensors as well as sudden changes in skin resistance indicate hand movements and repositioning. Fusion of all these data sources (e.g. by machine learning methods) could potentially allow for estimating the current uncertainty of measurements. However, lot more data than acquired in these studies would be necessary to develop such algorithms. Overall, a major challenge for using these devices in applied contexts is related to the usage of adequate signal analysis methods for maximising signal-to-noise ratio. An important strategy to deal with the strong individual component of all physiological parameters is extracting relative changes within one person. This was performed in the KomfoPilot studies by applying z-transformation and could be achieved as well in real-time algorithms by, e.g., calculating individual z-scores in sliding time windows and comparing current measures with these scores.

For z-scores of HR as well as HRV, a consistent decrease could be observed in most of the uncomfortable situations when the HR sensor was located at the inner side of the wrist. This effect was contrary to the initial expectations and prior findings of increased HR in situations with higher workload and stress. A possible explanation for the observed HR decrease could be the effect of "preparation for action", which means an anticipatory deceleration of HR prior to actions (Cooke et al. 2014; Schandry 1998). This effect was reported for sport actions, but as well for simpler reaction time paradigms (Andreassi 2000).

For *z*-transformed SCL, a general linear increasing trend could be observed in all studies, which is mainly related to the fact that people got warm during driving. However, in the truck approach situation, the increase tended to be steeper than before and after the discomfort interval, which could be interpreted as situation-related. However, the effect was generally weak (stronger when wearing the SCL-sensor at the inner side of the wrist) and hardly identifiable in other situations on the longer test route.

Eye tracking measures showed situation-specific increases of *z*-transformed pupil diameter as well as interblink interval time (meaning less eye blinks). Thus, eye-related measures could contribute to identify discomfort in situations that are visually monitored by the user. A problem of pupil diameter is the fact that it is primarily dependent of ambient light conditions and not only of mental states. Despite cameras/sensors could capture current light conditions and correction algorithms for ambient light exists, the exact association between ambient light and pupil diameter is rather complex and also influenced by other factors like age (Watson and Yellott 2012).

First preliminary results on video-based analyses of facial expressions showed situation-specific changes in facial action units related to tension and surprise (Beggiato et al. 2020). However, due to the late and originally not planned usage of the facet racking software, further detailed analyses are still ongoing. As a consequence of rapid technology development and easy application, video-based face tracking seems a promising tool for assessing driver state parameter such as head/body position and movements, eye-related measures as well as emotional states.

Body movements captured by the motion tracking system as well as the seat pressure mat from the project partner FusionSystems GmbH showed a situation-related push-back movement in the truck approach situation. However, this reaction is specific for such approach situations and the effects could not be identified for other uncomfortable scenarios such as overtaking an obstacle or merging into a highway. For real-world driving, seat pressure mat data mainly represent forces related to driving manoeuvres. To extract body motion independently from these forces, sophisticated algorithms would be required.

As a remark for all results, it must be noted that all physiological effects were assessed by aggregating data across participants, drives and situations. Individual discomfort diagnostics requires sophisticated data fusion of all parameters, including vehicle kinematic and situation information. All physiological parameters are sensitive to a variety of stimuli and are therefore per se not a clearly interpretable measure of specific processes without knowing stimulus conditions and the context (Andreassi 2000; Backs and Boucsein 2000; Cowley et al. 2016). However, the presented results including relevance, direction and strength of physiological effects as well as timings and variability provide valuable knowledge as a basis for developing individualised discomfort detection algorithms.

2.6 Algorithm Development for Discomfort Detection

2.6.1 Measurement of Discomfort

There is no consensus in the literature regarding the measurement of comfort or discomfort (Hartwich et al. 2018). However, it is very clear that comfort plays an important role in the interaction of humans with objects or systems. Depending on the specific application or research area, the measurement methods, which are usually only indicators of comfort or discomfort, are determined according to the respective model approach. Examples from the area of classic automotive comfort definitions are the measurement of seating comfort (Cieslak et al. 2019), e.g. through measuring forces and accelerations as well as the indoor climate or noise pollution. With the introduction of automated driving functions, this definition extends to motion sickness (Yusof 2019) and the most natural driving styles possible (Elbanhawi et al. 2015). The latter raises the question of whether there is a comfortable driving style that drivers perceive as natural. To answer this question, discomfort must be measureable and, when considering an adaptively acting system (driver-vehicle-team), it must be individually measurable.

In addition to the results on uncomfortable driving situations presented in previous section, a further research aim of KomfoPilot was the development of an algorithm for the real-time detection of discomfort during automated driving. Main objective was the quantitative operationalisation of the drivers' subjective discomfort perception. The integration of the handset control (Sect. 2.2.2.2) during the automated drives, which provides a direct feedback from the participants on their perceived discomfort in the current situation, was a first step of quantising discomfort in the research context. However, using the handset control still requires active and accurate action from the user. Hence, based on the previously presented analyses (Sects. 2.4 and 2.5), vehicle kinematic signals as well as physiological data seem applicable for further quantising. Subsequently, a machine learning approach is proposed to estimate discomfort considering the aspects described above.

2.6.2 Estimation of Discomfort

From a system-theoretical perspective, the driver's personal perception can be assumed as a state that cannot be measured directly. However, input and output variables can be used to infer this state. According to Slater (1985) and de Looze et al. (2003), the feeling of comfort is a reaction to external factors, accompanied by physical and physiological effects. Thus, the current driving situation represents an important reason for feeling uncomfortable. If an automated system does not react adequately to these given circumstances or if this reaction does not correspond to the intent of the passenger, the passenger will not be in harmony with the environment (Engelbrecht 2013; Slater 1985). This condition can become apparent in physical or physiological reactions. These assumptions offer two possible information sources for algorithmically detecting discomfort. On one hand, analysing the current driving situation could help to identify a potentially uncomfortable situation. On the other hand, physical or physiological reactions of the driver/passenger could provide additional relevant input.

The proposed estimation of discomfort is based on machine learning by an artificial neural network (ANN) approach. Multi-layer perceptrons (MLPs) were generated in order to recognise characteristics in physiological data with relation to discomfort affected by the current driving situation. These elemental neural networks fulfil all set requirements, such as possible data fusion from different sources, e.g. eye tracker and smartband, and the ability of pattern recognition while not requiring the immense amount of training data compared to deep learning approaches. With regard to vehicle kinematics, aspects of driving style or driving context causing discomfort shall be identified through involving patterns in physiological data at moments when the driver feels uncomfortable in the current situation.

Based on the results of the two driving simulator studies (Sect. 2.5), heart rate, pupil diameter and interblink interval time were used as input features for the MLPs. As a first step, the standardised truck approach situation (Sect. "Simulated Driving Tracks") was analysed. In contrast to other automated driving scenarios, this situation was very predictable by repeatedly heading slowly towards the truck. Hence, most participants reported a gradual increase in discomfort by the handset control, resulting in successively increasing values when coming closer to the truck. The handset control signal was discretised into two distinct classes: (1) no discomfort in case the handset value was below 10 and (2) discomfort when handset values were 10 or above. Based on these two distinct classes, which can be interpreted as a rating of the current situation, no discomfort was defined as 0 vs. discomfort = defined as 1. This discretised handset control signal was used as the reference output label for the MLP training (Fig. 2.37).

In order to obtain an effective but also efficient MLP, topology was iteratively extended until training converged. Beginning with a very simple net consisting of one hidden layer with two units, one unit per iteration was added to the last hidden layer. In case the added unit did not improve the training, another hidden layer was added. This procedure continued as long as adding units or layers resulted in

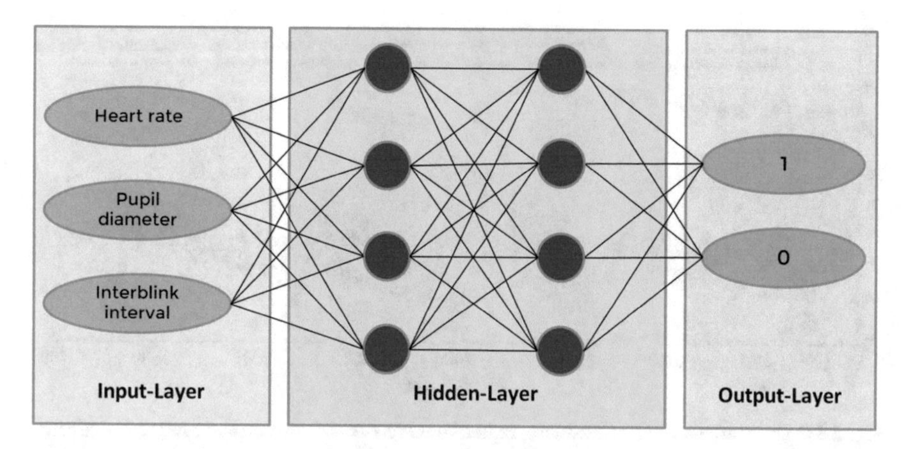

Fig. 2.37 Topology of the multi-layer perceptron; output: discomfort = 1, no discomfort = 0

improvements. Finally, the MLP possessed the necessary width and depth for the required pattern recognition without inflating parameter size and therefore reducing unnecessary long computation time especially during training. Thus, based on heart rate, pupil diameter and interblink interval time as input, the resulting topology of two hidden layers with four units respectively is shown in Fig. 2.37.

2.6.3 Data and Data Processing

In the driving simulator studies, several measurement data of the driver, such as head orientation, glance position, pupil diameter, interblink interval time or heart rate were asynchronously recorded with a common time base. The data were fed into a database for a centralised storage and generic, decentral access (Sect. 2.2.2.4).

Questionnaire data revealed that different driving styles could provoke discomfort to the driver (Sect. 2.3). These driving styles are objectively correlated to the given driving contextual situation, whereas physiological indicators are highly individual, e.g. differences between persons' heart rate at rest. However, statistical analyses with z-scores after transforming raw values into standard normal distribution showed promising indicators with regard to discomfort detection (Sect. 2.5). Data analysis shows specific effects in physiological data located at the same regions where participants indicated their individual discomfort using the handset control. These effects were most pronounced for pupil diameter, heart rate and interblink interval time (Sect. 2.5).

In offline analyses, z-scores of physiological responses (e.g. individual increase or decrease of heart rate or pupil diameter) could be calculated and aggregated over trips and persons. For a real-time application with continuous measurement data input, z-scores were calculated over a ten-second running window of past values. Figure 2.38 points out the differences between raw data, offline and online approaches illustrated

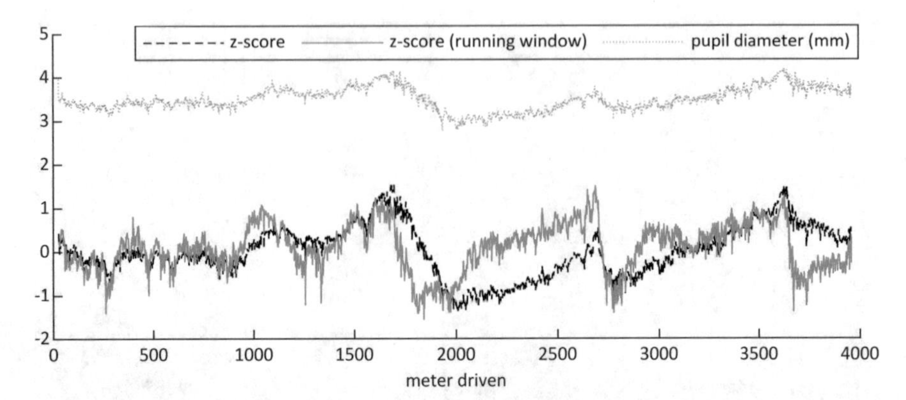

Fig. 2.38 Comparison of offline z-score, online running window z-score and raw pupil diameter

by the average pupil diameter during the standardised truck approach situations. While raw data variations in pupil diameter were partially levelled by the differences between participants, the z-scores highlight individual increases and decreases from the mean, resulting in stronger distances between minima and maxima. Using the running window approach, the z-scores possessed more fluctuation due to the reduced number of records for calculation. In consequence, this approach led to a more instable distribution. However, curve shape is still similar to the offline z-scores with clear peaks that could feature discomfort.

To ensure that the physiological reactions of the participants were related to the watched driving situation, measurements during glances into other areas then the front window or the street were filtered out for the discomfort detection algorithm. In doing so, also influences regarding adaptation and accommodation mechanisms could be reduced by taking out areas with darker light conditions in the interior, e.g. when the speedometer was focused.

For a supervised training of the described ANN in Fig. 2.37, a training data set had to be created. Therefore, the above-mentioned measurement data of pupil diameter, interblink interval time and heart rate were used as input. As label, the handset control values were divided into two defined classes (no discomfort = 0 versus discomfort = 1), by a threshold value of 10 out of 100. This means that the current situation was regarded as uncomfortable when the person pressed the handset control more than ten per cent.

At this point, two different approaches on how to use the label information were investigated. For the first approach, the individually indicated discomfort by each participant was used as label information during the training. The second approach tried to take into account occurring inaccuracy of direct use of handset control values based on the following observation: the handset control values are dependent on individual interpretation as well as the awareness for the current situation. In situations of brief discomfort, the handset control was either not activated at all or at least with some delay due to reaction time. In addition, the handset control represented a subjective indicator of discomfort with regard to decision making of when, whether

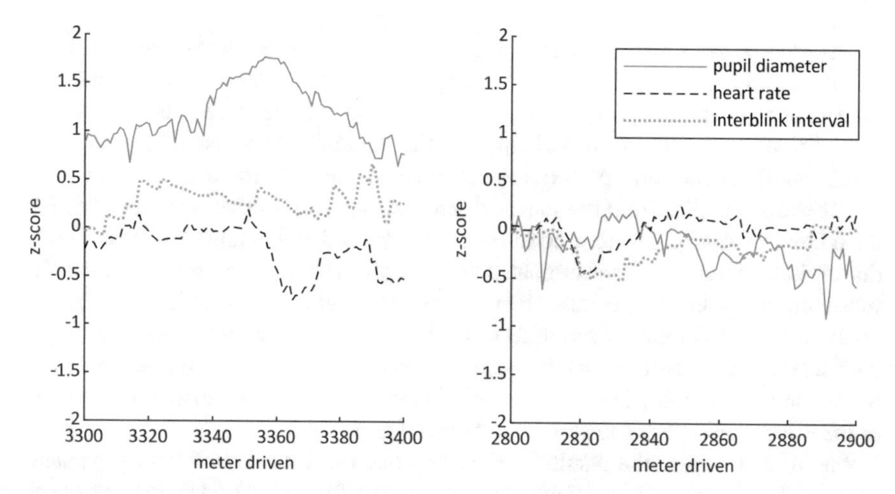

Fig. 2.39 Comparison of physiological data in z-scores calculated over a ten-second running window within a discomfort interval (left) and non-discomfort interval (right)

or how much to press the handset control, depended on individual habits of a participant regarding the usage of the handset control. Nevertheless, it can provide hints or can be used as orientation to identify discomfort in a differentiated manner. That is why for the second approach of labelling the output reference of class discomfort, accumulated data of the handset control were used. When several participants classified the situation as uncomfortable, the situation was labelled as discomfort section. An example is shown in Fig. 2.39 (left) between 3300 m and 3400 m driven. The selected discomfort interval was part of the intersection scenario (DS1) with the highest cumulated discomfort over all participants. In this situation, variations within pupil diameter, heart rate and interblink interval time were visible in z-score values based on the ten-second running window. Other situations with no indication of discomfort were labelled as non-discomfort intervals; see Fig. 2.39 (right). Between 2800 m and 2900 m driven, no participant indicated discomfort using the handset control.

2.6.4 Results

In order to evaluate the proposed approaches for the estimation of discomfort, different traffic situations of the data sets from the driving simulator studies were used. On the one hand, single situations such as the intersection situation and standardised truck approach situation were used as data set for training and testing the network described in Fig. 2.37. On the other hand, these trained networks were also applied to the entire drives.

In order to use as much data from the database as possible, a leave-one-out cross-validation, which is described below, was applied in the validation experiments. This method allowed the usage of the complete data set for training and evaluation. In each step, data of one person were left out from the training set and used for evaluation. Thus, iterating over every person, the calculated mean gives the overall result.

Measurement data from the standardised truck approach situations (Sect. "Simulated Driving Tracks") were used in testing the first labelling approach (Sect. 2.6.3) that divided segments of the automated drive by given handset control values into the two defined classes of discomfort by threshold value of ten per cent into the class of no discomfort = 0 versus discomfort = 1. These driving situations were very simple ones with no other distracting road users, e.g. vehicles and pedestrians, nor complex infrastructure. Thereby, the effect of the driving situation on the discomfort feeling of the driver could be demonstrated as isolated as possible.

Figure 2.40 shows the results of the leave-one-out cross-validation of the standardised truck approach situations. It contains data from all 43 participants of both driving simulator studies that wore the eye tracker to provide pupil diameter and interblink interval time. The dashed curve in Fig. 2.40 shows the relative frequency of how often the class discomfort (label = 1) occurred, based on the values from the participants during the automated drives using the handset control. Values predicted by the neural net (Fig. 2.37, solid line) are indicating the average result of all 43 iterations of the leave-one-out cross-validation. Values towards 1 can be interpreted as high discomfort. Values towards 0 are equivalent to situations without discomfort. It can be noted that the ANN can generally identify situations with discomfort, as the statistical tendency over all participants shows a similar shape regarding to increases and decreases compared to the given handset control values. Areas in which drivers indicated discomfort with higher frequency are also rated as uncomfortable by the

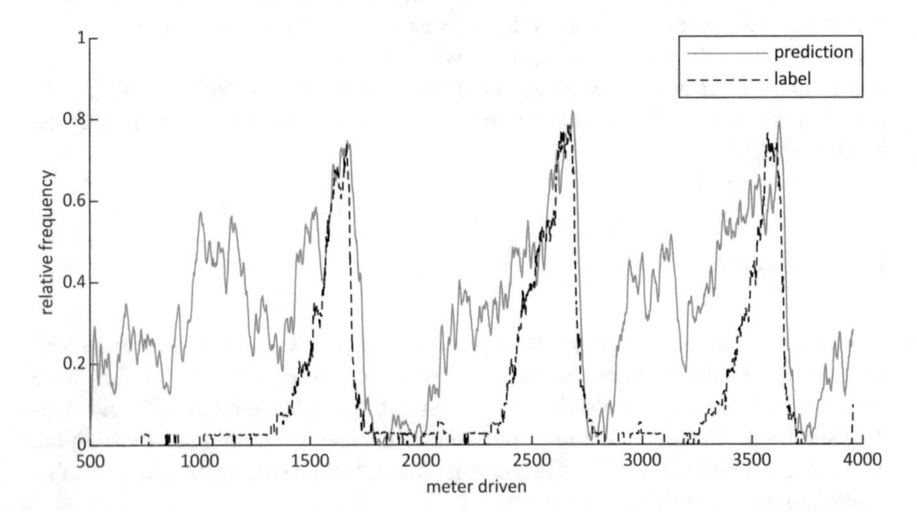

Fig. 2.40 Leave-one-out cross-validation of the standardised truck approach situations (solid: MLP prediction; dashed: relative frequency of label class discomfort)

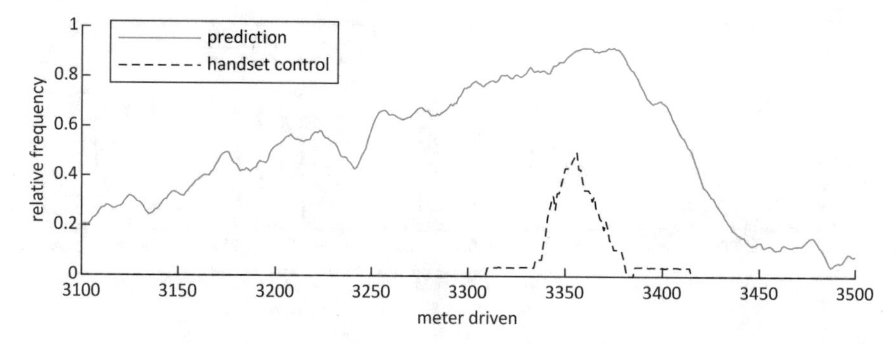

Fig. 2.41 Relative frequency of prediction (solid) and handset control signal (dashed) of class discomfort during the intersection situation

ANN, such as the areas around 1500, 2500 and 3500 m driven, where the ego vehicle approached closer to the truck.

The results of the leave-one-out cross-validation from applying the described network on measurements of the intersection scenario (DS1) are shown in Fig. 2.41. In this case, the second approach of labelling was applied. As described in Sect. 2.6.3, only the identified intervals of no discomfort and discomfort sections were used in the training step and the whole intersection situation to validate the learned parameters. This driving situation was more complex because of the non-signalised intersection and the approaching vehicle from the right side.

The average predicted values in Fig. 2.41 are increasing around the highest peak of indicated discomfort through the handset control. It is visible that the area of discomfort indicated by the participants can be identified at around 3350 m driven, as the tendencies with regard to increase and decrease of the prediction are matching the ratings given by the participants.

For comparison purposes, the MLP trained on intersection scenarios was also applied on other isolated situations, such as the approach to the red traffic light at 4563 m driven (Fig. 2.42). The classified discomfort based on the selected feature

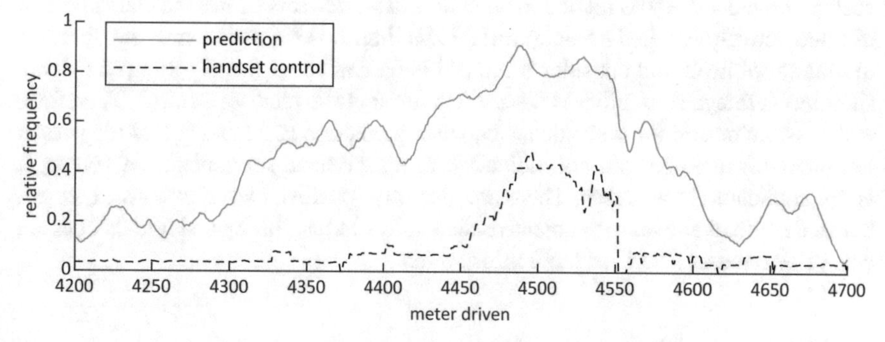

Fig. 2.42 Relative frequency of prediction (solid) and handset control signal (dashed) of class discomfort during the traffic light approach situation

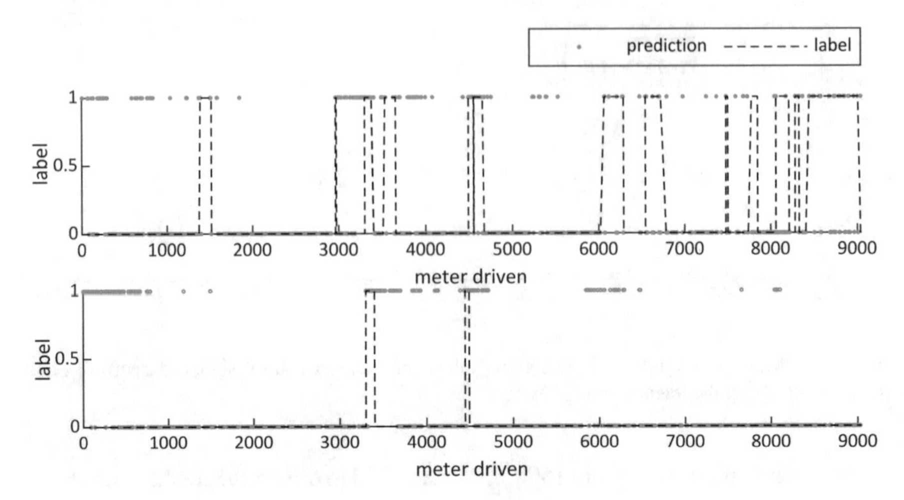

Fig. 2.43 Predicted discomfort labels of two single participants

signals (pupil diameter, heart rate and interblink interval time) increased in the relative frequency of predicted class discomfort during the interval from 4400 m driven to 4550 m driven. In general, the participants rated this traffic light approach situation as second most uncomfortable situation after the intersection situation (Sect. 2.5.4.2). In both situations, the predicted values show similar statistical tendencies as the rating of participants based on the handset control values.

A validation result using driving data of two single persons is exemplary depicted in Fig. 2.43. It characterises one processing step in the leave-one-out cross-validation method, which means that all other persons, except the one to be validated, were included in the training set in this step. The main discomfort situations indicated by the average handset control signal, i.e. intersection (3300 m to 3400 m), traffic light (4200 m to 4600 m), highway ramp (5300 m to 6200 m) and lane narrowing (7700 m to 8200 m), were classified as discomfort. By comparing the classification results with the individually indicated discomfort from these two persons, it is visible that not each predicted value was related to an indicated value. This shows the characteristic of a subjectively labelled non-comfortable feeling. It is the decision of the observer to monitor himself and consider if her or his feeling about the experienced driving situation is relevant to indicate discomfort through the handset control. Therefore, one possible reason for mismatches between prediction and label is that the person felt uncomfortable but did not indicate it using the hand set control. Of course, it is not applicable in all cases. There are also false-positive classifications expected, but as described above, the subjective discomfort indication by a single test person cannot be rated as a verified label information.

2.6.5 Discussion and Conclusions

Estimation of discomfort by an artificial neural network approach using MLPs was validated in different automated driving situations: the standardised truck approach scenario, a non-signalised intersection and an intersection regulated by traffic light. The MLP classifies discomfort by revealing patterns in physiological data. The validation results demonstrate that in general it is possible to estimate the perceived discomfort of drivers that monitor an automated drive in a driving simulator environment based on the physiological features heart rate, pupil diameter and interblink interval time. The usage of the handset control allowed a direct continuously and individual indication of discomfort, which opens up the possibility of including this information into the labelling process of the training data directly. Nevertheless, several uncertainties concerning an adequate label had to be handled, starting with the indication of discomfort, which is a highly subjective and individual process. The synchronicity between indicated discomfort and occurring of discomfort features is a critical aspect. The second synchronicity aspect is related to the individual specificity of the discomfort features between each driving situation sequence, meaning that one driver perceives discomfort in a situation not exactly at the same moment as another driver does. These issues were partly circumvented by creating discomfort and non-discomfort intervals based on the highest peaks in the averaged handset control value of all participants.

Pointing out these uncertainties about the label, it is also difficult to judge and interpret the prediction results. For example, there are predictions for a single person where she or he itself doesn't indicate individual discomfort, but it is clearly one of the main discomfort situations. However, the used approach to estimate discomfort is able to reproduce statistical tendencies, similar to participant ratings in questionnaires as well as direct responses by the handset control. Therefore, the interpretation of resulting estimation from developed MLPs can be considered a rather relative rating than an exact measurement of discomfort that is able to compare given situations and identify critical segments with regard to discomfort during an automated drive.

More research is required, especially regarding the detection algorithm and the acquisition of a naturally and realistic data set. Further measurement studies are necessary to acquire a larger data set. This must not necessarily be a wider range of persons. An essential step will be to acquire more data from one single driver. The results and analyses showed that in detail there are clear differences between drivers. Thus, it could be expected that an individually personalised classifier will have a higher performance for detecting discomfort. On sensor level, further investigations are also crucial. Especially a non-invasive measurement method for the pupil diameter and interblink interval time is required to use such an algorithm in a real vehicle. Results of the KomfoPilot algorithm development for discomfort detection can be used as basis towards discomfort detection algorithms that can be practically used in the strict sense of direct measurement to reflect driver's perception of discomfort during an automated drive.

2.7 Legal Aspects of Camera-Supported Driving Comfort Recording

The main concept of the project KomfoPilot is to adapt the vehicle behaviour during automated driving processes in a context-sensitive and personalised manner to the individual comfort and safety perception of the respective driver. This interaction between vehicle and driver requires continuous sensory acquisition of driver information. Within the scope of the project, this was done on one hand camera-based by recording of the driver's gaze, facial expressions and body positions via camera systems (eye, face and motion tracking) and on the other hand by using wearable devices (smartband), which continuously measured physiological data such as pulse frequency and skin conductance. The collected data, which capture driver's tension, search behaviour and attention in certain traffic situations, were finally transferred to a central database, so that the vehicle behaviour during automated driving could potentially be individually adapted to the driver. In this way, the vehicle could, e.g. automatically increase the distance between vehicles in dense rush-hour traffic if driver discomfort was detected.

Real-time comfort measurement with its effects on vehicle behaviour raises a number of legal questions, for example on the handling of the personal data, on the guarantee of device and IT security, on the liability in event of device malfunctions and measurement errors, and generally on the spheres of responsibility of the various parties involved in the overall system. The legal questions have to be raised at an early stage at the point of conception in order to ensure legal conformity by means of tailor-made technical and/or contractual solutions in the design of the systems as well as in their transfer to real traffic.

Only part of the project-related considerations can be presented due to the limited nature of such a book contribution. Therefore, only the data protection law and image law aspects of camera-based real-time comfort measurement will be discussed here.

2.7.1 *Camera-Based Real-Time Comfort Measurement*

The camera-based real-time comfort measurement via camera systems will be used for the multimodal capture of driving experience, inter alia the head movement, facial expressions, the gaze behaviour and body movements of the driver. However, it cannot be ruled that also information about the vehicle's surroundings will be recorded. When using these optical systems, a large amount of data are collected and processed, partly including personal data. Therefore, the data protection law and the art copyright law got a major importance in our project.

2.7.2 Data Protection Law

The starting point of all considerations is that every form of processing personal data is first of all an encroachment on the general right of personality in its special form as a basic right of informational self-determination as well as the basic right to guarantee the confidentiality and integrity of information technology systems (8 CFR[1]; Article 1 (1) GG[2] in connection with Article 2 (1) GG).

Everyone has the right to decide of collecting, using and processing of their own personal data. The entry into force of the European General Data Protection Regulation (GDPR) did not alter this situation. The principle of protecting personal data was merely placed on a European base, whereby the already known concept of data protection law was retained. Wherever data relates to a specific person or even allows conclusions to be drawn about their individual behaviour, normally[3] the scope of the GDPR will be opened. This is linked to a close-meshed catalogue of obligations for the protection of data in all processing operations.

2.7.2.1 Personal Data

Therefore, the decisive preliminary question is whether the data which are collected in the project are personal data (Metzger 2019; Erbguth 2019). Only these categories of data are subject to the GDPR protection regime.

"Personal data" means any information relating to an identified or identifiable natural person ('data subject'); an identifiable natural person is one who can be identified, directly or indirectly, in particular by reference to an identifier such as a name, an identification number, location data, an online identifier or one or more factors specific to the physical, physiological, genetic, mental, economic, cultural or social identity of that natural person (Article 4 No. 1 GDPR).

The information obtained from sensor-based camera systems—here video and image recordings, which record people's behaviour and/or biometric data—regularly have a personal reference in their original form,[4] because they show a person or capture information about a person.

This opens up the scope of application of the GDPR, even if the image no longer clearly identifies a person, but only traces him or her in the form of a thermal image, the image is available in a highly pixeled form or image sections merely show significant facial features (Haustein 2016). According to the GDPR, the characteristic of a personal reference is interpreted broadly; it is sufficient if conclusions about persons even indirectly are possible (Gola 2018, Art. 4 Rn. 16)—by whomever and however (e.g. because meta data, such as GPS information, image labels etc., which are also

[1] Charter of Fundamental Rights of the European Union.

[2] Basic Law for the Federal Republic of Germany.

[3] Exceptions are regulated in Art. 2 (2) GDPR.

[4] Judgement of the European Court of 11.12.2014–C-212/13, NJW 2015, S. 463 Rn. 22.

recorded, enable conclusions to be drawn about a person or methods of artificial intelligence convert a heavily pixelated image back into an original image[5]).

For that matter, it is irrelevant how much time is available for such a (re)identification process. Rather, the prevailing view is that even if an image of a person was created and buffered for biometric analysis purposes and has been deleted from the system within a few milliseconds, it can still be considered as a personal date, because it was—however briefly—an expression of the physical and physiological identity of a person within the meaning of Article 4 No. 1 GDPR (Simitis et al. 2019, Rn. 43; Auer-Reinsdorff and Conrad 2019, Rn. 271).

The GDPR does not know an exception for intermediate processing steps—in contrast to § 44 a UrhG,[6] which declares temporary and volatile acts of reproduction in the cache memory permissible and excludes them from copyright protection; according to the intention of GDPR, personal data can still be misused even in the case of rapid processing. On the other hand, it is equally irrelevant whether inference about a person is laborious and time-consuming. The only decisive factor is the possibility of identification.

The image and video material from the camera-based real-time comfort measurement can also be "particularly sensitive data", e.g. data concerning health (Article 4 No. 15 GDPR, because the health impairment or handicap of a person becomes apparent) or biometric data (Article 4 No. 14 GDPR, for example because physiological characteristics or behavioural features of a person become visible), which are subject to increased requirements (Article 9 GDPR).

It would now seem reasonable to assume that photos/images of persons regularly contain characteristics of persons and therefore should always be considered as biometric data and therefore as "particularly sensitive". However—and in this respect does the GDPR impose a restriction—according in recital 51, the processing of "biometric data for the unique identification of a natural person" that is particularly worthy of protection should only be assumed if it is processed through a specific technical mean allowing the unique identification or authentication of a natural person and if there is an intention to do this. These two legal requirements must be fulfilled cumulatively, which will be more often the case in research and real-life operations. It is quite conceivable that in the processing process, significant characteristics of a person are first filtered out and processed into a template in order to enable the recognition of the person with data set at a later point in time (e.g. so that a service robot in his field of operation could locate its reference subject between many others on the basis of the processed biometric image information by means of a system-immanent data comparison).

Whether such a biometric analysis is already planned at the time of recording or only during further processing of the image and video material must be determined in each individual case, because in every specific data processing operation (recording, saving, saving, deleting and publishing), a breach of data protection law could be

[5]An image processing software from the research department Google Brain calculates pixelated images to be sharp again with a success rate of 83% (Trinkwalder 2017).

[6]German Copyright Act.

occurred (Article 9 GDPR). As a result, it can be stated in any case that especially image and video material in human-technology interaction is suitable for qualification as a "special category of personal data".

On the other hand, "anonymised data", which do not relate at all to an identified or identifiable natural person (e.g. pure machine data), or personal data which have already been anonymised in the recording process itself in such a way that the person concerned cannot or can no longer be identified, may exclude the scope of the GDPR. The processing and use of anonymous data is not subject to any data protection regulations.[7]

While this may be rather difficult to realise or design at the recording of image and video material in the first processing step (usually location or metadata are stored here and the personal reference could be restored), the anonymisation could take place for the further processing of the recordings. This would rule out the applicability of the GDPR for the subsequent processing steps (e.g. sorting, storing, archiving). Such pseudonymisation should already be considered in the experimental design.

Anonymisation must be distinguished from pseudonymisation. In the case of pseudonymisation, the personal data is changed by a classification rule (e.g. assignment to an identification number) in such a way that the individual details to a person can no longer be assigned to that person without knowledge or use of the classification rule. Basically, three types of pseudonymisation and their reversal are conceivable, namely that the personal reference can only be restored by the person concerned himself (self-generated pseudonyms) or by means of a reference list (reference pseudonyms) or by using a so-called one-way function with secret parameters (one-way pseudonyms). Since the reference lists and parameters may also be known to third parties and therefore there is a risk of re-identification, the GDPR usually remains applicable if personal data is pseudonymised.[8]

All in all, it must be noted that in the era of big data and artificial intelligence, it will be increasingly difficult ensuring anonymity in the processing of video and image material, because data can be collected more and more "in real time" and put into relation to each other for new findings and connections. If the sum of apparently remaining "image and video fragments" or the inherent information of these sequences still allows conclusions to be drawn about a specific person, this is sufficient for the personal reference, and the actions would then again be measured against the GDPR.

---

[7]According to Recital 26 GDPG data are considered as anonymised, if "the data subject is not or no longer identifiable".

[8]Pseudonymisation has a different objective than anonymisation. Pseudonymisation is intended to exclude the direct knowledge of the full identity of the data subject during the processing and usage processes, if a personal reference is absolutely not necessary (Gola 2018, Art. 4, Rn. 36).

2.7.2.2 Prohibition with Reservation of Permission

If a personal reference would be possible during the recording and processing of image and video material, data protection law intervenes and ensures with its "prohibition subject to authorisation" structure that the personal rights of the data subjects, i.e. the persons to whom the information relates (Article 4 (1) GDPR), are protected. According to this provision, the processing of personal data is prohibited, unless there is a legal circumstance which justifies the processing (Article 6, 9 (2), (3) GDPR) or the data subject has expressly consented it.

For each processing step (recording, processing in the form of arranging, storing, adapting, reading out, querying, making available up to the deletion and destruction of the information), it must be determined whether such legitimising fact exist and—if not—what further measures must be taken to protect the personal rights.

At this point, the time has come to differentiate between the various groups of persons, connected to the research project, since the conditions of permission and justification are not the same for all involved persons.

In general, it will be possible to differentiate between (1) test persons, (2) users and (3) third parties: The test persons are those persons who act as experimental or test persons within the framework of the research project at the research institution or in the field. The users apply the systems in real operation. And finally, there are the third parties who are recorded or filmed by the sensor-supported camera systems (e.g. co-drivers, passers-by) in scientific and real-life situations, rather randomly. For all these stakeholder groups, the data protection concept must be taken into account right from the design of the systems ("privacy by design").

2.7.2.3 Test Person

Privilege of Research

In the scientific world, the misconception very often prevails that the "privilege of research" fully justifies any recording and processing of image and video material. However, this is not the case.

The research privilege does not create a general authorisation in the sense of a "carte blanche". On the contrary, Article 89 GDPR makes it clear that the general principles of data protection must also be observed in the scientific community and their activity. In order to achieve freedom of research, they can be relativised to a certain extent by the national legislator (Article 89 (2), (3) GDPR), whereby Article 89 (2) GDPR and also Article 9 (2) GDPR (i.e. on particularly sensitive data) explicitly ensure that the need for technical and organisational measures must always be guaranteed.

The latter concerns—just to name a few areas—the storage of information on cloud storage services in the form of commissioned data processing, which even in the case of a research privileges has to include the requirements of Art. 28, 32 or 44 GDPR (guarantee of processing security in third countries as well). Likewise, technical and organisational measures must also ensure an adequate level of protection to secure personal data in research environments, e.g. with regard to data encryption, the pseudonymisation, the continuous checking of the functionality of the systems, the reliability of research partners and so on. Finally, with regard to particularly sensitive data (health data, biometric data), the processing requirements of Article 22 (2) GDPR must be observed (general profiling ban).

The Federal Government and also the federal states have made use of the possibility for research privilegation and have made provisions for data protection for research purposes, which now work as legal permission through the opening clauses (e.g. § 27 BDSG[9]; § 12 SächsBDSG[10]).

According to state laws, the processing and further processing of personal data (also particularly sensitive data according to Article 9 (1) GDPR) are permissible for scientific research purposes, as far as the processing of previously non-anonymised (image) data serves scientific research purposes and is itself part of a serious and planned attempt to determine findings, interpretation and establishment of the truth (purpose limitation).[11] However, if the research purpose could be achieved in the same way with anonymised data, the necessity would be missing and the privilegation had to be dropped off—this must to be checked carefully in the case of video and image material, especially with regard to further processing.

Furthermore, the interests of the controller[12] (researcher) in the processing of the (image) data must outweigh the interests of the data subject in excluding the processing. This also must be determined on a case-by-case approach, but it is often unproblematic in relation to the test persons because on side of the researcher such a predominance normally exist.

In addition to these rather general statements in many federal state regulations the duties of anonymisation and pseudonymisation are further specified, e.g. by the storage image data the separation of main and auxiliary marks is regularly arranged, i.e. all identifying (image) characteristics must be stored separately in order to make it difficult to draw conclusions about the affected person.

On the other hand, the rights of the affected persons of Article 12 ff. GDPR—such as the enforcement of rights of rectification and opposition—are restricted or excluded in interest of the research purpose, because their use could hinder the achievement of the research purpose. These particularities under federal state law must be identified and taken into account in the conception phase.

While the research privilege also legitimises cooperation and data transfer to non-public bodies, provided that these bodies are involved in the research project, the entire field of research communication[13] is not covered by the research privilege. Therefore, the propagation of image and video material to a wider audience always requires a special consent of the data person according to Article 6 (1a), 9 (2a) GDPR.

[9]Federal Data Protection Act.

[10]Federal State Data Protection Act of Saxony.

[11]BVerfGE 35, 112 ff. (1973).

[12]Article 4 No. 7 GDPR.

[13]Research communication is the communication of the findings to a broad audience.

As a result, the research privilege represents a well-functioning basis of legitimacy in the purely scientific environment, also with regard to recording and further processing the image and video material of camera-based real-time comfort measurement. However, it does not release the operators from the data-protecting processing principles (Article 5 GPDR, e.g. data economy, storage limitation, confidentiality) and the necessary technical and organisational measures (Article 25, 30, 35 GDPR, e.g. data-protecting pre-settings, keeping a processing directory, data protection impact assessment, etc.).

Consent of the Test Person

Instead of the research privilege, the consent of the test persons can also be obtained for justification (Article 9 (2a) or Art. 6 (1) GDPR), whereby the test person must be precisely informed in advance of the nature, purpose and extent of the processing of his/her personal data and of the scope of the intervention. Furthermore, his or her consent must observe the formal requirements of Article 7 (2), (4) GDPR, i.e. it must also precede the collection of data, be voluntary and specific. For the specially protected data categories (biometric data), the consent must also explicitly refer to the "release" of these (biometric) data. The further processing operations must therefore be communicated very precisely so that the data subject can fully trace the path of his/her data (mandatory information and transparency requirement) and then expressly declare his/her consent with regard to each processing step (Wolf and Brink 2019, Art. 7, Rn. 52; Gola 2018, Art. 7, Rn. 21 and Art. 9, Rn. 16ff). So-called opt-out solutions are not sufficient here.

However, the exact description of the projects is rather difficult at the beginning of a project. Often, the type, purpose and scope of scientific processing with regard to the image data collected at the beginning cannot be completely specified by those responsible either at the time of recording or during their first further processing, as the use of divergent research methods suddenly leads to new research questions/purposes or unforeseen considerations. In accordance with the 33rd recital to the GDPR, for scientific research it should be sufficient to obtain consent for an abstract research area or for a possibly still unspecific research project (so-called sectoral consent or broad consent). However, giving consent for "any research" (so-called open consent) is not regarded as permissible.

Via the "declaration of consent", it is even possible—in contrast to the privilege of research—to include research communication and thus also the publication of image and video material at events, trade fairs, etc. (Article 6 (1a) GDPR). In return, however, the federal state law restrictions no longer apply, which are proclaimed as part of the research privilege, such as the restriction of the right of revocation, with the further consequence that the consent of the test person always stands under the reservation of revocation for those responsible, i.e. a previously granted consent would be revocable at any time and without justification, with the further consequence that the previously lawfully obtained recording could no longer be used and have to be deleted or the processing actions based on it would suddenly become illegal. In this respect, research projects should carefully weigh up which legal basis (research privilege or declaration of consent) would to be based for the data processing (and also the data protection declaration). It is also conceivable to use both together.

Privilege of Media

In connection with the use of media tools—such as the processing of personal image and video recordings—the media privilege is also often mentioned, according to which the processing of personal image recordings, including the stored EXIF files, should be lawful even without the consent of the data subject. This is reasoned with the freedom of expression and information (Article 85 (1) GDPR, Article 5 GG). However—and insofar this is not quite correct—the media privilege only covers the processing of personal image and video recordings for journalistic, literary and artistic purposes—for example, in the sense of the federal state press laws, the federal state broadcasting treaty or the telemedia law. In the context of the research project, the media privilege is not addressed in this form and therefore it cannot provide any justification for the processing of personal data.

2.7.2.4 User in Real Operation

In real operation, camera-supported comfort measurements will collect personal data of the users in order to process them further on the vehicle side. If the data cannot or should not be made anonymous because the automobile manufacturers want to make the data usable for further business models, the informed consent of the users alone[14] is suitable to legitimise data processing (Article 9 (2a) or Article 6 (1a) GPDR in conjunction with Article 7 and 12 GDPR).[15] In practice, the user (e.g. the automobile manufacturer, supplier) will have to define exactly what is to be done with the personal data that may already be generated by a "log file".[16] He will also be responsible for the use of the data in accordance with data protection regulations, because he decides on the purpose and means of processing personal data (Article 4 No. 7 GDPR).

With regard to the consent that will be required, it should be mentioned at this point that it must always be given voluntarily. Art. 7 (4) GPDG prohibits the linking of consent and performance of the contract (so-called linking prohibition). As the consent and the concerned data are not absolutely necessary for the performance of the contract, the performance of the contract or the provision of the respective

[14]It will be difficult to justify an overriding legitimate interest in the processing of comfort data.

[15]As not all driving functions and business models have been developed on the way to autonomous driving, the anonymisation of personal data may limit innovation. In this respect, further use of the data should at least be considered in the design phase, which would require the consent of vehicle users. When and how this consent could be given is to be considered during the development of systems, even if the concrete purposes are ultimately determined by the user of the systems.

[16]Registration of the vehicle user via a personal account in the vehicle.

service may not be made dependent on the consent. This is particularly problematic in constellations in which the vehicle user has booked the journey in the autonomous vehicle or has even purchased the vehicle and can only ride or drive it, if he previously consented to the use of his data ("sim or sink" situation) (Metzger 2019; Albrecht 2016). Since the decision in fact obtained, there is now no voluntary consent (Steege 2019). Furthermore, it must be considered in which way the necessary data protection information and declaration should reach the user and/or should appear in the vehicle (for various models, see Metzger 2019).

2.7.2.5 Uninvolved Third Parties

While the research privilege or the obtaining of consent for the recording and further processing of video and image material in the immediate project environment may be helpful towards the test persons and/or the users, the question remains how to deal with situations in which suddenly uninvolved third parties are detected by the sensor-supported camera systems—because they are in the direct vicinity of the test vehicle in road traffic or in the test environment of the image acquisition system.

Their personal rights are affected in the same way as those of the test persons or users, because the usage of other technologies (e.g. RFID/Beacon) could also determine their identity. If the complete anonymisation of personal data (possibly also particularly sensitive data) cannot be ensured during the recording process, the scope of application of the GDPR would be opened again for both the recording itself and for its further processing.

For the justification of the particular data-processing operations, it would be relevant again whether the encroachment on personal rights could be legitimised.

The research privilege does not apply here, because the picture information about the third parties (passers-by, etc.) are not necessary or required for the realisation of the research purpose (e.g. comfort recognition). Although the purpose of the research is broadly defined (recital 156), the DGPR leaves the precise details—as already mentioned—to the Federal Government and the Federal States.

However, both the BDSG and the regulations of the federal states do not contain any regulation regarding the handling of personal data outside the specific project, i.e. regulation which only appear "in the context" of the scientific research purposes. In this respect, the focus must be on the general legal justification.

First of all, the consent for processing under Article 6 (1a), 9 (2a) GDPR could be obtained. However, to seek such consent from uninvolved third parties would be an undertaking that is virtually impossible to achieve in practice. Furthermore, there are also a number of minors in the public area who would not be able to give their consent at all (Article 8 (1) GDPR).

In this respect, only the legal elements of the consent under Article 6 (1b)—(1f) GDPR can be considered, which could legalise the processing of image and video material, especially Article 6 Abs. (1f) GDPR. According to this rule, the processing of personal data is permitted even without an explicit consent, if the processing is necessary to protect the legitimate interests of the controller or a third party and if the interests of the data subject in protecting his/her personal data do not outweigh.

An example are research projects in connection with rear-view cameras for passenger cars, where the cameras used also image people who are in the rear camera field of view. Here, it is directly in the interest of the uninvolved third party to be seen in order to avert danger to his limb and life. In this respect, the data protection interests of the uninvolved third party take second place to the legitimate interests of the researcher or future user. As clear as the interest situation is in the event of a risk to life and health, such an overriding safety interest of the uninvolved third party cannot be assumed in comfort measurements, so that in this case his/her data protection interests must also be taken into account in the conception.

The capture of uninvolved third parties should therefore be avoided as far as possible, e.g. by placing the cameras in the vehicle so that the image/video recording extends to those groups of persons from whom consent can be obtained. If the recording of uninvolved third parties cannot be avoided, efforts should be made to ensure that the image data of third parties in the "black box" will be deleted or made anonymous during image recording, so that technical or manual access is no longer possible.

2.7.3 Art Copyright Law

Particularly with regard to the uninvolved third party, the art copyright act (KUG), which regulates the right to one's own image and thus pursues a similar objective to that of the GDPR, can also be of importance via the opening clause of Article 85 (1) GDPR. The KUG is a simple legal regulation to protect the right of visual self-determination. The design of the KUG is similar to the GDPR. The dissemination of photographs requires the prior consent of the photographed person (§ 22 KUG), although—in contrast to the GDPR—this can also be tacitly or impliedly declared and can only be revoked again for important reasons.

The KUG therefore allows an exception to the basic rule "dissemination only with consent". According to § 23 KUG, under certain conditions a personal image may also be permitted without the consent of the person concerned, e.g. if uninvolved third parties appear merely as "decorative attachment". These persons do not have to agree the dissemination of their photographs.

Whether this rule-exception system of the KUG may also influence the weighing decision within the framework of Article 6 (1 f) GDPR is currently disputed (Reuter and Schwarz 2020, 31ff). Some argue quite rigorously with the priority of the higher-ranking GDPR and the final provision of Article 6 (1f) GDPR (Weberling and Bergmann 2019; Lauber-Rönsberg and Hartlaub 2017), which completely replaces the KUG in producing a picture/video (in particular with regard to the consent). Others consider the KUG and its regulations to be a special legal exception to the GDPR, at least with regard to the use of the image material (Ziebarth and Elsaß 2018,

583) (even without a corresponding reference in the KUG[17]), which would then also remain applicable in the context of the opening clause under Art. 85 (2) GDPR.

The Higher Regional Court of Cologne[18] also recently decided in this direction—albeit with regard to a journalist—and confirmed that the GDPR merely states that the practical concordance between data protection and freedom of information and opinion had to be brought about, but did not itself make any stipulations under substantive legal. Therefore, the KUG must be incorporated as a special provision into the GDPR—at least for journalistic photojournalism. What the outcome of this dispute will be in the end is currently open. However, if one assumes the correctness of the opinion of the Higher Regional Court of Cologne, it is quite possible, in view of the initial situation "inclusion of uninvolved third parties" and the weighing decision "exemption from the requirement of consent" (Article 6 (1f) GDPR), to include the ideas of § 23 KUG. Consequently—in the research environment—the legitimate interest of the person takes a back seat to the research interest, because the affected person plays only a subordinate role according to the overall impression, the context of the recording and is only regarded as a "decorative attachment" to the main motive. Then, a consent would no longer be required.

This also has an impact on the area of research communication. If uninvolved third parties are recorded on pictures/videos, they must agree to the dissemination of their personal data (data protection declaration), unless they are merely classified as "decorative attachment"—in this case, the previous remarks can be taken up again.

In addition, it should only be pointed out that the video material from human–technology interactions cannot be qualified as "video surveillance", which would be subject to special regulations, among other things, special transparency and information obligations (camera symbol, communication of the identity of the person under surveillance, mandatory information at the place of surveillance) and organisational obligations (data protection impact assessment, data protection officer). Furthermore, the interests of the third party would be classified as "particularly important". Assuming that the video surveillance regulations would apply, the balancing of interests would be in favour of the uninvolved third parties. However, the special rule in § 4 BDSG, in addition to Article 6 (1f) GDPR, was expressly created for highly frequented public spaces (e.g. multi-storey car parks) and only for this area is the protection of the affected party classified as "particularly important interest". In the cases described here, the special regulations on video surveillance do not regularly play a role and the "normal" balancing of interests remains, in which § 23 KUG could also become relevant.

[17] Different in numerous federal state press laws.

[18] OLG Köln ZUM-RD 2018, 549; attested by OLG Köln 2019, 382.

2.7.4 Project-Related Solutions

Within the framework of the joint project, the above-mentioned principles of personal and data protection were already taken into account in the design of the experimental and test set-up. In doing so, great importance was attached to obtaining the necessary declarations of consent. Furthermore, in order to exclude profiling, image and video data were made anonymous in the central database, so that access in all subsequent processing steps only referred to aggregated information, and it was no longer possible to establish a personal reference.

With regard to the recording of uninvolved third parties, the dispute about the priority of the GDPR over the KUG was noted. Since the competition between the GDPR and the KUG has been disputed and unresolved to date, the uninvolved third parties were classified as affected parties within the meaning of the GDPR. In order to circumvent the problem of obtaining a declaration of consent, the cameras in the trial and test set-up were placed and focused in such a way that they did not permit detection of the surroundings (Art. 25 GDPR).

Acknowledgements The research project KomfoPilot (2017–2019) was funded by the Federal Ministry of Education and Research (BMBF) under grant no. 16SV7690K. More information can be found at https://bit.ly/komfopilot. We are very grateful to Konstantin Felbel, Marty Friedrich, Maximilian Hentschel, Paul Marcion, Isabel Engelhardt, Madeleine Bankwitz, Laura Heubeck, Daron Arto Kreutzer, Philipp Prokop, Nensy Le Thu Ha, Melanie Wutzler, Marla Schönecke and Vivien Lehmann for their assistance with data collection and analysis.

References

Albrecht JP (2016) Das neue EU-Datenschutzrecht — Von der Richtlinie zur Verordnung. Computer und Recht 32(2):88–97. https://doi.org/10.9785/cr-2016-0205

Andreassi JL (2000) Psychophysiology: human behavior and physiological response, 4th edn. L. Erlbaum, Mahwha, N.J

Auer-Reinsdorff A, Conrad I (2019) Handbuch IT—und datenschutzrecht, 3rd edn. C.H. Beck Verlag, München

Backs RW, Boucsein W (2000) Engineering psychophysiology: issues and applications. Lawrence Erlbaum, Mahwah, N.J

Beggiato M (2015) Changes in motivational and higher level cognitive processes when interacting with in-vehicle automation (Doctoral Dissertation). University of Technology Chemnitz. Retrieved from https://nbnresolving.de/urn:nbn:de:bsz:ch1-qucosa-167333

Beggiato M, Krems JF (2013a) The evolution of mental model, trust and acceptance of adaptive cruise control in relation to initial information. Transp Res Part F: Traffic Psychol Behav 18:47–57. https://doi.org/10.1016/j.trf.2012.12.006

Beggiato M, Krems JF (2013b) Sequence analysis of glance patterns to predict lane changes on urban arterial roads. Paper presented at 6. Tagung Fahrerassistenz—der weg zum automatischen fahren, Munich, 28–29.11.2013. https://mediatum.ub.tum.de/node?id=1187197

Beggiato M, Hartwich F, Schleinitz K, Krems JF, Othersen I, Petermann-Stock I (2015) What would drivers like to know during automated driving? Information needs at different levels of automation. 7. Tagung Fahrerassistenz, Munich, 25–26.11.2015. https://doi.org/10.13140/RG.2.1.2462.6007

Beggiato M, Hartwich F, Krems J (2017) Der Einfluss von Fahrermerkmalen auf den erlebten Fahrkomfort im hochautomatisierten Fahren. At—Automatisierungstechnik 65(7). https://doi.org/10.1515/auto-2016-0130

Beggiato M, Pech T, Leonhardt V, Lindner P, Wanielik G, Bullinger-Hoffmann A, Krems J (2017) Lane change prediction: from driver characteristics, manoeuvre types and glance behaviour to a real-time prediction algorithm. In: Bengler K, Hoffmann S, Manstetten D, Neukum A (eds) UR:BAN human factors in traffic. approaches for safe, efficient and stressfree urban traffic. ATZ/MTZ-Fachbuch. Wiesbaden: Springer Vieweg, pp 205–221. https://doi.org/10.1007/978-3-658-15418-9_11

Beggiato M, Hartwich F, Krems J (2018a) Using smartbands, pupillometry and body motion to detect discomfort in automated driving. Front Hum Neurosci 12:3138. https://doi.org/10.3389/fnhum.2018.00338

Beggiato M, Hartwich F, Krems J (2018b). Discomfort detection in automated driving by psychophysiological parameters from smartbands. In: Van Nes N, Voegelé C (eds) Proceedings of the 6th HUMANIST conference. Lyon: Humanist Publications, pp 8–13. ISBN 978–2–9531712–5–9

Beggiato M, Hartwich F, Krems J (2019) Physiological correlates of discomfort in automated driving. Transp Res Part F: Traffic Psychol Behav 66:445–458. https://doi.org/10.1016/j.trf.2019.09.018

Beggiato M, Rauh N, Krems J (2020) Facial expressions as indicator for discomfort in automated driving. In: Ahram T et al (eds) Intelligent human systems integration, IHSI 2020, AISC 1131, Springer Nature Switzerland. https://doi.org/10.1007/978-3-030-39512-4_142

Bellem H, Schönenberg T, Krems JF, Schrauf M (2016) Objective metrics of comfort: developing a driving style for highly automated vehicles. Transp Res Part F: Traffic Psychol Behav 41:45–54. https://doi.org/10.1016/j.trf.2016.05.005

Bellem H, Thiel B, Schrauf M, Krems JF (2018) Comfort in automated driving: an analysis of preferences for different automated driving styles and their dependence on personality traits. Transp Res Part F: Traffic Psychol Behav 55:90–100. https://doi.org/10.1016/j.trf.2018.02.036

Berntson GG, Quigley KS, Norman GJ, Lozano DL (2017) Cardiovascular psychophysiology. In: Cacioppo JT, Tassinary LG, Berntson GG (eds) Handbook of psychophysiology 4th ed. Cambridge University Press, pp. 183–216

Bocklisch F, Bocklisch SF, Beggiato M, Krems JF (2017) Adaptive fuzzy pattern classification for the online detection of driver lane change intention. Neurocomputing https://doi.org/10.1016/j.neucom.2017.02.089

Brookhuis KA, de Waard D (2011) Measuring physiology in simulators. In: Fisher DL, Rizzo M, Caird JK, Lee JD (eds) Handbook of driving simulation for engineering, medicine, and psychology. Boca Raton, FL: CRC Press, pp 17–1 to 17–10

Cacciabue PC (ed) (2007) Modelling driver behaviour in automotive environments: Critical issues in driver interactions with intelligent transport systems. London: Springer. https://doi.org/10.1007/978-1-84628-618-6

Charles RL, Nixon J (2019) Measuring mental workload using physiological measures: a systematic review. Appl Ergon 74:221–232. https://doi.org/10.1016/j.apergo.2018.08.028

Cieslak M, Kanarachos S, Blundell M, Diels C, Burnett M, Baxendale A (2019) Accurate ride comfort estimation combining accelerometer measurements, anthropometric data and neural networks. Neural Comput Appl 1–16. https://doi.org/10.1007/s00521-019-04351-1

Constantin D, Nagi M, Mazilescu C-A (2014) Elements of discomfort in vehicles. Procedia—Soc Behav Sci 143:1120–1125. https://doi.org/10.1016/j.sbspro.2014.07.564

Cooke A, Kavussanu M, Gallicchio G, Willoughby A, McIntyre D, Ring C (2014) Preparation for action: psychophysiological activity preceding a motor skill as a function of expertise, performance outcome, and psychological pressure. Psychophysiology 51(4):374–384. https://doi.org/10.1111/psyp.12182

Cowley B, Filetti M, Lukander K, Torniainen J, Henelius A, Ahonen L, Barral Mery de Bellegarde O, Kosunen IJ, Valtonen T, Huotilainen MJ, Ravaja JN (2016) The psychophysiology primer: a guide to methods and a broad review with a focus on human–computer interaction. Found Trends® Hum–Comput Inter 9(3–4):151–308. https://doi.org/10.1561/1100000065

Cropley M, Plans D, Morelli D, Sütterlin S, Inceoglu I, Thomas G, Chu C (2017) The Association between work-related rumination and heart rate variability: a field study. Front Hum Neurosci 11:217. https://doi.org/10.3389/fnhum.2017.00027

Dawson ME, Schell AM, Filion DL (2017) The electrodermal system. In: Cacioppo JT, Tassinary LG, Berntson GG (eds) Handbook of psychophysiology 4th ed. Cambridge University Press, pp. 217–243

de Looze MP, Kuijt-Evers LFM, van Dieën J (2003) Sitting comfort and discomfort and the relationships with objective measures. Ergonomics 46:985–997. https://doi.org/10.1080/001401303 1000121977

Ekman P, Hager JC, Friesen WV (2002) Facial action coding system: the manual. Res Nexus, Salt Lake City

Elbanhawi M, Simic M, Jazar R (2015) In the passenger seat: investigating ride comfort measures in autonomous cars. IEEE Intell Transp Syst Mag 7(3):4–17. https://doi.org/10.1109/MITS.2015. 2405571

Engelbrecht A (2013) Fahrkomfort und fahrspaß bei einsatz von fahrerassistenzsystemen. Disserta-Verl, Hamburg

Erbguth J (2019) Datenschutzkonforme verwendung von hashwerten auf blockchains. Wann sind kryptografische hashwerte von personenbezogenen daten selbst wieder personenbezogene daten? Zeitschrift für IT-Recht und Digitalisierung (MMR) 2019:654–660

ERTRAC (2019) Connected automated driving roadmap. european road transport research advisory council. Retrieved from https://www.ertrac.org/uploads/documentsearch/id57/ERTRAC-CAD-Roadmap-2019.pdf

Festner M (2019) Objektivierte bewertung des fahrstils auf basis der komfortwahrnehmung bei hochautomatisiertem fahren in abhängigkeit fahrfremder tätigkeiten. Retrieved from https://due publico2.uni-due.de/receive/duepublico_mods_00070681

Field AP (2013) Discovering statistics using IBM SPSS statistics, 4th edn. SAGE, Los Angeles

Gola P (2018) Datenschutz-grundverordnung. Kommentar. 2. Aufl. 2018, München: C.H. Beck Verlag

Gratton G, Fabiani M (2017) Biosignal processing in psychophysiology: principles and current developments. In: Cacioppo JT, Tassinary LG, Berntson GG (eds) Handbook of psychophysiology 4th ed. Cambridge University Press, pp 628–661

Griesche S, Nicolay E, Assmann D, Dotzauer M, Käthner D (2016) Should my car drive as I do? What kind of driving style do drivers prefer for the design of automated driving functions? 17. Braunschweiger symposium automatisierungssysteme, assistenzsysteme und eingebettete systeme für transportmittel (AAET), ITS automotive nord e.V., pp. 185–204

Hartwich F, Beggiato M, Dettmann A, Krems JF (2015) Drive me comfortable: individual customized automated driving styles for younger and older drivers. In: VDI (eds) Der Fahrer im 21. Jahrhundert. VDI-Berichte 2264. Düsseldorf: VDI-Verlag, pp. 271–283

Hartwich F, Pech T, Schubert D, Scherer S, Dettmann A, Beggiato M (2016) DriveMe: Fahrstilmodellierung im hochautomatisierten Fahren auf Basis der Fahrer-Fahrzeuginteraktion. https://doi. org/10.2314/GBV:870302329

Hartwich F, Beggiato M, Krems JF (2018) Driving comfort, enjoyment and acceptance of automated driving—effects of drivers' age and driving style familiarity. Ergonomics 61(8):1017–1032. https://doi.org/10.1080/00140139.2018.1441448

Hartwich F, Witzlack C, Beggiato M, Krems JF (2019) The first impression counts—a combined driving simulator and test track study on the development of trust and acceptance of highly automated driving. Transp Res Part F: Traffic Psychol Behav 65:522–535. https://doi.org/10. 1016/j.trf.2018.05.012

Haustein B (2016) Datenschutzrechtskonforme ausgestaltung von dashcams und mögliche ableitungen für den autonomen PKW. In: Taeger J (ed) Smart world—smart law? Weltweite netze mit regionaler regulierung. Edewecht, pp 43–59

ISO—International Organization for Standardization (1997) ISO 5805: Mechanical vibration and shock—human exposure—vocabulary. Genève, Switzerland

Jennings JR, Allen B (2017) Methodology. In: Cacioppo JT, Tassinary LG, Berntson GG (eds) Handbook of psychophysiology 4th ed. Cambridge University Press, pp 583–611

Jentsch M (2014) Eignung von objektiven und subjektiven daten im fahrsimulator am beispiel der aktiven gefahrenbremsung. eine vergleichende untersuchung. PhD-Thesis, Chemnitz University of Technology, Chemnitz

Jian J-Y, Bisantz AM, Drury CG (2000) Foundations for an empirically determined scale of trust in automated systems. Int J Cogn Ergon 4(1):53–71. https://doi.org/10.1207/S15327566IJCE0401_04

Johnston P, Rodriguez J, Lane K, Ousler G, Abelson M (2013) The interblink interval in normal and dry eye subjects. Clin Ophthalmol 253. https://doi.org/10.2147/OPTH.S39104

Klein G, Woods D, Bradshaw J, Hoffman R, Feltovich P (2004) Ten challenges for making automation a "team player" in joint human-agent activity. IEEE Intell Syst 19(06):91–95. https://doi.org/10.1109/MIS.2004.74

Ko B (2018) A brief review of facial emotion recognition based on visual information. Sensors 18(2):401. https://doi.org/10.3390/s18020401

Lauber-Rönsberg A, Hartlaub A (2017) Personenbildnisse im spannungs-feld zwischen Äußerungs—und datenschutzrecht. Neue Juristische Wochenschrift (NJW) 70(15):1057–1062

Leonhardt V, Pech T, Wanielik G (2017) Fusion of driver behaviour analysis and situation assessment for probabilistic driving manoeuvre prediction. In: Bengler K, Hoffmann S, Manstetten D, Neukum A (eds) UR:BAN human factors in traffic. Approaches for safe, efficient and stressfree urban traffic. ATZ/MTZ-Fachbuch. Wiesbaden: Springer Vieweg, pp 223–244. https://doi.org/10.1007/978-3-658-15418-9

Marquart G, Cabrall C, de Winter J (2015) Review of eye-related measures of drivers' mental workload. Procedia Manuf 3:2854–2861. https://doi.org/10.1016/j.promfg.2015.07.783

Metzger A (2019) Digitale mobilität—verträge über nutzerdaten. Gewerblicher rechtsschutz und urheberrecht (GRUR), 129–136

Meyer G, Beiker S (eds) (2018) Road vehicle automation 4. Cham: Springer International Publishing. https://doi.org/10.1007/978-3-319-60934-8

Michon JA (1985) A critical view of driver behavior models: what do we know, what should we do? In: Evans L, Schwing RC (eds) Human behavior and traffic safety. Plenum Press, New York, pp 485–524

Morris DM, Erno JM, Pilcher JJ (2017) Electrodermal response and automation trust during simulated self-driving car use. Proc Hum Factors Ergon Soc Annu Meet 61(1):1759–1762. https://doi.org/10.1177/1541931213601921

Pech T, Lindner P, Wanielik G (2014) Head tracking based glance area estimation for driver behaviour modelling during lane change execution. In: 2014 IEEE 17th international conference on intelligent transportation systems (ITSC) pp 655–660. https://doi.org/10.1109/ITSC.2014.6957764

Qatu MS (2012) Recent research on vehicle noise and vibration. Int J Veh Noise Vib 8(4):289. https://doi.org/10.1504/IJVNV.2012.051536

Reuter W, Schwarz J (2020) Der Umgang mit Personenbildnissen nach Inkrafttreten der DSGVO. Zeitschrift für Urheber- und Medienrecht (ZUM) 31–38

Riener A, Boll S, Kun AL (2016) Automotive user interfaces in the age of automation (Dagstuhl Seminar 16262): Schloss Dagstuhl — Leibniz-Zentrum fuer Informatik; Dagstuhl reports, vol 6(6). Retrieved from https://drops.dagstuhl.de/opus/volltexte/2016/6758/pdf/dagrep_v006_i006_p111_s16262.pdf

Roßner P, Bullinger AC (2018) Hochautomatisiertes fahren — Welche Fahrmanöver- und Umgebungsmerkmale beeinflussen erlebten Diskomfort?. In: VDI Wissensforum GmbH — VDI Berichte 2235 (Hrsg.), 34. VDI/VW Gemeinschaftstagung Fahrerassistenzsysteme und Automatisiertes Fahren. (S. 331–344). Düsseldorf: VDI Verlag

Rossner P, Bullinger AC (2019) Do You Shift or Not? Influence of Trajectory Behaviour on Perceived Safety During Automated Driving on Rural Roads. In: Krömker H. (eds) HCI in Mobility, Transport, and Automotive Systems. HCII 2019. Lecture Notes in Computer Science, vol 11596. Springer, Cham. https://doi.org/10.1007/978-3-030-22666-4_18

Roßner P, Bullinger AC (2019b) How do you want to be driven? Investigation of different highly-automated driving styles on a highway scenario. Proceedings 10th International Conference on Applied Human Factors and Ergonomics (AHFE 2019) and the Affiliated Conferences. 24.07.2019 bis 28.07.2019, Washington, USA. https://doi.org/10.1007/978-3-030-20503-4_4

Roßner P, Dittrich F, Bullinger AC (2019c) Diskomfort im hochautomatisierten Fahren — eine Untersuchung unterschiedlicher Fahrstile im Fahrsimulator. In: Gesellschaft für Arbeitswissenschaft e.V. (Hrsg.), Arbeit interdisziplinär analysieren — bewerten — gestalten, 65. Frühjahrskongress der Gesellschaft für Arbeitswissenschaft. Dortmund: GfA-Press

Schandry R (1998) Lehrbuch psychophysiologie: körperliche indikatoren psychischen geschehens (Studienausg.). Weinheim: Beltz, Psychologie Verlags Union

Scherer S, Dettmann A, Hartwich F, Pech T, Bullinger AC, Wanielik G (2015) How the driver wants to be driven—modelling individual driving styles in highly automated driving. 7. Tagung Fahrerassistenz, München, 25.-26.11.2015 https://mediatum.ub.tum.de/doc/1294967/1294967.pdf

Schmidt E, Decke R, Rasshofer R (2016) Correlation between subjective driver state measures and psychophysiological and vehicular data in simulated driving. In: 2016 IEEE intelligent vehicles symposium (IV) pp 1380–1385. https://doi.org/10.1109/IVS.2016.7535570

Silva MCG (2002) Measurements of comfort in vehicles. Meas Sci Technol 13(6):R41. https://doi.org/10.1088/0957-0233/13/6/201

Simitis S, Hornung G, Spiecker gen. Döhmann I (2019) Datenschutzrecht. Kommentar. 1. Aufl. 2019. Nomos

Simon K (2017) Erfassung des subjektiven erlebens jüngerer und älterer autofahrer zur ableitung von unterstützungsbedürfnissen im fahralltag, (Dissertation). TU Chemnitz. Retrieved from https://nbn-resolving.org/urn:nbn:de:bsz:ch1-qucosa2-318664

Slater K (1985) Human comfort. Springfield, Ill., U.S.A.: C.C. Thomas

Smith S, Koopmann J, Rakoff H, Peirce S, Noel G, Eilbert A, Yanagisawa M (2018) Benefits estimation model for automated vehicle operations: phase 2 final report. U.S. Department of Transportation. Retrieved from https://rosap.ntl.bts.gov/view/dot/34458/dot_34458_DS1.pdf?

Steege H (2019) Ist die DS-GVO zeitgemäß für das autonome fahren? Da-tenschutzrechtliche aspekte der entwicklung, erprobung und nutzung automati-sierter und autonomer fahrzeuge. Zeitschrift für IT-Recht und Digitalisierung (MMR), 509–513

Tran C, Trivedi MM (2009) Driver assistance for 'Keeping hands on the wheel and eyes on the road'. In: 2009 IEEE international conference on vehicular electronics and safety (ICVES) pp 97–101. https://doi.org/10.1109/ICVES.2009.5400235

Tran C, Trivedi MM (2010) Towards a vision-based system exploring 3D driver posture dynamics for driver assistance: issues and possibilities. In: 2010 IEEE intelligent vehicles symposium (IV) pp 179–184. https://doi.org/10.1109/IVS.2010.5547957

Trinkwalder A (2017) Bildbearbeitung mit hirn—künstliche intelligenz macht bildbearbeitung intuitiv. c't, p 76. Retrieved from https://www.heise.de/select/ct/2017/11/1495733265623730

Van der Laan JD, Heino A, De Waard D (1997) A simple procedure for the assessment of acceptance of advanced transport telematics. Transp Res Part C 5(1):1–10. https://doi.org/10.1016/S0968-090X(96)00025-3

Wade (2017, November 15). Wearable technology statistics and trends 2018. Retrieved from https://www.smartinsights.com/digital-marketing-strategy/wearables-statistics-2017/

Watson AB, Yellott JI (2012) A unified formula for light-adapted pupil size. J Vis 12(10):12. https://doi.org/10.1167/12.10.12

Weberling J, Bergmann J (2019) Aktuelle fragen der umsetzung des medienprivilegs der DSGVO. Zeitschrift für das gesamte medienrecht (AfP), 293–298

Wolff St, Brink HA (2019) BeckOK Datenschutzrecht. Kommentar. 30. ed, München: C.H. Beck Verlag

Yusof NBM (2019) Comfort in autonomous car: mitigating motion sickness by enhancing situation awareness through haptic displays (Doctoral Dissertation). Eindhoven: technische universiteit eindhoven. Retrieved from https://research.tue.nl/files/127355785/20190703_Yusof.pdf

Ziebarth L, Elsaß L (2018) Neue maßstäbe für die rechtmäßigkeit der nutzung von personenbildnissen in der unternehmenskommunikation? Zeitschrift für urheber- und medienrecht (ZUM), 583–585

Chapter 3
KoFFI—The New Driving Experience: How to Cooperate with Automated Driving Vehicles

Rainer Erbach, Steffen Maurer, Gerrit Meixner, Marius Koller, Marcel Woide, Marcel Walch, Michael Weber, Martin Baumann, Petra Grimm, Tobias Keber, Judith Klink-Straub, Julia Maria Mönig, Jakob Landesberger, Ute Ehrlich, and Volker Fischer

Abstract Imagine you are at the beginning of a journey from Stuttgart to Munich. It will take you almost three hours because of heavy traffic. It is quite warm outside; you have just had lunch and feel a bit tired. This is sure to be a long, exhausting, and boring trip. The good news is that your car can drive automatically and that you have KoFFI (in German: "Kooperative Fahrer-Fahrzeug-Interaktion") on board—the new intelligent driver assistance system for collaborative driving in both manual and automated driving modes. In this chapter, we describe how KoFFI supports you in typical traffic situations during that drive. On the one hand, there is the so-called guardian angel function, which helps you to survive critical traffic situations but also offers some convenient features during manual driving. On the other hand, you will learn how KoFFI can assist the driver at system boundaries and vice versa in various cooperative driving scenarios. In addition, we explain how to apply ethics-by-design during system development and how to take care of your personal data required for automated driving (e.g., driver monitoring video streams or data needed for personalization). KoFFI communicates with the driver via its innovative speech dialogue system, which can even distinguish between priorities and a user-centered

R. Erbach (✉) · S. Maurer
Robert Bosch GmbH, Renningen, Germany
e-mail: Rainer.Erbach@de.bosch.com

G. Meixner · M. Koller
UniTyLab, Heilbronn University, Heilbronn, Germany

M. Woide · M. Walch · M. Weber · M. Baumann
Universität Ulm, Ulm, Germany

P. Grimm · T. Keber · J. Klink-Straub · J. M. Mönig
Hochschule der Medien, Stuttgart, Germany

J. Landesberger · U. Ehrlich
Daimler AG, Ulm, Germany

V. Fischer
EML European Media Laboratory GmbH, Heidelberg, Germany

G. Meixner (ed.), *Smart Automotive Mobility*,
Human–Computer Interaction Series,
https://doi.org/10.1007/978-3-030-45131-8_3

human-machine interface. The results of and lessons learned from several user tests show that the cooperative assistant KoFFI is able to ensure a convenient, pleasant, and safe drive in either manual or automated driving mode.

3.1 Starting Your Journey/Introduction

The dream of automated driving is slowly coming true—very slowly, or at least slower than it was thought years ago. Although many consortia have now been formed comprising vehicle manufacturers, suppliers, and entirely new players such as Waymo (Alphabet/Google daughter) with Fiat Chrysler, JLR, and Lyft (Bloomberg 2018), or Aptiv with Mobileye, or Bosch with Daimler, the task is still very complex and it is still necessary to adapt the legal framework (Gleiss 2017). Automated driving on fixed and short routes, which are mostly available exclusively for automated vehicles, has already been mastered in part. One example of this is automated parking that works both for on-street parking spaces and in specially equipped parking garages (Bosch 2019).

Waymo recently announced that the company has now driven more than 20 million miles fully automatically with its vehicle fleet since 2009, half of them in the last year (Autoblog 2020). In addition, Waymo has been operating a robot taxi project in Arizona for the past year, but for safety reasons the company still engages a security driver on board. Even if he/she were to be eliminated soon, monitoring by telemetry and operator is still necessary in order to intervene in the event of errors or traffic situations which are unsolvable by automation (Techcrunsh 2019).

Nevertheless, in the next few years, more and more vehicles will be capable of highly automated driving in real road traffic, at least in special areas. However, people remain responsible in the lower levels (up to SAE L2, SAE = Society of Automotive Engineers) of automation (SAE 2014). Even with the next higher level (SAE L3), the driver must quickly take control again at the system's request. At the same time, conventional and automated vehicles will be on the road side-by-side. In practice, this will lead to many traffic situations that neither automation nor humans can solve alone. This is where the KoFFI project comes into play. In that public founded project, researchers of Bosch (head of consortium), Daimler, European Media Lab, University of Ulm, HdM Stuttgart and Hochschule Heilbronn have worked together for 3 years on HMI concepts for automated driving.

Cooperation between the vehicle and the driver is intended to deal with such complex traffic situations quickly and safely. At the same time, the project examines and evaluates ethical and legal aspects that play an important role in this context (BMBF 2019).

In a user-centered research and development process, a software framework has been developed with a multimodal user interface. The innovative speech dialogue system can even recognize urgencies based on the language style. This is the next step of natural language communication between humans and machines.

Situations were examined in which humans and automation are overwhelmed, but which can be solved jointly. The guardian angel function is of particular importance here. Interpretation thereof (when and how should the guardian angel intervene?) was optimized with test subjects in a gamification approach.

Various expert opinions on legal issues were drawn up during the project. In addition, ethical monitoring was carried out continuously and ethical recommendations and guidelines worked out for automated driving.

The results form the basis for continuous further development of cooperative driver–vehicle interactions. Later generations of KoFFI-like systems should be able to recognize new traffic situations independently, and to propose their own actions.

Summary and Lessons Learned/Status January 2020: To date and continuing into the near future, highly automated vehicles are still unable to cope with every individual complex traffic situation. The intelligent assistance system KoFFI can recognize many such situations and solve them through trustworthy driver–vehicle cooperation. To optimize this, technical, ethical, and legal experts worked closely together on KoFFI right from the start. The main statement is that the KoFFI project contributes to close the gap between manual and automated driving and will thus contribute to greater traffic safety.

3.2 Data Protection and Selected Legal Issues of Automated Driving

3.2.1 Introduction

The technical side of highly and fully automated driving appears to be very advanced and sophisticated. Ethical aspects have been discussed more and more intensively for some time now, even in non-scientific contexts, and have been causing a stir beyond national borders. In contrast, the legal development and therefore the evaluation and (case-by-case) examination of the phenomenon is despite the (national) legislator's best efforts, still in their infancy.

Opaque legal questions address both the whether (is automated driving generally permissible?) as well as the how of "automated driving." The first issue concerns traffic and liability law, the second (among others) data protection law. Automated vehicles are dependent on data. Systems have to identify, prioritize, monitor and exchange certain data to ensure functional and safe operation. In Germany, data protection law was traditionally linked to informational self-determination. The direct applicability of the General Data Protection Regulation 2016/679 (GDPR) since May 2018 has rather complicated the legal framework and raises difficult questions regarding the regulation of human–technology interaction in general and automated driving in particular (Keber et al. 2018).

Due to content guidelines and formal restrictions applicable to this compendium, the topic and subject matter of this piece are, based on a general consideration of the current situation in road traffic law, together with some selected examples within a comparative legal context (see Sect. 2) limited to a focused account of the complex of data protection law (Sects. 3, 4, 5, and 6), with particular attention paid to the voice assistance system included and used in the KoFFI project.

3.2.2 A Short Comparative Law Trip[1]

Due to the associated risk, it is essential to provide sufficient legal framework conditions for automated driving. The permits for testing and operation in road traffic, the duties of the motor vehicle driver, liability issues, and specifications for data storage have to be considered. Following Society of Automotive Engineers (SAEs) definition, automated driving is understood to mean highly and fully automated as well as autonomous driving according to SAE levels 3–5.

The operation of automated vehicles up to and including level 5 on public roads has so far only been legally permitted in parts of the USA. As there is no federal law with a uniform framework yet, the federal states still have extensive legislative leeway, which they have used very differently. While, in California, the legal requirements are very high and the technical specifications are particularly detailed, the picture in Arizona is quite different: Here, for example, the vehicles are not even required to have an accident data recorder, as it is the case in all other countries and federal states considered. Neither requirements for an emergency plan, nor for the training of test drivers are stated. Not even an increased sum insured is prescribed. In Arizona, self-certification is already sufficient for the operation of vehicles with automation level 5.

In contrast to the USA, EU member states are bound by the Vienna Convention on Road Traffic and the United Nations Economic Commission for Europe (UNECE) regulations, which is why they are currently unable to make regulations on autonomous driving according to level 5. Germany is the only European country where regular operation has been fully approved for levels 3 and 4. Austria allows regular operation as well, but only for parking and highway driving assist systems. However, in both countries the driver has to be present during operation. While, in Austria, there must still be a driver inside the vehicle during tests, requirements in the Netherlands, Great Britain, and France are much weaker. In the latter, the only requirement is to ensure that the vehicle can permanently be controlled remotely.

In Germany and Austria, there are similar liability systems, providing liability for fault and strict liability. Neither country currently sees any particular need for

[1]This section summarizes our paper comparing the legal situation for automated vehicles in several countries (Klink-Straub and Keber 2020).

legislative action with regard to liability in the event of an accident involving an automated vehicle. The Netherlands and France also have comparable liability systems, but these are structured differently. UK's "Automated and Electric Vehicle Act" of 2018 contains the remarkable rule that the owner's insurance company is *always* liable in the event of an accident involving an automated vehicle.

In Germany, the 2017 amended Traffic Act (Amendment Regulating the Use of "Motor Vehicles with Highly or Fully Automated Driving Function" from July 17, 2017) in § 1a Section 2 Nr. 3 provides for an technical solution which makes it possible for the driver to manually override or deactivate the automated driving mode. This approach was also proposed in the Report of the Ethics Committee "Automated and Networked Driving" (EC AND 6/2017, para. 19). The concept, as an expression of the fundamental assessment that the driver retains control of the vehicle at all times and thus retains his status as a driver (Vienna Convention on Road Traffic, Art. 8(5) bis)), is consistent in this respect, but—strictly speaking—is diametrically opposed to a system feature which (in certain cases) may (mandatory) "override" the driver ("guardian angel function").

Harmonising the regulations remains a major challenge for the legislator if it is to be avoided that a driver must first inform himself about the legal situation in the respective country before each border crossing and change his usage behavior of the automated vehicle. The abovementioned British liability rule may serve as a blueprint, as it concentrates claims by the injured party, who may also be the insured driver himself, in one place.

Country	Tests	Regular operation	Drivers in motor vehicles	Liability	Accident data memory compulsory
Europe					
Germany	1–4	1–4	Yes[a]	Liability for fault and strict liability	Yes
Austria		1–4[b]	Yes[c]		
Netherlands			No[d]	Liability for fault, strict liability for "vulnerable road users"	
United Kingdom			No[e]	No strict liability, since 2018 insurance of the owner	
France			No[f]	Liability for fault and extensive strict liability	

(continued)

(continued)

Country	Tests	Regular operation	Drivers in motor vehicles	Liability	Accident data memory compulsory
US					
California	1–5	1–5	No[g]	Liability for fault and product liability	Yes
Arizona	1–5	1–5	No		No

[a]Driver must be perceptive during regular operation and—depending on the individual case—during test drives
[b]Limited to parking and highway driving assist systems
[c]In normal operation, the obligation to hold the steering wheel with at least one hand is waived, in test drives the driver must take up position as intended (except for parking assist system)
[d]However, driver intervention must be possible
[e]Must not turn his attention to other things
[f]But driver must control vehicle at any time
[g]But driver must be in control from outside

3.2.3 Project-Related Data (Types): Personal Data?

In the academic discourse, some progress has been made with regard to the structuring and categorization of the data (types of data) generated in automated vehicles (instructive Schwartmann and Jacquemain 2018; Klink-Straub and Straub 2018; Metzger 2019).

It is possible to differentiate between:

1. Maintenance and repair data (e.g., tire pressure, wear and tear, fill levels),
2. Data on the use of emergency systems (e.g., eCall system),
3. Data on the evaluation of driving behavior (e.g., distance measurements),
4. Data on driver monitoring (e.g., Volvo system “Alcolock”),
5. Traffic data (e.g., “Cooperative Intelligent Traffic System,” known as C-ITS),
6. Data on journey analysis (e.g., location, routes),
7. Accident-related data (e.g., steering and braking behavior, speed, vehicle occupancy),
8. Data on the use of media and communications (e.g., search queries, radio programmes),
9. Data on the environment and surrounding area (e.g., weather, lighting conditions, traffic volume).

Although this list could be expanded further and divided into further subcategories, and although distinctions are often difficult to make, it was possible to define and confine problem analysis within the context of the legal data protection examination in the KoFFI project insofar as it primarily concerns data within the passenger compartment, particularly in the form of voice recordings.

Superficially, this concerns mere content data (text content), which could, however, also be used as data for driver monitoring or journey analysis if necessary, or even as data on the use of media and telecommunications, and designated as such. Here, the additional purpose of application and use would be a decisive factor for well-founded qualification and classification. The indication and (prior) definition of it would, in practical terms, not just be necessary for legally compliant data collection, but would also be a necessary part of informing the user properly.

For the above reasons, it is also unnecessary to perform an in-depth examination of whether all data related to the (KoFFI) vehicle would be deemed "personal" within the meaning of the GDPR (Art. 4 Nr. 1) if, ultimately, a personal reference to the identity of the keeper (vehicle identification number, registration number) could always be drawn for each piece of data. This issue, which appeared to have first been clarified since the fundamental decision by the ECJ (ECJ C-582/14, 19.10.2016) and the finding made there on the application of a so-called relative personal reference, but which continues to entail significant legal uncertainty and practical problems, does not arise here. This is applicable at least when the keeper drives their vehicle themselves and is therefore the notified subject of the data processing.

This is because such a connection is regularly made in the case of personal audio recordings and should—with increasing differentiation and perfection of the technical possibilities—at least apply if the person concerned is recorded directly and up close, so that he or she can be (re-)identified clearly and unambiguously on the recording, either audibly or visibly, possibly also by further technical means (e.g., speech analysis, biometric measurement; for (initially unidentifiable) sound, image and video recordings see ECJ ruling C-212/13, 11.12.2014).

Furthermore, it is frequently assumed that this type of data collection concerns "special categories" of personal data (e.g., Schwenke 2018) which are subject to special protection in accordance with Article 9 GDPR and generally require the explicit consent of the person recorded. Voice and sound recordings may represent "health data" within the meaning of Art. 9 para. 1 DSGVO if, for example, a physical/psychological infirmity or illness can be inferred, e.g., stuttering, alcoholisation, depression. That means that voice data may be especially sensitive regardless of the issue of whether the recordings also contain (special) content-related information.

3.2.4 *Personal Data: Who is Who?*

"Personal data" within the meaning of the GDPR would therefore be processed, in the case of both the vehicle driver's voice and audio recordings and, if applicable, their images and video recordings. This relates to both stored and, if applicable, encrypted raw data (i.e., source data) and the specific content-related data of the voice recording. What is known as "household privilege" (Art. 2, para 2 lit. c GDPR) does not apply here, even though the enclosed, protected space inside the vehicle can be fully considered a personal area (for video-takes: Schwenke 2018).

This is when data leaves this area and is transferred and processed outside this enclosed space (manufacturer, lessor, employer, provider, trustee, if applicable), unless only internal vehicle data processing is performed (so-called black box). The latter means that there is only an on-board data unit (OBD) interface in the vehicle itself, which can be used, e.g., for the purpose of later readout by car repairers, public authorities, manufacturers, etc. (Balzer and Nugel 2016).

However, this does not take into account passengers and other occupants whose voices or comments may be recorded by the voice recognition system under certain circumstances (whether technically intended or not). This poses the question of the legitimation of a (joint) recording. This would have to meet the conditions for legality of Article 6(1), as far as not even concerned by Art. 9 GDPR (see point III). This would involve taking into consideration the alternatives of (a) (consent), (b) (contract performance), (d) (the data subject's vital interests) or even (f) (the controller's legitimate interests).

In this case, however, (d) (the aspect of "traffic safety") would not really apply because the voice recognition system would be intended to work in relation solely to the driver to avoid mixed, unclear, different or even conflicting "commands." Regarding Article 6(1)(f) GDPR (the controller's legitimate interests) practically all economic or non-material and not simply illegal interests of the person responsible may be taken into acount. However, the interests of the person concerned, i.e., his fundamental rights and freedoms (in particular Art. 8 GRCh, privacy) may not outweigh. The "reasonable expectation" of the same must be assumed (Recital 47 GDPR).

With regard to any consent that may be required (lit (a) of Article 6), it would also have to be taken into account that passengers are often minors. The protection for children and young people provided by Article 8 GDPR, which lays down conditions for consent from individuals under the age of 16 years, applies here. This means that, in principle, consent would have to be obtained from the individual with parental responsibility on a case-by-case basis (in general, the parents must exercise the parental custody in mutual agreement for the best interests of the child; see §§ 1627, 1687 German Civil Code BGB). This gives rise to huge practical problems, which is why, in this regard, implementing the provisions of the GDPR in a legally compliant way could be difficult in voice-controlled vehicles (Schwenke 2018).

Another problematic area in terms of practical application could create situations in which the drivers of an automated vehicle often change (e.g., lease or hire cars, use of a company vehicle by an employee). Here, the developers and manufacturers of systems are also required to establish an authentication procedure that works reliably but minimizes data at the same time, and which guarantees the secure assignment of the data being processed and a low-risk deletion procedure. One step may be achieved by (biometric) templates, which guarantee the (re)recognition of already registered users by storing reference values in a permanent data set (Art. 29 Data Protection Working Party 2012).

More generally, the legal data protection principles of Article 5(1) GDPR must be taken into account even in the development phase of these systems, as must the provisions of Articles 24 and 25(1) ("Privacy by Design") and (2) ("Privacy by Default") GDPR. Although the wording of Art. 24 et seq. obliges (just) the data controller, Recital 78 GDPR also "encourages" manufacturers to take these principles into account when developing the product and "to ensure that they take due account of the state of the art."

In respect of the KoFFI project, authentication of the driver should be available through voice input(s), including (additional) the simple specification of a code or password, if necessary. As in the case of a simple driver authentication (password), passengers and co-drivers cannot be reliably excluded, the actual identification of persons—including subsequent release of the system or vehicle and the (exclusive) reaction of the system to the respective person—should be carried out by means of reliable, individual speech, and voice recognition and should therefore be implemented primarily in the system with respect to Art. 6/9 GDPR.

Reflecting the recently published Guidelines 1/2020 on processing personal data in the context of connected vehicles and mobility related applications by the European Data Protection Board (adopted on January 28, 2020), all processes concerning (personal) speech and voice recognition data should operate (and stored) **inside the vehicle** as this guarantees "the user the sole and full control of his/her personal data and, as such, it presents, "by design," less privacy risks especially by prohibiting any data processing by stakeholders without the data subject knowledge" (EDPB, Guidelines 2020).

3.2.5 *Potential Recipients of the Data and Data Mobility*

The key question in respect of the general "passion for collecting" data remains, however, one regarding the data controller and (potential) recipients of the data (Krausen 2019). The new Section 63a of the German Road Traffic Act appears to give a clear, unambiguous response from the legislator (in-depth analysis and critical: Schmid 2017; Berndt 2018).

According to the new law, "data concerning location, date and time obtained using a satellite navigation system" must be stored if control of the vehicle switches from the driver of the vehicle to a highly or fully automated system, if the driver of the vehicle is required by the system to take over control of the vehicle or if a technical error occurs within the system (§ 63a Section 1 German Road Traffic Act, i.e., StVG). Hence, there is an obligation for implementing a "driving mode" or "accident data memory" as a requirement for the vehicle design (e.g., Lutz 2019).

The data stored in this way may be transferred to the "authority responsible for the prosecution of road traffic offenses under state law" at the authority's request and may be stored and used by it. When doing so, the scope of the data must be restricted to the extent necessary (§ 63a, Subsection 2, StVG). This provision does not grant an independent power of access in favor of authorities. Instead and according to the “two doors model” of the German Constitutional Court (BVerfG, judgement 1 BvR 3541/13 of 6.3.2014) only a right to data transmission is constituted (on the possibilities of criminal prosecution authorities to (independently) read out the “black box”: Berndt 2018).

Basically, Data according to § 63a StVG must be erased after 6 months (§ 63a Abs. 4 StVG). However, the keeper of the vehicle must arrange such transfer to third parties (insurance providers, manufacturers, etc.) if this is necessary in order to make, satisfy or defend legal claims in the event of an accident (see Section 7 StVG) and the automated vehicle concerned was involved in this accident. The time limit for erasure in this case is 3 years. There is legal debate concerning the calculation of the beginning of the three years period (recording or transmission) and regarding the addressee of the obligation to delete.

While, therefore, the scope and group of recipients of data collection limited in this sense in a highly or fully automated vehicle could be viewed as clarified, there is the other question of the legal nature and the regulatory effect depending on such a definition and associated with Section 63a StVG.

Reading § 63a StVG together with the provisions of the GDPR, one may argue that the processing of data mentioned there would constitute justification within the meaning of lit. (c) (legal obligation) or even lit. (e) (public interest) of Article 6(1) GDPR. § 63a paragraph 1 StVG does not address data about the (exact) identity of the driver, his driving style, the speed of the motor vehicle or route logs as a combination of position data. Against this background could be argued, that further data collection in an automated vehicle—and therefore other groups of recipients too—may be admissible and permitted.

However, the regulation should reflect the (beyond legal permission) generally recognized principle of reserving the right to give consent under data protection law (instructive Rüpke et al. (2018). Hence, any collection, storage, and transfer of the data found in a (highly or fully automated) vehicle beyond the data processing governed by Section 63a StVG would, without consent, be inadmissible (for the discussion during the legislative process see BT-Drs. 18/10192; BT-Drs. 18/11300).

Anyway, it is not readily apparent from the requirement of the application of “general regulations” (i.e., the GDPR) whether data processing beyond that specified in Section 63a StVG is admissible. Instead, further data processing would still be (case by case, data category by data category) subject to the requirement for grounds to justify it within the meaning of Article 6(1) GDPR (i.e., consent, contract performance, and legitimate interests in particular).

Another fundamental problem in relation to the question of the group of (potential) data recipients concerns the data controller, i.e., the subject of the retention obligation. In respect of Section 63a (1) StVG, this has not (yet) been clarified by the legislator.

A wide concept regarding data controllership (and hence responsibility) is reflected in the “Joint Declaration by the independent data protection authorities of the Federal and State Governments and the German Association of the Automotive Industry (VDA)—Data Protection Aspects in the Use of Networked and Non-Networked Vehicles” (2016), according to which the person who “reads” personal data is also the responsible party.

Also, based on recent ECJ case law regarding joint responsibility (ECJ, C 210/16, 05.06.2018) it can be said with some legal certainty that the manufacturer, should it have direct access to data streams in the vehicle, would be considered the data controller even if it did not provide its own storage media (e.g., Cloud, for details see Wagner and Goeble 2017).

3.2.6 *The Road Ahead*

Regarding the exercise of "data sovereignty," so-called data custodians are currently under discussion. Details are still controversial, e.g., which relevant data in this regard should initially (exclusively) be collected and stored and only released to those entitled or to the requesting party as necessary (e.g., on request in accordance with Section 63a (2) or (3) StVG). In particular, discussions have covered who should be the responsible party (also for the costs) of such an institution (for details see Brockmeyer 2018; Steinert 2019; Krausen 2019a, b).

With strong impetus from the automotive industry, the question of data ownership is currently (still) being raised in legal policy (de lege ferenda), which, depending on its legal design, could favor the manufacturer as the provider of infrastructure, the driver as the data producer or the owner of the vehicle as the holder of exclusive rights. The associated questions, including whether such a concept should only refer to personal data and/or also to non-personal machine data, are still open. The German Data Ethics Commission has rejected the concept of data ownership in its report in October 2019.

As to certain functions of a market-ready KOFFI an important question regarding existing law (GDPR) concerns the right to data portability. The right was newly introduced with Art. 20 GDPR. It supplements the right of access which has existed for some time and aims to enable the data subject to change providers (the legislator had social networks in mind, for example) while taking his or her own personal data with him or her. In the context of automated driving, the challenges are also in this context to clarify responsibility (since manufacturers, suppliers, dealerships, providers, sharing companies, fleet managers, car repairers, insurance companies, etc., are involved in the use of data), to take account of the data protection rights of third parties or business secrets (since they may constitute a reason for exclusion) and to check whether the condition of provision within the meaning of Art. 20(1) GDPR is met. The intention of the legislator was to develop interoperable formats to support data portability, which has not yet been achieved by the automotive industry (Klink-Straub and Straub 2018).

3.2.7 Conclusion

As shown, it is not just the technical processes of highly and full automated vehicles that are complex, but also their inclusion and classification under (data protection) law.

This begins with the question of which types of data to define and categorize and how precisely to do so, and specifically how much protection they (should) have, before moving on to the problem of the users (or non-users!) concerned and the definition of this group of people, finally ending with the in fact rather alarming admission by the German legislator that it itself does not know exactly who the controller, and therefore the subject of the regulations it has imposed, actually is.

Therefore, in respect of the KoFFI project, in which speech-based driver–vehicle interaction with content-related communication has been at the center of the investigation, there are—besides the implementation of the problem of legal liability issues and the corresponding legal requirements also significant difficulties related to data protection law. These lie, in particular, in limiting the recording and further processing of voice content, its sole focus on the driver and, last but not least, its particular need for protection, as so-called sensitive data in accordance with Article 9 GDPR.

In this regard, manufacturers and suppliers of such products (and their components) are recommended to produce a system specification that is worded clearly, concisely and comprehensively. This should then also form the basis for any contractual declarations or declarations of consent by users (buyers/keepers) that are required in this regard; these would preferably be obtained in writing, and also be worded clearly and concisely.

A problem could arise here in respect of obtaining consent when the driver and/or passenger differs or changes. A simple but reliable authentication procedure should therefore be carried out for the driver (at least) at the beginning of each journey. In contrast to non-registered drivers, declarations of consent and notices that are standardized and/or obtained in advance can be produced for this, for example, if the keeper/buyer and driver are the same person, with these defining the content of the contract or even making explicit reference to legitimate or even vital interests (see Article 6(1) GDPR). If, due to the objectionable processing of "sensitive data" (Article 9 GDPR), consent is not obtained and, for example, legitimate interests are taken as a basis instead, it must be ensured by the manufacturer, even in the development stage (*Privacy by Design*, Article 25(1) GDPR), that appropriate technical and/or organizational protective and circumvention measures are taken, for example, to guarantee erasure plans and procedures or even prompt anonymization or pseudonymization procedures by the system for the purpose of data minimization. Pending further legal regulations, voluntary use of the system, including the easy and personal deactivation of individual categories of data at any time (*privacy by Default*, Article 25(2) GDPR), should be a matter of course, not just because of current general legal conditions but also for reasons of autonomous action on the part of the data subject and a huge comparative advance in result at least.

Legal requirements are still vague, ambiguous and therefore inadequate. Additional sector-specific clarifications and the closure of normative legal loopholes would be required here. For reasons of legal certainty and therefore also the prompt attainment of road safety, this should preferably lie with the legislator—and not, for example, rather slow case law that does not keep up with rapid technological development—and could definitely be implemented at national level for the first time. In the medium-term to the long-term, however, it is not an issue that can be addressed single-handedly on a national scale due to the nature of the matter (e.g., limitless flow of data, vehicles and goods). National regulations should therefore be drawn up in further consideration of the international or, in any case, European context and its legal framework.

3.3 Design of a Guardian Angel

This chapter explains the function and design of an innovative driver assistance system that acts like a guardian angel. The first section provides motivation of what the system is about and why its invention could help to reduce the number of car accidents. The second section describes the proposed HMI of the system and presents the results of corresponding user studies. In the last part of this section, a framework for collecting unbiased driver information on situations and behavior is presented and a study associated with it is explained.

3.3.1 Why Do We Need a Guardian Angel in the Car?

The WHO proclaimed that "road traffic injuries are the eighth leading cause of death for all age groups" (WHO 2018). To help reduce the number of fatal accidents, the German Road Safety Council presented the "Vision Zero" (DVR 2012) in which it emphasizes the development of new technologies and interfaces between the car and the driver. The European Union issued an order that all new vehicles must be equipped with advanced safety features, such as emergency braking assistant systems, by 2022 (EP 2019).

Automated and autonomous driving is another emerging technology that will help reduce the number of accidents dramatically. But even in automated cars, the driver can and must take control of driving in some cases. A car that is capable of automated driving does not need to disable all of its sensors while being controlled by a human. Therefore, it is very likely that the car can detect dangerous situations and human errors. In such situations, the car could come up with a strategy to mitigate the risk and to avoid (or at least attenuate) an accident. If this is the case, the car could act like the driver's guardian angel as it takes control.

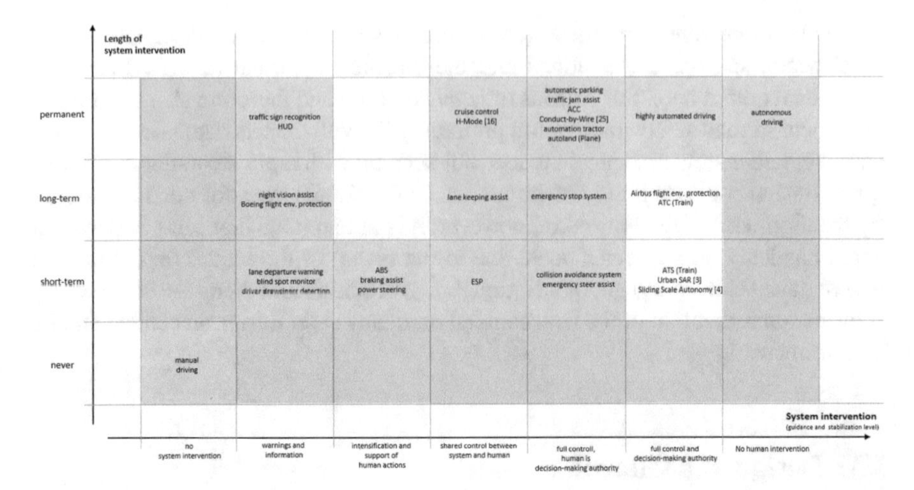

Fig. 3.1 Taxonomy of various assistance systems in various transportation modes. Green highlighting indicates that at least one automotive technology is present in the respective category (Maurer 2018a).

Available assistance systems can be classified using a taxonomy presented by Maurer et al. (2018a). Systems are classified into two dimensions based on an intervention level in accordance with work by Sheridan and Verplank (1978) and the duration of such intervention. Systems with functionalities similar to the proposed guardian angel are already present in other modes of transportation such as trains and planes by the European manufacturer Airbus (see Fig. 3.1).

In the automotive domain, such systems are prohibited by current legal regulations originating in international conventions on road traffic from 1949 and 1968 (UN 1949, 1968). These regulations originally required every vehicle to have a (human) driver. By amending the "Vienna Convention of Road Traffic" (UN 1968) in March 2016 (UNECE 2016), automated driving was legalized under the condition that the human driver can always take control and oversteer the system. In contrast, a guardian angel is "hard automation" (Young 2007) that cannot be overridden by the driver without being rendered useless and therefore needs additional adjustments in current traffic legislation.

3.3.2 Designing the Guardian Angel

A system that interferes directly with the control of a human driver needs to communicate its actions immediately and unambiguously to the driver. However, a guardian angel may act at the last possible moment, which means there may not be enough time for a warning in advance. The human–machine interface needs to be designed in such a way that the actions of the system are transparent and understandable for the driver, even in a stressful situation.

The systems that can be found in current vehicles have "visual and auditory display modalities" (Lee et al. 2004) as interfaces to the driver. The more advanced the advanced driver-assistance system (ADAS) is, the more "car makers need to attend not only to the design of autonomous actions but also to the right way to explain these actions to the drivers" (Koo et al. 2015). One key factor is facilitating trust in the system, while Nothdurft et al. showed that justifying and explaining the reasons for the action "are the most promising ones for incomprehensible situations in human–computer interaction (HCI)" (Nothdurft and Ultes 2015).

In order to learn more about preferred feedback modalities for the guardian angel in various situations and to gather deeper insights, a first user study was conducted by Bosch. Another goal of that study was to collect general feedback from the participants concerning the system, and to observe how people react during a system intervention.

The driving simulator used in the study consists of a BMW 3-series cockpit, mounted on a 2DOF platform. Monitors in front of the simulator provide the outside view, while smaller monitors behind the shell are used as image sources for the shell's mirrors. SILAB software (WIVW 2020) is used to create the virtual environment. The simulator is located at Robert Bosch GmbH in Renningen, Germany (Maurer 2018b) (Fig. 3.2).

24 participants were invited to experience two different conditions in a driving simulator. Half of the participants were females, ranging in age from 20 to 59, and average driver's license ownership of 12.2 years (SD = 9.45).

Fig. 3.2 Driving simulator (Maurer 2018b)

Participants in the study had to complete a series of test tracks to familiarize themselves with handling of the simulator before they were allowed to complete the study's objective.

The participants had to drive on two different routes to a predefined town. One route consisted of seven T-junctions with a longer rural road between them. At each junction, participants had to follow the road signs and turn either left or right. The last junction had a programmed mechanism, which would trigger a police car crossing at high speed, as soon as the participant drove to the stop line. The guardian angel system then stopped the car to avoid a collision. The other route entailed a longer drive to the same town as in the previous condition, but with the difference of no turns at the junctions passed during the drive. Shortly before the last junction, the participants were given several objectives to use their phones while driving, e.g., texting a friend. Due to this distraction, all of the participants missed the sign indicating they should turn left at the next intersection. Another mechanism was programmed to activate if the vehicle was not in the turning lane of the intersection, triggering the guardian angel system to turn at the last possible moment and prevent a detour for the participant. The sequences of these two conditions were balanced among the participants.

For both conditions, two methods of system feedback were provided. One method involved a beeping sound, repeated once. It was designed similar to current systems in cars, e.g., emergency braking systems. The other method was also an auditory feedback, but with a spoken explanation of the action and the reason for it. Both feedbacks were started as soon as the guardian angel intervened. Again, the two methods were balanced among the participants and situations.

After each drive, the participants had to complete a questionnaire. The first part consisted of the NASA TLX (Hart and Staveland 1988) questionnaire in its raw version (Hart 2006). This questionnaire provides the task-load of a person for a given task. The higher the value, the higher the perceived effort to obtain the respective result in the respective task.

Figure 3.3 depicts the results as box plots for both safety-critical intervention by the system and non-safety-critical (comfort) intervention. In the safety-critical condition, the task-loads for the two feedback methods do not differ very much. In the comfort condition, the mean task-load (marked with "X") of the beeping sound feedback was higher, although not significantly so (Maurer 2018b).

When asked if they would like to have such a system in their own car, all but one participant answered with "yes" regarding the safety-related condition. Where the comfort-related condition was concerned, participants had mixed feelings about the system. Half of them answered with "yes," the other half with "no," for example, "because you have the feeling of being in an emergency situation" (Maurer 2018). Most participants experiencing the beeping sound feedback wanted to have a spoken cue.

The video observation of the study situation revealed interesting behavior by participants in the comfort-related condition. While all participants in the safety-critical situation remained calm when the system took control, this was not the case for participants in the other situation. All participants were surprised by the unexpected behavior of the vehicle. The group experiencing the beeping sound feedback tried

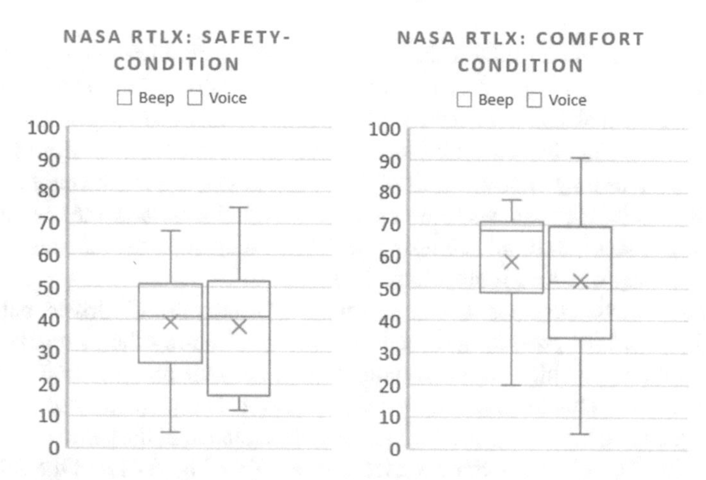

Fig. 3.3 Results of the NASA TLX questionnaire (Maurer 2018b)

to actively counteract the system and looked frustrated when they were unable to take back control. The group with the spoken feedback instead removed their feet from the pedals and their hands from the steering wheel. Afterward, this group was smiling and, in some cases, started interacting with the system, saying "thank you" or asking "may I now drive again?" (Maurer 2018b).

To summarize, a system that can impeach the human driver of a vehicle in a certain situation needs to communicate this action immediately. Spoken feedback is the favored type of auditory feedback, and it helps to reduce the mental effort involved in grasping the system's actions. Interfering with driving in a safety-critical situation seems to be acceptable, while in non-critical situations the acceptance rate is only at 50%.

3.3.3 Changing the Perspective of the Driver

As described in the previous section, there is no mutual consent as to which situations in the car should activate the guardian angel. Asking people in which situations they would see a risk or danger during driving or in which situations they need help from the automation system might not result in a valid definition of situations. The majority of people think they are doing better than "the average driver" (McCormick 1986). This bias is particularly prevalent in dangerous and risky situations. Drivers tend to overestimate their own skills and underestimate their reaction times and physical boundaries, such as braking distances. However, it is crucial for the acceptance of an intervening system to find a set of situations that is perceived as dangerous by a majority of drivers and to base the actions of the system on those findings.

One way of eliminating the bias is to not let people judge their own decisions and skillsets but other people's actions instead. In order to change the perspective, a game was designed for that specific task, where participants could take the place of the guardian angel and try to guard a virtual driver during the drive. As an incentive to only intervention when absolutely necessary, a gamified element was added. Participants were awarded points for the time they refrained from intervening with the virtual driver. As soon as they intervened, they did not get any points for the duration of the intervention. A leader board was presented where participants were ranked according to their performance and could compare themselves with other participants. This was a simple yet very effective motivational tool.

The game environment not only provided a "standardized" driver, but also a safe environment for provoking dangerous situations. During the game, the virtual driver would drive on highways and rural roads to keep the road geometries simple. Different weather situations and driver states such as being "distracted" or "tired" were included in the game. In total, 15 levels were included in the game representing 31 situations. The game provided a user interface (see Fig. 3.4) and the following information to the player:

1. Current speed and current speed limit
2. Driver state
3. Surroundings of the ego vehicle on the road
4. Environment (rural road or highway and if oncoming traffic was to be expected)
5. Current weather conditions
6. Automation status.

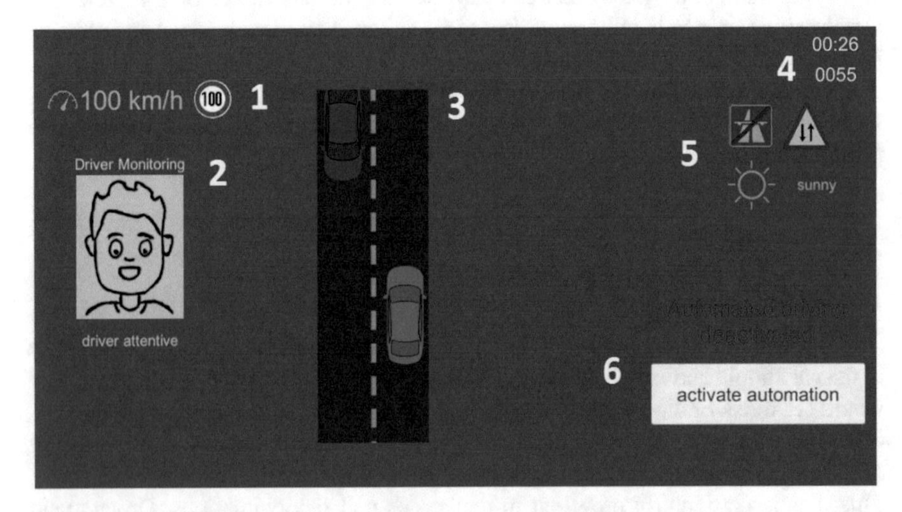

Fig. 3.4 Screenshot of the game screen used in the user study. 1—speed and speed restriction; 2—driver monitoring; 3—top-down view of the car and the surroundings; 4—environmental information; 5—weather information; 6—automation status and button for activating/deactivating automated driving; 7—current playtime and points (Maurer 2018b)

In a study, 25 participants (11 female, 14 male; aged between 20 and 65) with a valid driver license played the game. A tutorial was provided to explain the functionalities to the participants and a test level to get used to playing the game had to be completed by all participants. It took an average of 17 min to complete all levels of the game. All of the participants' actions were logged in a text file for later evaluation.

Only two of the 25 participants experienced a crash in the game, at a level where the ego vehicle would veer off the road if not corrected by the participant. One common example of intervention involved tailgating behavior, i.e., both the ego vehicle tailgating the road user in front of it and the ego vehicle being tailgated.

Most participants were not bothered by fast driving, regardless of the weather—only the driver going very fast during snowfall seems to be viewed as risky, as nearly half of the participants activated automation in this case. In every situation where the driver was driving faster than allowed by the speed limit while not being attentive, the majority of the participants intervened. In those cases, all participants deactivated automation as soon as the driver's state was displayed as "attentive" again.

When asked about their own driving, the vast majority of participants only wanted the system to intervene in the event of changing lanes with another car in the blind spot, involuntary fishtailing, involuntary tailgating, or veering off the road. A small majority wanted intervention in the event of unintentional speeding or being distracted in general.

When combining the data from the logfile with the data questioned, the previously mentioned bias was still present. One very obvious case of this bias was driving much too slowly, where all but two people intervened in the game, but only one participant wanted a guardian angel system to intervene in that case in real life.

Some cases where noticeable where the opposite was true: some participants did not intervene in a situation in the game, yet they wished for intervention by the system in this situation if they were to experience it while driving. One example involved driving more than 20 km faster than allowed. Presumably, this happened as a result of gamification where people were rewarded for enduring a risky situation and suppressing their urge to intervene. All participants displaying this distinctive behavior ultimately reached top positions on the leader board.

3.3.4 Summary

A system which is capable of automatically intervening when a human being is driving when it senses a dangerous situation or an imminent accident can help to reduce the number of fatalities and accidents in general. One requirement involves accurate representation of the environment of the vehicle and reliable prediction of actions by the driver and other road users. This might certainly be possible when highly automated driving becomes available. As soon as automation is able to compute a way of mitigating an imminent incident, it should activate, even if overriding the human driver is necessary. Although a feature already exists in some airplanes or trains, artificial intelligences and machines can also fail in their decisions.

The action carried out by the system needs to be communicated to the driver immediately. As evaluated in a driving simulator study, a spoken explanation of the intervention works best and is widely accepted.

Dangerous situations are a clear use case for a guardian angel-like system, whereas situations without a clear threat to safety result in mixed feedback from participants. In this case, a possible solution would be to give the driver the option to activate or deactivate the system in such situations.

3.4 Cooperation at System Limits of Automated Cars and Human Beings

Cooperation between the automated vehicle and the driver will play a central role in future automated driving. Furthermore, for future mixed traffic, the interaction between the driver and another automated vehicle or vice versa has to be considered. The first cooperative concept for human–machine interaction was made by Hollnagelt and Woods (1983), outlining that the joint functioning of human and machine needs explicit consideration. For this purpose, a complete task should be dynamically divided into subtasks, which are then integrated and flexibly distributed among the automation and the human operator. In general, adding an automated agent is similar to adding a new team member that requires new coordination demands. The more complex the task and automated the support, the more additional cooperative effort is needed to coordinate the joint team (Christoffersen and Woods 2002; Hoc (2010). According to Hoc (2010), a cooperative situation is present, if each agent "strives toward goals and can interfere with the other on goals, resources, procedures, etc. (and if both try to) manage the interference to facilitate the individual activities and the common task when it exists." (p. 515). As a result of this definition, the cooperation partners not only have to manage their specific tasks, but also those resulting from the joint cooperation. Another meaning of cooperation is the definition by Bengler et al. (2012), "Cooperation is considered a relation between human and machine where the interaction or interference between the two (or more) partners occurs with shared authority in dynamic situations." Besides those theoretical definitions of cooperation in human–machine interaction, several works have summarized principles facilitating human–machine interaction in highly interdependent activities (Christoffersen and Woods 2002; Endsley 2017; Klein et al. 2004). For an overview, Walch et al. (2015) summaried four essential concepts that must be addressed for a successful human–machine interaction:

Mutual Predictability: Mutual predictability is essential for practical cooperation. Through shared knowledge of current actions or status of both cooperation parties can the prediction of future activities of the partner and the planning of own actions be successful (Klein et al. 2004; Walch et al. 2017b). Additionally, both interaction partners should be supported in collaborative planning (Klein et al. 2004).

Directability: To intervene or change strategies according to a new situation, team players should be able to modify the actions of the other interaction partner. To achieve this directability, machines must be flexible and easy to handle with respect to the human's goals, state, and capabilities (Christoffersen and Woods 2002; Dekker and Woods 2002; Klein et al. 2004; Walch et al. 2017b).

Shared Situation Representation: A common situation representation, in which relevant information about the situation is shared between humans and the automated system, serves as the basis for practical cooperation between two actors (Christoffersen and Woods 2002). Each the driver and the automated system develops their specific situation representation, e.g., due to different sensor systems, processing procedures, and available resources (Walch et al. 2017b). In a collaborative approach, sharing of relevant information regarding the current problem state and the motivation to communicate with each other is necessary to reach a common situation representation (Christoffersen and Woods 2002; Klein et al. 2004).

Trust and Calibrated Reliance on the System: Trust in an automated system influences human–machine cooperation in various ways (Walch et al. 2017b). In addition to system factors (e.g., comprehensibility and predictability of a system), individual (e.g., willingness to trust) and situation-related aspects (e.g., time constraints) influence adequate trust-building (Endsley 2017). In this context, adequate trust-building means an appropriate level of trust in automation, without the occurrence of its non-use or misuse. Since such an adequate level of trust in the system is necessary for successful cooperation between two parties, these factors have to be considered in the design process of cooperative systems (Walch et al. 2017a, b).

For a better understanding of trust, Kraus et al. (2015) proposed the model of efficient driver–vehicle cooperation, which is shown in Fig. 3.5. According to the authors, beliefs mediate the effects between initial interaction and five variables on level 3-attitude (trust, acceptance, intention to buy, satisfaction, intention to use) and four variables on level 4-cooperation (allocation, safety, well-being, usage). Besides, other factors are relevant for the development of trust, which is described in more detail in Kraus et al. 2015.

Besides those key concepts for successful cooperation, several authors have already worked on conceptualizations of cooperation, such as shared control (Steele and Gillespie 2001; Abbink et al. 2012) the H-Mode or conduct-by-wire.

Shared Control: Shared control permits the driver to be in the control of the vehicle all the time, even when the car is driving autonomous mode, for example, through a haptic control interface (Steele and Gillespie 2001).

H(-orse) Mode: The central idea of the H-Mode is that the system and the driver can both influence the behavior of the vehicle during the driving task. In doing so, the grade of the driver's influence on the behavior of the vehicle can be varied. One resulting benefit is that this mechanism keeps the driver in the loop. Another

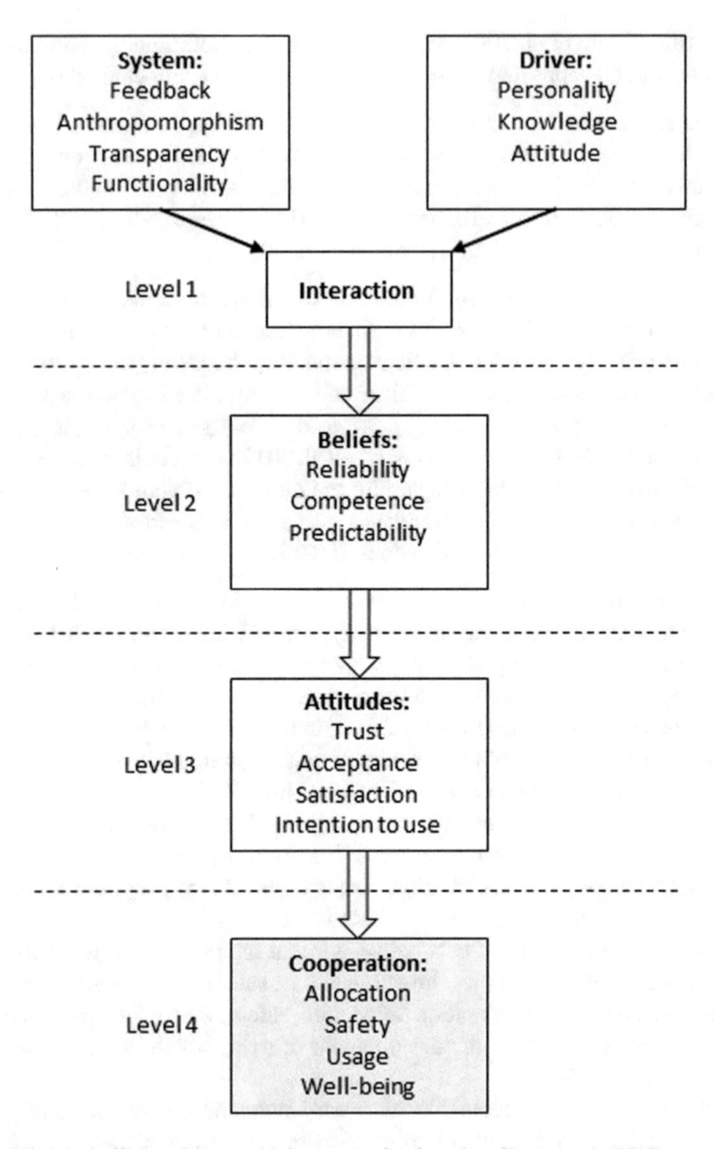

Fig. 3.5 Model of efficient driver–vehicle cooperation based on Kraus et al. (2015)

advantage of this concept is the bidirectional feedback. Both cooperation partners have the opportunity to give the other partner feedback about their activities through, for instance, a haptic feedback interface (Kienle et al. 2009).

Conduct-by-Wire: Similar to the H-Mode is the concept Conduct-by-Wire in which the driver is kept in the loop. Therefore, the driver has to select maneuvers provided on a regular basis by a maneuver selection interface. Those maneuvers are then performed by the automated car.

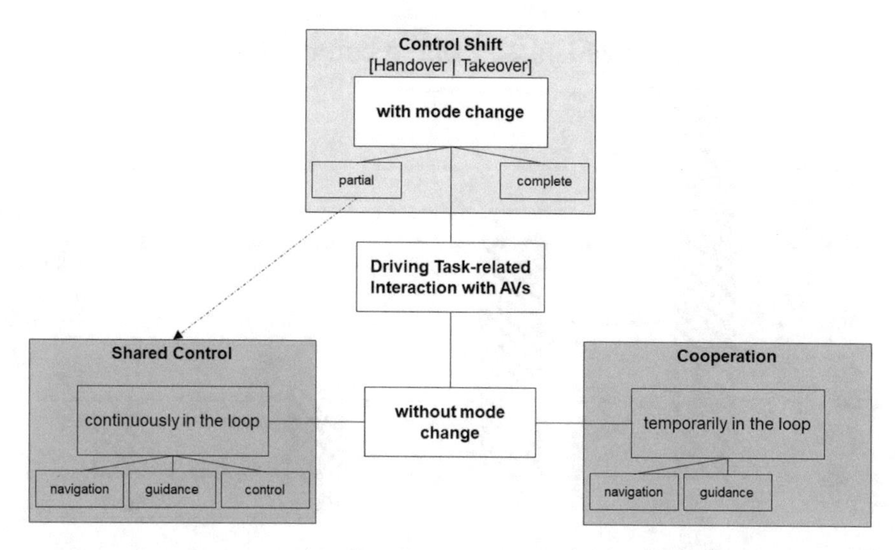

Fig. 3.6 Overview of concepts for automated driving based on Walch et al. (2019a, b, c, d)

Walch et al. (2019a, b, c, d) made the first attempt to structure the different conceptualizations of how automated driving can be realized in one overview (see Fig. 3.6). According to this model, the driver is just temporarily in the loop, and the cooperation takes place on the level of navigation and guidance.

3.4.1 Driver Limitation/Cooperation

In a driving simulator study, Woide et al. (2019) identified a situation that can be handled by the automation and usually cannot be handled in manual control by the driver. This leads to a conflict between the driver and the automated system because the driver normally would not perform this maneuver. As a result, drivers try to intervene, which leads to safety-critical situations. To avoid such situations, a system is required, which provides the correct information, and additionally, the opportunity for the driver to intervene without shifting the complete control to the driver.

The study was conducted in the static driving simulator at Ulm University. Participants were introduced to the functions of the automated vehicle with a focus on the capacity of the sensors of the automated vehicle, which were able to see through fog. In order to generate a situation that creates a conflict between the automated vehicle and the driver due to the perceived limits of the driver, participants drove a simulated autonomous vehicle on a rural road during foggy view conditions. The moment when the vehicle overtook another vehicle in the fog represents the conflict. Three different fog conditions that limited the view of the driver to 50 m, 100 m, 150 m, and free visibility, but not the view of the automated vehicle, were conducted

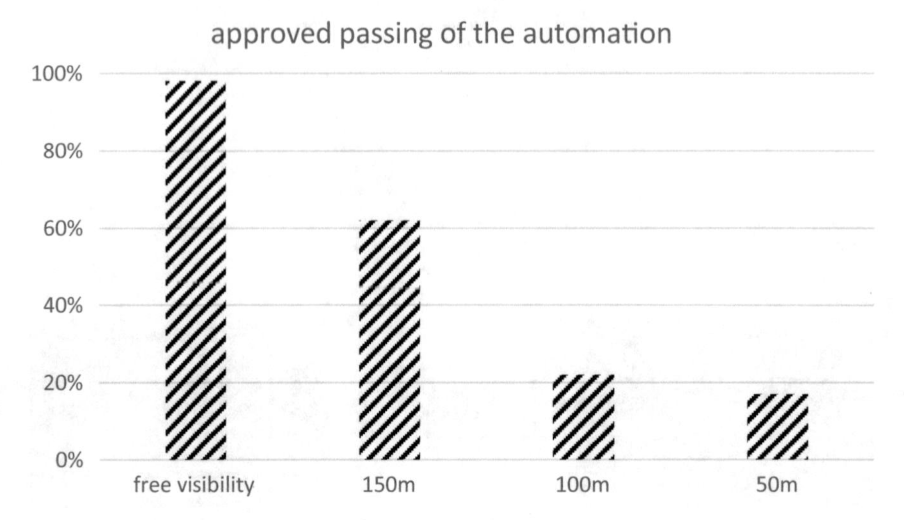

Fig. 3.7 Approved passing of the automation (ratio of conflict)

in this study. Woide et al. (2019) showed that with decreasing visibility, the conflict between driver and automated system increases, even if the participants know that the automated car can see through the fog by using its sensors. The driver's intervention in the automated overtaking maneuver was considered a conflict. Figure 3.7 shows the amount of approved overtaking maneuvers by the driver. When the driver had free visibility, nearly all overtaking maneuvers by the automation were admitted. In contrast, the drivers canceled more than 80% of the overtaking maneuvers by the automation when the visibility was 100 m or 50 m.

The results indicate that the driver needs the appropriate information to understand the behavior of the automation. In addition, it is also becoming apparent that even with SAE level 4, the driver needs the ability to influence the driving behavior of the automation without taking full manual control of the vehicle. Approaches to how such an intervention can look like without a complete rollover, are described in the following.

3.4.2 Cooperation at System Limits

Cooperation at system limits was investigated in different areas: object recognition, pedestrian intention prediction, and maneuver approval.

Object Recognition

Automated vehicles' high definition maps likely contain information about the traffic infrastructure like speed limits and other traffic signs. Nevertheless, the recognition and classification of objects like signs, traffic lights and the state of traffic lights are very important for these systems; for instance, there can be temporary road signs and traffic lights at construction sites. One circumstance that makes object recognition difficult is occlusion, for instance by dirt, snow, or stickers. The recognition of a temporary or partially occluded road sign is not hard for a human driver, and in contrast, it can be to an artificial system. Consequently, the system can ask the user to classify unrecognized objects.

On-the-fly object labeling (Walch et al. 2019a) allows conditional (SAE Level 3) and highly (SAE Level 4) automated vehicles to stay in automated mode when they cannot classify an object. They can ask their user in such situations to help them out and classify objects. This allows to avoid handovers of control and provides information to improve object recognition algorithms and to revise maps.

An empirical exploration was conducted with two different implementations of the system (Walch et al. 2019a): a free text system allowed the users to name an unrecognized object while the choice system provided a set of suggested traffic signs to choose among. Both implementations were rated similarly demanding regarding workload, highly usable and as suitable for object labeling by the study participants. On average, the participants needed more time to tap a microphone button and then express the information verbally than to select a suggested object on a touchscreen in the center stack. The majority of participants reported that they did not at all or only rarely shift their visual attention to the traffic scene in front. In contrast, they looked at the screen in the center stack that displayed the unrecognized object. Using the choice system, they selected in 87.5% of cases the correct suggestion of the system. Using the free text system participants sometimes provided only information regarding the class of the object, for instance, indicating that it is a traffic sign, or a description of how the vehicle has to behave. Consequently, in some cases they provided information that was valuable for labeling of the data but that was not necessarily sufficient for the system to react appropriately.

Pedestrian Intention Prediction

Even harder than recognizing static objects is to recognize vulnerable road users and to predict their intentions. Again, automated vehicles can profit from the help of a human user who is likely better in predicting the behavior of, for instance, a pedestrian next to the road. A particular scenario of interest in which driver–vehicle cooperation can be implemented is crosswalk scenarios with pedestrians on the sidewalks who potentially can cross (see Fig. 3.8). A preliminary evaluation (Walch et al. 2019b) has shown, that users of automated vehicles sometimes oversee pedestrians or do not act responsibly and do not look at both roadsides when asked whether a crosswalk is clear. Consequently, a cooperative system should implement driver monitoring and check if the user has looked at all relevant objects or areas. If a user does not behave as expected, the system should incorporate this information in the dialogue to establish the responsibilities and to guide the user's attention toward relevant entities.

Maneuver Approval

In the scenario described in the previous section, where the user helped the system out recognizing if a crosswalk is clear, the user triggered implicitly an action of the

Fig. 3.8 In cases when an automated vehicle cannot predict whether pedestrians at the sidewalk want to cross, it can ask the human user

vehicle—waiting until the pedestrians have crossed the street or continuing driving. There are other scenarios in which automated vehicles can ask their users explicitly to decide which and when to execute a maneuver, for instance overtaking a slower or standing vehicle or merging faster at a T junction onto a priority road (Walch et al. 2017a). Two different maneuver approval interactions with safety mechanism to avoid erroneous inputs have been evaluated and compared to a single tap baseline condition without any safety mechanisms (Walch et al. 2017a): The first safety mechanism was a confirmation dialog, the second was to enable the maneuver approval button only after a certain time preventing the users to just approve a maneuver instantly. Additionally, speech output was used to inform the users regarding their responsibilities and which actions the system expects them to perform, e.g., "If you would like to overtake, you are responsible for the maneuver" (Walch et al. 2017a, p. 208) or "You can speed up the continuation of the journey by checking the lanes at the intersection" (Walch et al. 2017a, p. 208). Most participants (78.94%) did not prefer to execute the maneuvers manually in contrast to use the cooperative system. Participants preferred the one-tap baseline condition; however, the study results imply that they were aware that error prevention mechanisms like the confirmation dialogue are reasonable in overtaking scenarios regardless of the additional required effort. Some erroneous maneuver approvals were observed which emphasizes the importance of such mechanisms.

Another touch screen maneuver approval mechanism has been investigated and compared to a single tap baseline (Walch et al. 2018, 2019c). Again, an overtaking scenario was investigated: a slower vehicle in front blocks the sensor range of the ego vehicle as shown in Fig. 3.9. Users had to hold down the maneuver approval button to start the maneuver and let the finger rest on the button until the cooperative part of the

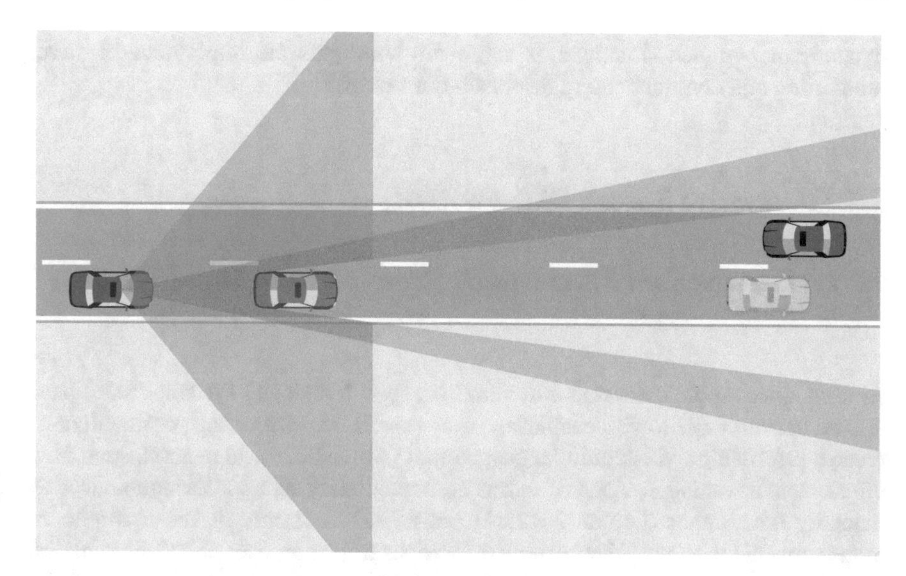

Fig. 3.9 Slower driving car ahead of the blue automated vehicle blocks the sensors and thus hinders an automated overtaking. Adapted from Walch et al. (2019c), Fig. 1

maneuver is finished. In the scenario under investigation the cooperation ended, when the ego vehicle is driving on the oncoming lane as the vehicle has no blocked sensor range anymore. In case users lift off their fingers earlier, the system would cancel and the car would move back to the old lane. During the cooperative phase (lane change to the oncoming lane), system and user can cancel the maneuver; in case, there are surprisingly oncoming vehicles. This cooperative interaction can be classified as *blended decision-making* (Endsley and Kaber 1999): the system generates options (offers to perform overtaking maneuver), the user can make a selection (approve the maneuver), the vehicle implements the selected option (executes the maneuver), during the execution system and user monitor and can select the cancel maneuver option.

A driving simulator study (Walch et al. 2019c) has proven the feasibility of the concept and revealed that both maneuver approval techniques provide good usability. The majority of participants preferred the cooperative approach overtaking over control and driving manually. Oncoming cars were intentionally very hard to see for participants because the ego vehicle kept only a short distance to the vehicle in front. This design decision allowed to study erroneous approvals and maneuver cancellations. In scenarios with oncoming traffic, the system had to intervene in 62.5% of trials. However, it was observable that system and user complement each other. When the overseen contraflow was far away, the user canceled. In contrast, when the car was close, the system reacted quicker. The hold down system showed its advantages in these situations: The users were able to cancel quicker when they just had to lift off their finger from the touch screen. A gaze analysis revealed that users tend to neglect safety precautions (i.e., looking in the rear-view mirror) in more dynamic or complex situations, which again highlights the importance of driver monitoring and communicating responsibilities clearly.

3.5 Innovative Speech Dialogue System

3.5.1 Introduction: Trust and Explanations with Highly Automated Vehicles

At least since Apple's assistance system Siri, speech dialogue systems (SDS) have arrived in everyday life. Siri enables the user to perform a wide range of functions or receive information via natural language input (Apple 2020). In general, such SDS offers great advantages, since speech is an immediately accessible communication modality for humans (Lemon 2012). However, SDS not only makes smartphones more convenient to use, they can also serve security purposes. Used in vehicles, they enable the driver to interact with the car without having to take his eyes off the road. In modern vehicles, the driver is supported by a variety of assistance systems. If these systems intervene in the driver's driving action, the intervention is usually not communicated or only communicated by cryptic warning symbols. SDS has the potential to reduce distraction and increase safety by making important information intuitively accessible via natural language (Weng et al. 2016). Especially explanations of interventions in driving actions by assistance systems can increase the understanding and acceptance of such systems. Koo et al. (2015) showed that when driving with an automatic brake assistant, short verbal explanations of why the intervention happened were preferred and led to better driving performance. A further explanation of how the obstacle was overcome, in addition to steering the vehicle, was cognitively already too demanding for many participants.

As written before, the focus of the KoFFI project is not only on autonomous driving, but also on the dynamic change between autonomous and manual driving. As soon as the task of driving is no longer solely in the hands of the driver, he will be able to devote himself to other activities and turn his attention away from the traffic. This inevitably leads to new situations and challenges for man and machine. In order for the driver to be able to devote his attention to other tasks, he must be able to trust the automation to handle a safe journey, because relinquishing control always requires a certain degree of trust. Several models of trust in machines have already been presented in Sect. 4. Most are based on definitions of trust in the interpersonal sphere. Mayer et al. (1995) define trust in human–human interaction to be "the extent to which one party is willing to depend on somebody or something, in a given situation with a feeling of relative security, even though negative consequences are possible." This reflects the situation during autonomous driving very well, as the driver has to rely on the vehicle, not to cause an accident or other negative consequences. However, to understand more precisely how interaction influences trust with a machine, more precise structures are required. According to a model by Madsen and Gregor (2000), an important part of trust in a machine is cognition-based. This "Cognition-Based Trust" consists of the perceived "Understandability, Technical Competence and Reliability." The user therefore assigns characteristics such as technical competence to the system and suspects certain patterns of action. This mental representation of the system's behavior patterns does not necessarily correspond to reality. It is formed by previous experience with similar systems, expectations, and observation of behavior (Norman 2013).

If—in the case of automated driving—the system behaves different than the driver would expect, a discrepancy between expectation and action arises. Caused by a break in the mental model, this can lead to a loss of trust in the automation. The "Cognition-Based Trust" is particularly affected. This loss of trust can be reduced or even eliminated by a suitable explanation (Lim et al. 2009) However, explanations can pursue different goals, all of which allude to different aspects of information. Sørmo and Cassens (2004) suggest four goals of explanations, namely justification, transparency, relevance, and learning. These can be described as follows if transferred to the context of autonomous driving or automated actions:

- Justification: Explain why the action is/was a good action.
- Transparency: Explain how the system arrived at the decision to take this action.
- Relevance: Explain why the action is important.
- Learning: Explain general automation behavior.

Explanations with the goal of justification or transparency showed a particular positive effect regarding trust in a system (Nothdurft et al. 2014). In order to explain information to the driver intuitively, the modality of spoken language can be used. This makes it possible to communicate information regarding automation, vehicle states or road conditions in a simple, natural and easy way to understand (Weng et al. 2016). At the same time, it is possible for the driver to actively ask for information or to ask follow-up questions for further details.

3.5.2 What Does the Driver Say?

The dynamic change between autonomous and manual driving poses new challenges for the driver and for the interaction with the vehicle. In order to obtain the appropriate information during demanding driving tasks or to cope the additional possibility of permitted secondary activities, a dialogue must be designed efficiently. A useful mechanism also used in human–human communication is the ability to make several requests in one utterance. In everyday conversations, humans tend to speak about more than one topic and even switch to a new topic within one utterance. They do so especially when giving orders to an assistant, e.g., "first do this, and then that."

If each intent contributes to different tasks or activities, we call an utterance multi-intent (MI) (Landesberger et al. 2019a, b). In order to find out what multi-intent user requests look like in the context of highly automated driving, we carried out two data collections, among other things.

In an Internet study, it was investigated how participants formulate utterances to a fictitious in-car speech dialogue system. The participants had to solve seven different types of tasks in an unsupervized fashion on their own devices via a web interface. The tasks all aimed at interacting with a highly automated vehicle. The instructions ranged from everyday scenarios, such as entering a navigation destination, to less familiar tasks, such as prompting the vehicle to search for a parking lot automatically. In order to not influence the linguistic style, the test persons were instructed exclusively with pictograms. A fictional story in which the participant acts as the protagonist and experiences the journey from his home to the workplace with a highly automated vehicle helped to interpret the pictograms correctly (Landesberger et al. 2017).

The data collected showed a mainly natural speaking style for both multi-intent and single-intent utterances (Braunger et al. 2016; Hofmann et al. 2012). Expressing several intents in one utterance to a system seems to be as natural as to express single intents in isolation. Furthermore, the analysis of the data showed a high proportion of interrogative sentences (11.2%) despite the tasks being almost exclusively orders. This highlights the need for explanations in the context of highly automated driving.

A second study investigated how MIs are used to introduce additional topics during an ongoing dialogue. Each participant conducted six dialogues with a Wizard-of-Oz speech dialogue system (SDS) of an autonomous car. To keep the study controllable, the system tried to clarify the participant's need by asking closed questions. While the system was asking a question, a picture regularly appeared on the screen in front of the participant. This picture represented one out of four user conditions likely to occur during a car ride such as the driver feels cold. Participants were instructed to answer the question and to respond to the shown picture. A regularly simulated misunderstanding occurred after the participant used a MI utterance. The participants received instructions to correct possible errors, and no matter which strategy they chose, the wizard ensured that resolving the misunderstanding was successful (Landesberger and Ehrlich 2018).

Users again had no difficulties using MIs while talking to a simulated SDS. They even used MIs to resolve misunderstandings and talk about other things in one turn. Despite the usefulness of MIs, it seems that the entire dialogue becomes very demanding, if the system uses MIs as well in order to add topics or to clarify on multiple topics at once (Landesberger et al. 2019b, CHIRA). Instead, a system-side prioritization of one intent could be useful to reduce the high cognitive load imposed on the user, if the prioritization is logical and comprehensible.

3.5.3 Explanations

In order to improve acceptance and trust, the system must be transparent to the driver. Therefore, it must be able to explain what it is doing in the respective situation and why. This can be done proactively, i.e., based on its assumptions about the driver's knowledge, the system itself decides what and when explanations are given to the driver. However, many persons would not want this, because the system may become annoying and less transparent.

Explanations can also be triggered by driver's questions or orders. We analyzed four different types of syntactic realizations of utterances which possibly require explanations in the context of taking over and handing over the steering wheel:

- Questions about the current situation:
  "Warum fährst du so langsam?" ("Why are you going so slowly?")
  "Können wir nicht schneller fahren?" ("Is it not possible to speed up?")
- Orders or requests:
  "Bitte übernimm jetzt das Steuer!" ("Please, take over!")
  "Jetzt nicht überholen!" ("Don't take over!")
- Questions about a previous situation:
  "Warum hast du vorhin nicht überholt?" ("Why didn't you take over a moment ago?")
- General questions:
  "Wie funktioniert das, wenn wir die Autobahn verlassen. Muss ich da übernehmen?" ("How does it work when we leave the highway? Do I have to drive on my own?")

These types are not always unique. Utterances, which are syntactically realized as questions addressing the system ("you") or the system and the driver together ("we"), are often ambiguous and can be interpreted as different intents within the context of one task:

"Warum überholst du hier?" ("Why do you overtake here?")
"Warum überholen wir hier nicht?" ("Why don't we overtake?")
"Kannst du jetzt (nicht) überholen?" ("Can (can't) you overtake now?")
"Warum hast du vorhin (nicht) überholt?" ("Why did you (not) overtake a moment ago?")

Why questions or sentence questions in the present tense can be interpreted as a question for an explanation of the actual situation (type 1) or, especially in the negated form, as an order to do so (type 2). On the other hand, why questions or sentence questions in the past tense in our context are always questions for an explanation of a passed situation (type 3).

The design of a dialogue strategy must deal with this ambiguity of orders and questions for explanations. Therefore, no matter of the syntactic realization of an order or request, the system's reaction should contain an explanation. For example, the instruction to drive faster can be given in different ways:

- Direct order: "Fahr schneller!" ("Speed up!")
- Question: "Warum fährst du so langsam?" ("Why are you going so slowly?")
- Statement: "Ich hab's eilig" ("I am in a hurry!")

Dependent on the feasibility of the request the system gives the following information:

- A confirmation containing the description of the action if the action is executed: "Ok, ich fahr schneller!" ("Ok, I speed up!")
- An explanation of what to do if the system is not allowed or able to execute the action without the help of the driver, i.e., a cooperative action is required: "Ok, drück auf den Knopf (dann werden wir schneller.)" ("Ok, push the button to speed up!")
- An explanation of why the action is not possible: "Es sind nur 100 Stundenkilometer erlaubt." ("Speed limit is 100 km per hour.")

If the driver can change the premise himself, the system can propose it in addition to the explanation:

> "Du wolltest langsamer fahren. Soll ich wieder Gas geben?" ("You told me to go slower. Should I speed up again?")

3.5.4 Prioritization and Urgency

In cases where the users utters a MI to the system, the prioritization for processing is not necessarily given by the sequence (e.g., "do this, but first that"). Models like the sequential prioritization model (SPM) (Fig. 3.10) define criteria for the prioritization of one task over the other. The model consists of six steps. Each step defines criteria for the prioritization of one task and must be considered before going on to the next one (Landesberger et al. 2019a, LondonLogue).

In the context of highly automated driving, the aspect of urgency plays a special role. If the driver is "out of the loop" (Walch et al. 2019a, b, c, d) due to a permitted secondary activity, there is often no time to grasp complex driving situations in order to be able to correctly assess the situation. In this case, the driver should be able to request the urgently needed information intuitively and quickly by voice. If the driver is already in a dialogue or has previously made additional requests to the system, the urgent intent must be detected and prioritized.

By urgency or urgent dialogue topics, we mean topics that must be clarified in a certain, very short time frame, otherwise they lose relevance. This topic does not necessarily have to be important; it is a matter of time. It is therefore crucial to recognize such urgent issues in user statements and to react accordingly, for example by

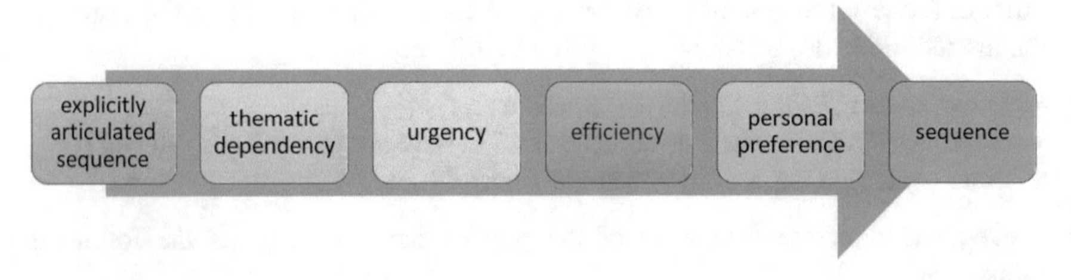

Fig. 3.10 SPM: a six-step sequential prioritization model

postponing other issues and prioritizing the urgent topic. This prioritization becomes particularly relevant when already talking about several topics or more than one topic are addressed in one statement.

The sense of urgency and the ability to quickly grasp and interpret complex situations are highly subjective factors and vary accordingly from driver to driver. In order to map these aspects and at the same time consider objective factors such as the driving situation and the addressed task, we propose a detection of urgency based on four factors.

Acoustic Features. Various acoustic features obtained directly from the audio signal have already been successfully used in the detection of stress, emotions or the Lombard effect (Fernandez et al. 2003; Tawari et al. 2010; Boril et al. 2011). We investigate the change of pitch, intensity, mel-frequency cepstral coefficients or articulation rates in different speech situations where urgent and non-urgent utterances occur.

Deictic Expressions. Deictic expressions gain their meaning only with respect to the situation in which they are uttered. Certain expressions can give indications of urgency. Local or temporal deixis are particularly relevant here. Temporal deixis describes an expression—like, for example, "now," "tomorrow," and "then"—that refers to the temporal dimension of the speech situation. Accordingly, a direct indication of urgency can be recognized by a temporal deictic expression. Local deixis describes expressions—like, for example, "here" and "there"—that refer to the spatial dimension of the speech situation. Such expressions may contain an indication of urgency, but are dependent on the situation: An utterance like "What is that greenhouse there?" becomes urgent if, for example, the house is only briefly visible because the vehicle in which the speaker is located is moving at high speed.

Dialogue Domain. Each utterance, if it can be processed by the system, is assigned to a certain domain or task in a task-based system. In general, knowledge of the current situation is required to classify an utterance as more urgent than others. For example, requests during an automatic overtaking manoeuver in fog are usually more urgent than a telephone call.

Situation. Knowledge of the current situation is also needed to correctly assess deictic expressions or the context in which a dialogue is taking place. A planned or actually carried out automated overtaking manoeuver becomes more or less critical for the driver depending on the speed of the vehicle, traffic density, road conditions, and visibility conditions. Accordingly, the likelihood for urgent user utterances regarding certain tasks increases or decreases.

3.5.5 Architecture

General SDS Architecture. Figure 3.11 shows the SDS architecture within the KoFFI system. Spoken words are recognized by the automatic speech recognition

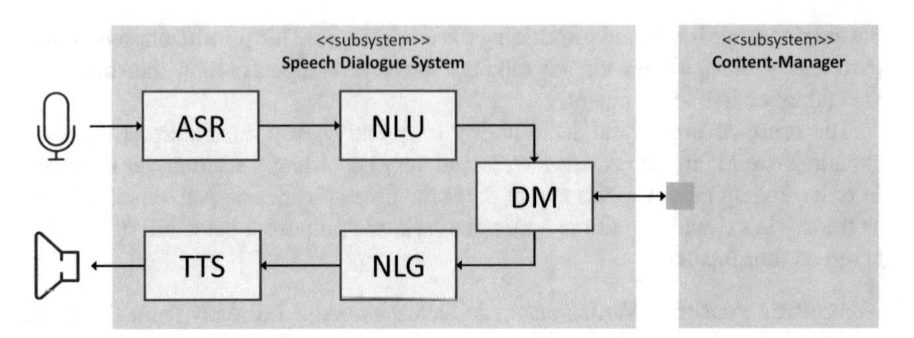

Fig. 3.11 SDS architecture within the KoFFI system

(ASR) module and are passed to the natural language understanding (NLU) component, which determines the meaning of the spoken words. The dialogue manager (DM) is then responsible for classifying the utterance into the current dialogue and making a decision about the next step. This can be an answer to the speaker, a request to the content-manager for more information or a trigger for additional visual feedback. Regardless of the type of answer, the language generation (NLG) component is responsible for generating the text of the answer, which is then synthesized by the text-to-speech (TTS) unit into an acoustic speech output (McTear 2004).

Automatic Speech Recognition. High accuracy and low latency are the two most desirable properties for any automatic speech recognition engine. Both can increase the users' driving experience and thereby trust in technology. High accuracy also facilitates the linguistic processing by other components of a dialogue system, and in the context of autonomous driving both properties have a critical impact on safety. With driving becoming highly automated or autonomous, automatic speech recognition must be even more robust and flexible, since it must deal with an increased number of use cases that differ with respect to user, environment, or driving situation. In vehicle, speech recognition tasks may comprise standard navigation or infotainment control tasks, the time critical recognition of driving instructions, and large vocabulary speech recognition for non-driving related tasks like, for example, email dictation.

Like many other pattern recognition tasks, automatic speech recognition has tremendously benefitted from the application of deep learning techniques in the past 5–10 years (Hinten et al. 2012; Yu and Deng 2014). A similar tendency is observed for natural language understanding tasks, where end-to-end processing and deep learning are successfully replacing rule-based strategies and empiricism; see (Goldberg 2017; Deng 2018) for an overview. The availability of huge amounts of annotated training data is crucial for the development of robust and accurate systems. For acoustic modeling, such data is available in many languages (Moreno et al. 2000), and the simulation of an automotive acoustic environment—e.g., by adding car noise to studio recordings—has also been studied extensively (Fischer and Kunzmann 2001). To a lesser extend, there is also data that allows modeling of in-vehicle dialogues (Hansen et al. 2002), but—to the best of our knowledge—there is no data available that considers an autonomous driving scenario.

While end-to-end trained systems are already approaching the scene (Hannun et al. 2014; Amodei et al. 2020), today's commercially available speech recognition engines usually use rather sophisticated recurrent neural network architectures for both acoustic and language model (Graves et al. 2013; Mikolov et al. 2010).

The EML speech recognizer utilized by KoFFI's speech dialogue system makes use of a bidirectional long short-term memory (BLSTM) neural network as acoustic model (Zeyer et al. 2017). The network was trained on the German and English SPEECHDAT-CAR databases and some additional in-house data. In order to meet the tight latency requirements of the automotive scenario, pruning methods for the elimination of network parameters (He et al. 2014) were transferred from forward networks to BLSTMs, and an online-capable BLSTM variant was introduced (Fischer et al. 2018). Furthermore, a component for robust speech detection with a short decision horizon was developed (Ghahabi et al. 2018).

The recognizer uses a lexicon with approximately 1 million words and a conventional 4-g language model. The latter was chosen to support all kind of possible conversations in the car. It is based on a general language model for German and slightly adapted to the use cases considered in the project. If hardware permits, the recognizer is capable to run several recognition tasks—each defined by either a finite state grammar or a language model—simultaneously in separate search spaces. In this case, the dialogue system (more general: the respective application) is responsible for the proper selection and further processing of results.

For use with KoFFI's speech dialogue system, the spoken words are enriched with some additional features such as the amount of speech and non-speech, the speaking rate—measured in words or phonemes per minute—and the number of sentences in an utterance. They are accompanied by an indication for the urgency of a sentence that is computed based on acoustic features, see above. Recurrent neural networks and FastText (Joulin et al. 2016) have been used for the identification of the driving situation from the recognized words. However, while successful in principle, the obtained results need validation on larger corpora which—to the best of our knowledge—are not available at the time of writing.

Natural Language Understanding As pointed out above, in everyday conversations humans tend to speak about more than one topic, even switching to a new topic within one utterance. Therefore, KoFFI must detect utterances with more than one sentence, segment them into the single sentences and interpret them.

There are a few corpora with MI-utterances available. For example, Beaver and Freeman (2017) introduce a commercial customer service speech corpus containing MIs. Another method to get MI data is concatenating combinations of single-intent sentences like Kim et al. (2017) used to provide a two-stage method to detect MIs with only single-intent labeled training data.

None of these corpora is suitable as a base for training both segmentation of utterances into sentences and their semantic interpretation, especially not in the context of automatic driving. Therefore, we developed a three-stage approach to analyze and interpret multi-intent utterances, which can also be the basis for a future deep learning approach.

Phrase Spotting The result of the recognizer is chains of word hypotheses. The first step of the linguistic analysis is the search for syntactically and semantically correct and meaningful phrases in the word chain. The resulting sequence of phrases must not be connected, i.e., words which are not relevant can be left out. These phrases are defined by grammars modeling simple syntactic phrases like noun phrases, prepositional phrases, or verb phrases. Using unification of the semantic features of lexical entries, a phrase is interpreted semantically. This approach allows, similar to statistical methods, the understanding of natural or colloquial human speech.

Segmentation into Sentences The phrases detected in step one get as an additional feature the information if they can potentially be the beginning or the end of a sentence. For example, coordinations ("and") or questions ("where," "which kind of obstacle") mark the beginning of a sentence, pauses, or participles as part of verb phrases the end of a sentence. Using this information the utterance is splitted into single sentences containing each a sequence of phrases, which can be semantically interpreted together in the same context. To enable the analysis if a sentence is urgent or not (see Sect. 5.4), the sentence gets the feature "deictic" if it contains deictic parts (e.g., "*here,*" "*now,*" "*the car in front of us*"), and the feature "prosodically urgent," if it contains parts marked as such by the recognizer.

Contextual Interpretation In the last step, the sentences are mapped to different tasks (e.g., *driving* for "*Why did you not overtake?*" or *climate* for "*I am cold. Please switch on the heating.*"). In this step also contextual references and references to the actual situation are resolved.

Dialogue Management Dialogue management maintains continuity over turns in a conversation between human and computer. While many approaches have been developed to address this problem, it is believed that the essence of dialogue management resides in performing two functions: interpreting user inputs with respect to task(s) within the domain, and maintaining the coherence, over time, of the conversation (Bohus and Rudnickey 2003). Several SDS-architecture approaches are viable. This means that the boundaries of responsibility of the individual components are becoming increasingly blurred. Like in our system where the NLU component does most of the user input interpretation and the focus of dialogue management is more on a coherent dialogue flow. We identified the following requirements: Maintaining the coherence of the conversation includes that we bring together various factors that may indicate urgency and thus priorities and sequence the different intents when MIs occur.

If several intents are found in an utterance, the dialogue manager calculates an urgency rating for each intent. All factors are taken into account depending on the situation. To determine the influence of the factors correctly, each factor is weighted. For example, for new drivers the dialogue domain with the situation can be weighted higher until enough data is available for an additional reliable prosodic urgency classification. If the difference between several intents in the calculated urgency rating exceeds a certain threshold, it is assumed that one intent is more urgent than the other. Thus, the dialogue management prioritizes this topic and creates a sequence for processing the remaining intents.

In order to convey this sequence in a comprehensible and intuitive way, the dialogue management is able to use meta dialogue modules. For example, to signal that several intents have been detected during a MI, a response can be initiated with "first" or "first this topic." If the urgent matter has been sufficiently dealt with, the dialogue management will independently return to the deferred topic.

3.5.6 Conclusion and Future Work

Highly automated driving poses new challenges for the driver and the vehicle due to the dynamic change of driving modes. In order to establish and maintain trust in the technology of automation, adaptive user-centered interaction concepts are required. Language as a means of communication directly accessible to humans offers immense advantages as a modality. Information can be asked for directly, and needs can be expressed freely and in natural language. Therefore, speech dialogue systems are confronted with increasingly complex utterances and the challenge to find a suitable answer to them. Utterances such as "KoFFI, take me to my favorite night club, but first stop at Eric's place to pick him up, and during the trip I want to go along the ocean and see the sunset" will become more and more common (Eliot 2020). We have identified some challenges of complex statements like MIs and pointed out solutions. The recognition of urgency and the following sequencing of intents when MIs occur is a very useful and helpful mechanism. These mechanisms are not only crucial for MIs, but can be applied in any situation with parallel tasks. In addition, even with single intents and short utterances, a recognized urgency can lead to a better interaction, for example by using modified answers or by adding further modalities to guarantee a holistic and fast response. We will investigate these and other use cases more closely in the future to further improve voice interaction with highly automated vehicles.

3.6 The Making of an Automated Driving HMI

As already described, a central challenge of KoFFI is the cooperation between the driver and the autonomous vehicle. The design becomes even more challenging when we consider the change in the tasks while driving autonomous. With the introduction of fully automated driving, the former primary driving tasks are handed over to the vehicle. Other tasks come to the focus of the driver, which affects the driver's behavior (Hensch et al. 2020). This is influencing the complete HMI system as it needs to provide the right information for the driver in specific situations. Our focus was on the cooperation between a vehicle and a driver.

In a first step, we integrated the handover of the driving task (McCall et al. 2016). We are implementing a driver-initiated handover where the driver is actively triggering and holding the action. We chose to use two interaction elements that are located on the steering wheel. To hand over the driving task to the car, the driver must press both buttons and hold them for one second. Afterward, the car is driving autonomously.

Our main focus was on the development of the center stack that allows us to display many information. We chose for the project a 17″experience with the touch screen that is mounted in portrait mode. The screen is divided in three big parts (Fig. 3.12). The upper part displays more of general information such as time, status of the speech dialogue system, and status of automation. The central part contains relevant information for driving such as position of the car on the street or traffic rules (e.g., speed limit). The lower part contains elements for interaction.

Fig. 3.12 HMI representing the current driving situation

The key part of the UI is to provide the driver with relevant information on the current status of the autonomous car. As a first part, we integrated a representation that shows the status of the intelligent speech dialogue system (see Chap. 5). For this, we created a four-status image allowing to see the systems current status. We defined four status:

- *System available*: The speech dialogue system is available for use; however, it is not actively listening. It can be activated with the activation phrase.
- *Listening*: The speech dialogue system is listening to the statements by the driver and recording it.
- *Processing*: The speech dialogue system is processing the recorded statements.
- *Talking*: The speech dialogue system gives a response to the driver's actions.

In the middle of the center stack, we display the current driving situation. This contains the track on which the car is currently driving and possible other surrounding cars. We implemented this by showing the schematic track without other details, such as colors or terrain. Woide et al. (2019) showed the driver's vision is a central aspect. To visualize what the car's sensors are able to capture, we introduced a view 300 m to the front.

User-initiated touch interactions are grouped in one circular menu, which is placed in the lower left corner. Using this menu, the driver can select various actions that are relevant for the driving.

One among these interactions is the maneuver approval. With this, we are able to operate at system limits (see Sect. 4.2), we implemented, as indicated by Walch et al. 2019c. In our implementation, we used a cooperative overtaking scenario. There, the driver must trigger a possible overtaking maneuver because the car cannot see beyond the leading car. To approve the maneuver, the driver must check the surroundings and then hold the button until the car has changed lanes.

Virtual Reality Driving Simulator

To conduct studies, it is necessary to have a sophisticated driving simulator. This is of special importance as fully autonomous cars are not yet available or accredited for public streets. One solution, which is representing a high-fidelity modality, might be to use Wizard-of-Oz cars that are partly allowed in public traffic. This car is piloted by a driver that is seated in the co-driver seat (Fig. 3.13), which allows the driver to hand over control to the vehicle. Further possibilities to study autonomous HMIs are simulations that assemble the driving situation in various ways. One way is to use low-fidelity installations that use 2D screens (Allen et al. 2011).

Another way is to implement and use driving simulators, which are widely used in various use cases. These simulators span from simple desk installation to sophisticated full-car simulators. During the course of this project, we implemented a sophisticated driving simulator that is based on state-of-the-art technology. The hardware-base of our driving simulator is the front part of a real Audi S8. The car was cut behind the driving seat, which leaves the rear part open (Fig. 3.14). The front part is intact and can be integrated in the driving simulation.

Figure 3.15 depicts the currently used components in our driving simulator. For

Fig. 3.13 Use of a Wizard-of-Oz car

Fig. 3.14 VR driving simulator mounted on a motion platform

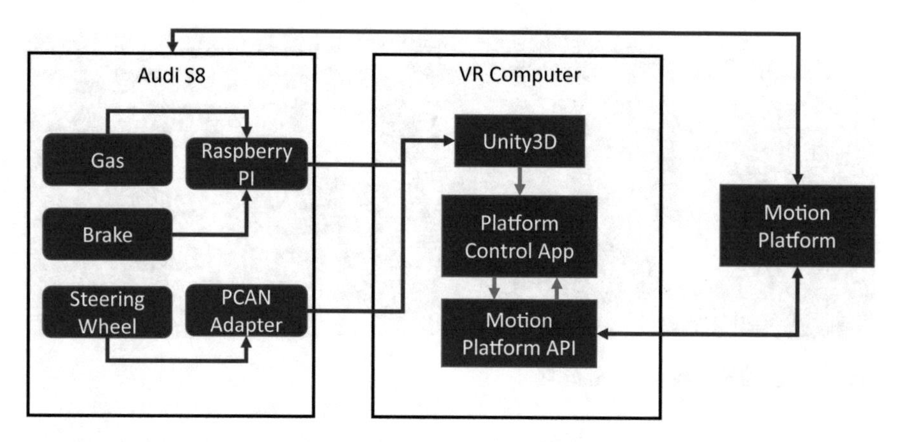

Fig. 3.15 Integrated components of the VR driving Simulator (Adapted from Martinez-Maradiaga et al. 2019)

driving, we use the components mounted in the car, i.e., gas, brake, and the steering wheel. We are able to read the intensity of the gas from the built-in BUS system. In contrast, the values for steering wheel and brake are not visible in the car's communication system. Therefore, we installed external sensors to gather the respective data. All data are then transmitted to the VR application, which is implemented using Unity3D. Afterward, the input variables are processed in the Unity3D application. The simulated physical behavior is transmitted to the control application of the motion platform. The implemented algorithm is then controlling the motion platform and moving the vehicle accordingly.

The motion platform used contains 6″ vertical actuators that allow 3-degree-of-freedom movements in the Z-direction. The motion platform is mounted directly to the chassis of the Audi S8 and is moving the whole car. The particular challenge is to install the actuators in a way that the weight of the car is distributed even. If not, this may cause a shut-down of the motion platform due to safety reasons, i.e., overheating in the example. Furthermore, we implemented an algorithm that allowed us to control the intensity of movements while the physics are still valid. By this, we expect to address the issue of motion sickness (Lucas et al. 2020) that may be reduced by physical movement.

A further set of complex components is the electronic installation. As already stated, the Audi S8 comes with a fully equipped cockpit that we integrated in our driving simulation. In particular, we integrated the steering wheel, blinker, gas, and brake pedal as input for the driving simulation. We were able to read the values of blinker out of the car's CAN-bus. The steering wheel is not sending the needed values (e.g., steering angle) over the car's CAN. Therefore, we integrated a force-feedback motor that allows us, first, to read the steering angle and, second, simulate the feeling of steering a real car.

The third big component is the Unity3D application that is simulating the virtual world where the driver is immersed in. We decided to use virtual reality (VR) as

Fig. 16 Virtual interior of the Audi S8

technology to immerse the driver in a virtual driving situation. We chose this technology due to the tremendous flexibility it provides. We implemented a virtual world containing a track with several degrees of complexity, e.g., driving situations with curves where the end is not visible. Furthermore, we have a full virtual representation of our Audi S8 containing the whole interior; see 3.16.

VR offers us the opportunity to change certain components of the HMI or interior. For example, in Fig. 16 we changed the content of the center stack with an interaction we tested.

To immerse the driver, we use a head-mounted display (HMD). As compared by Lê et al. (2017), the HTC Vive and Oculus Rift Consumer Version in order to find be most suiting HMD. Due to the setting, we found that the Oculus Rift currently is fitting the needs best. The easy installation of the camera sensor and its robust tracking lead us to this decision. Besides that, we did not implement any physical movement in VR, which would require a tracking in space. During the use of the HMD, the driver is seated and due to that relatively static, which is supported by both HMDs. The interaction with the (virtual) car was a challenge we addressed during the whole development. We implemented the whole interactions virtually, which means that the real car did not receive any messages back from the simulation.

One challenge we faced is the interaction with the car. In the VR, the driver is seeing the interior from the perspective of the head-mounted display (HMD). When we task the driver to interact with the car, they are controlling the movement of their hand on a visual basis, i.e., what they see in the VR. This challenges the design of the virtual environment (VE) that is mapping the real world, i.e., the real car's cockpit. One illustrative example is the steering wheel, which we always need in manual driving scenarios. The virtual and physical steering wheel must match; otherwise, this may cause a break in presence, which is undesired (Cummings and Bailenson 2016). For the interaction with the HMI, in some cases it is necessary to see the hands, e.g., when pressing a button. For that reason, we integrated the leap motion for the hand tracking. The leap motion is mounted to the front the HMD, which allows a hand tracking when the hands are in the field-of-view of the leap motion and, consequently, the HMD.

Using this set up, we are able to test and validate HMI concepts. The combination of a real car, that comes with all real advantages, with the opportunities of VR allows us to conduct studies in a close to real setting.

3.7 How to Apply Ethics-By-Design

Currently, we are not only experiencing an "ethics turn" (Grimm 2020) in the debate about AI and other state-of-the-art technologies, but also a "design turn" in ethics (van den Hoven et al. 2017). This means that the public debate is open for philosophical and ethical examinations and, possibly, their consequences. This also means that ethicists have understood that design is shaping our world and the way we perceive it, which, conversely, probably shapes us and our behavior. The challenge is therefore not only to *consider* "ethical and legal implications" of technology but to find ways to *apply* ethics by design.

3.7.1 Challenges and Questions

In 2016, we started the ELSI-Project with the idea to implicate ethics *by design* into the interdisciplinary research project KoFFI. The main deliverable of the ethical part of the ELSI-project is guidelines for the research and development of highly automated driving (SAE Level 3–4).[2] The guidelines can be found in the annex to this chapter.

Our approach is narrative ethics by design (Grimm and Kuhnert 2018b). We used empirical methods, i.e., from the field of narratology (Müller and Grimm 2016), to learn about people's opinions and preferences when it comes to highly automated driving. At the beginning of the project duration, we organized an expert round table to discuss KoFFI's concern to raise acceptance of and trust in highly automated vehicles. As a result, and as part of the ethics by design approach, the interdisciplinary perspective on a phenomenon such as "trust" became clear, as well as the necessity to be precise and aware about one's own terminology, especially in an interdisciplinary and transdisciplinary setting (Grimm and Kuhnert, Upcoming 2020). Other topics in the discussion were the relation between "privacy" and "trust," the difference between "trust and trustworthiness," "security," ethics by design as a concept and "the future of mobility" (Grimm and Kuhnert 2018a). Furthermore, we conducted narrative interviews ($n = 10$). One advantage of narrative interviews is, that the interviewer does not only ask a person about their opinion but learns about the—biographic—foundations of their beliefs. Therefore, it is possible, to see their answers and narrations in a larger context, which has consequences for the attribution of

[2]Other guidelines such as the ones by the German Federal Ministry of Traffic and Infrastructure have been taken into account (Ethik-Kommission Automatisiertes und Vernetztes Fahren 2017).

meaning to the interviewee's answers. We used this approach because we wanted to find out, which the experiences are that ground people's expectations considering highly automated driving. The results of the narrative interviews were, in addition, used to define the questions for further interviews. In three surveys,[3] we asked the persons who had just tested the KoFFI Software ($n = 54$) in the driving simulator, open-end "ethical" questions about trust in machines and highly automated cars. Most of the test persons were tech-savvy persons due to the recruitment process for the technical studies (Maurer et al. 2018a, b). Most of them considered themselves as good drivers and distinguished themselves from drivers who, according to them, should rather use highly automated driving functions than drive on their own. Not surprisingly, in general they would like to use automation when facing boring or annoying traffic situations or repetitive tasks such as stop-and-go-driving or long rides on the motorway. When it comes to ethical questions about human-machine interaction, most of the test persons are aware that trust in and cooperation with human beings is different from machines. Several answers are suggesting technical solutions to possible problems (e.g., the opinion that trusting means that the system has driven many kilometers without an accident and that the person themselves has not had any bad experience with the machine/car).

In terms of the project internal procedure, we regularly discussed our findings with our project partners. We questioned, e.g., which values were concerned, when it comes to highly automated driving and defined the values that we considered as important for the development of the KoFFI Software. According to the Ethics by Design-approach, we worked with feedback loops. This means that we didn't formulate ethical principles and expected our partners to follow them, but we included them into the discussion and reformulated also the values concerned in KoFFI.[4] This question is important because technology is never neutral (Friedman and Hendry 2019), since human beings always inscribe their ideas, assumptions, and (moral) beliefs into the technology they design and create. One way to make these visible is to ask which values play a role when designing technology and user interfaces in order to raise the engineer's[5] awareness for this fact.

Finally, we conducted an ethical study together with students of the Hochschule der Medien who produced a fictional film about highly automated driving.[6] As future

---

[3]After the first survey, we used the results to rethink the design of the questionnaire. We therefore worked with two different sets of questions. The first survey was conducted at Bosch ($n = 23$), the second at Daimler ($n = 21$), and the third at Ulm University ($n = 14$). The narrative interviews (six individual interviews and one group interview) were conducted at the Hochschule der Medien and at the University of Ulm.

[4]The values have at first been formulated in the "Smart Mobility Matrix" (Deliverable/Milestone 2.1). For the KoFFI Code, we reconsidered them (Deliverable/Milestone 2.3).

[5]By "engineer" we mean all professions concerned with the development of highly automated driving: computer scientists, software developers, programmers, designers, and alike.

[6]The film is called "StattLandFlucht" and was directed, filmed, and produced by students of the Audiovisual Media program at the Hochschule der Medien, course: "Studioproduktion Film & VFX" (studio production film and visual effects), winter term 2019/2020, supervized by Petra Grimm and Katja Schmid.

media professionals, they are responsible for science communication and will play a crucial role when communicating information about technological developments to society. The film, at the same, time also takes into account a central point uttered by the interviewees: the need to get information about highly automated driving functions through videos. It is a fictional story about two gangsters who use a highly automated car to escape after committing bank robbery. The film does not directly represent a tutorial about highly automated driving; it does, however, use a story, a narrative, that might possibly trigger (ethical) reflection about the topic.

3.7.2 What is Ethics (By Design)?

The development of new technologies always raises ethical and moral questions. Many technologies promise that they will facilitate people's lives. A dream, probably as old as mankind, is to automate repetitive or difficult tasks. One example for this is the invention of the escalator. But what about highly automated driving? Does its introduction mean that we need to do without the "joy of driving" promised to us by the car industry and its marketing? Most of the participants in our studies said they were enjoying driving in general. How does this comply with the idea that autonomous driving might make driving more safe, lead to less accidents, reduce the number of humans dying in car crashes or even to make driving more ecological and more efficient through reducing the number of cars? How can such conflicting interests be weighed against each other? With these examples we already see, why ethics have to play a role in the development of highly automated driving.

But what is ethics? Ethics means discussing and reflecting moral behavior. Moral is a set of "traditional" rules and conventions and passed on possibilities how to act. As an academic discipline "ethics" is part of "practical philosophy" and provides the opportunity to reflect on moral or moral standards. Ethics cannot only provide orientation in moral issues; i.e., it does not only reflect on how the individual can act good, and what the implications of one's actions are on others, but it also raises the question of how a good, successful life can look like.

In the media, a lot of attention goes to discussing the "Trolley problem" (Foot 1967)—the question which person should be overrun by an autonomous vehicle. Of course, this question is rarely being asked this directly (Matzner 2019), but it is rather asked into which direction a car should swerve when facing a dilemmatic situation in which hitting a person on the road or on the sidewalk has become inevitable. The MIT "moral machine experiment" is an online simulation that asked users exactly this question. In different scenarios, the user could decide which "character" should die in the upcoming accident. The results were not surprising: They make visible cultural differences in the opinion whose life is worth more, meaning, which characteristics of a person lead to the assessment that their life deserves protection (e.g., the respect for children vs. the respect for elderly people). They do, however, at the same time, show the universality of certain moral values (Awad et al. 2018). In KoFFI, we do assume that there are universal values, as the ones on which the Universal Declaration of Human Rights is based (United Nations General Assembly 1948). But this is only one example of a scenario that fully autonomous cars might encounter in the future. This extreme thought experiment, that we do need to consider, since cars might face it, does, however only play a minor role in the current development. Jeroen van den Hoven even argues that we have a moral obligation of a higher order to prevent dilemmatic situations (Van den Hoven et al. 2017). There are more realistic and maybe even more pressing questions that are important, when thinking about highly automatic cars, such as the general reflection about an individual's responsibility in their development.

Ethics by design includes the idea that problems can be avoided or at least discussed early on, when ethical issues or, e.g., potential ethical value conflicts are considered at an early point of the development of a product or a research and development process, or a process in general (Van den Hoven et al. 2017). In the current European (and German) debate, it is being argued that value-sensitive design—especially of artificial intelligence—might represent a "competitive advantage" compared to AI programmed and developed in other world regions (e.g., Hasselbalch and Tranberg 2016).[7]

The following aspects are part of the ethics by design approach as understood in KoFFI:

- Involvement of "ethics partners" right from the beginning of a project or process: Ethical monitoring should be undertaken right from the beginning and regular re-assessment should take place, even after the product is ready for the market.
- This includes feedback loops: Decisions should be reevaluated, values reconsidered.
- Conducting a stakeholder analysis: It is important that all stakeholders, indirect as well as direct ones, do have a say, but one has to be careful that this does not lead to so-called ethics washing.
- Paternalism should be avoided. This means, that the ethicists should not patronize the other researchers.
- The ethical theorists should act and argue, if possible, informed by empirical results.
- A value analysis should be undertaken to identify underlying convictions and assumptions, in order to avoid bias.
- A questionnaire, such as the one developed in KoFFI,[8] can be used to make visible possible conflicts. However, this does not mean that ethical issues can be resolved by ticking off a checklist.
- Narrative use cases can prompt ethical reflections and facilitate the interdisciplinary communication.

---

[7]See also the list of guidelines on AI ethics that have been published in the last few years: Algorithmwatch (2019).

[8]The 44 questions cover the six following topics: 1. About the project 2. Dealing with involved persons 3. Design 4. Programming 5. Ethical evaluation 6. Possible future consequences of the invention.

- The question whether we should program ethical machines needs to be discussed further. So far, programmable ethics is only possible to a certain degree.
- The voluntary commitment to an ethics code should not be used as a pretention to avoid legal obligations.[9]

3.7.3 Open Questions

Further research should be done on:

1. *Research ethics*: the question on how to actually imply ethics into the research and development process needs further inquiry. In automated driving, for instance, the "problem" of the so-called Turing deception. This means that one has to be careful when using Wizard-of-Oz technology and one has to choose carefully the moment at which and how to reveal to the test persons that the car was not driving on its own. While the fact that it is important to consider a test person's reaction and possible "deception" is well known in research ethics (Riek and Howard 2014), the engineers concerned need to be trained in handling this phenomenon.
2. *Awareness raising*: This leads to our second point: researchers, engineers, designers, and developers need training to understand and become aware of possible ethical issues. We suggest that this should be done like we did it in KoFFI: ethicists—and possibly lawyers or social scientists—should be involved early in the research process. This should not only be done in the sense of an ethics board "okay-ing" or passing the research as "ethical" but as a competent partner to whom one can turn with one's questions and concerns or to raise concerns that one hasn't seen in the first place. Ethicists therefore should be trained, too, not only in applied ethics but also in interdisciplinary and technological questions and interdisciplinary and transdisciplinary communication.
3. *Research funding*: Our third remark is that ethics committees should become reference persons and competent points of contact with the resources and competences to answer people's questions without being perceived of as "ethics police" or "ethics nanny." Thus, in short there needs to be more funding for ethical research and action and training for ethicists as well as for researches and developers in "technical" professions.

3.7.4 The KoFFI Questionnaire

In the first block of questions, we ask to provide general information such as the name of the project, a list of the project partners and a short description of the project to set the scene and to make visible what is seemingly obvious. This leads to questions that have the future users and therefore already ethical aspects in mind. To name the stakeholders shows that someone might be involved in a project or a

[9]For the relationship between law and ethics see for instance: Grimm et al. (2019), Schliesky (2019).

process that might play a role but in the sense of an indirect stakeholder (Friedman and Hendry 2019). The next question is about the "needs" of potential users with the questions after next aiming at a dimension beyond user-centered design: "Who could be affected by possible design choices? Whom does the technology serve? Who benefits from it? What is the purpose of the technology? Does the technology serve members of "vulnerable groups" (e.g., children, elderly people, or people with disabilities)?" The final point raises an explicitly ethical aspect: "Does the technology improve people's quality of life?"

The questionnaire continues with topic 2: Dealing with involved persons. If human beings are involved in the development of a product, specific (research) ethics questions can arise. To make, again, the obvious visible, the first questions echo questions from the first block but with a focus on the research and development process. It is therefore asked whether test persons are involved in the development of the product and whether the needs of different user groups are being assessed. This does overlap with legal questions, since the collection of user data has to comply with the General Data Protection Regulation (GDPR) (European Union 2016). The test persons need to have the possibility to consent to the collection and processing of personal data[10] and it has to be assessed whether it is possible to collect the data anonymously, or to render them anonymous (i.e., no possibility to (subsequently) attribute them to a specific data subject) or at least pseudonymous.[11] In case a Wizard-of-Oz vehicle is being used, that pretends to act "autonomously," precautions might have to be taken to avoid so-called Turing Deceptions (Riek and Howard 2014), which is the feeling of having been deceived about driving an autonomous car when it was actually steered by someone (cf. also above). In technology development, it is common to consider different use cases. We therefore ask whether the ones used represent different (moral) perspectives of the various stakeholders and whether also unlikely "edge cases" are being discussed.

The questions in paragraph 3 are concerned with the design that is being developed and deployed. We quote the questions directly, since these are the immediate result of the ELSI part in KoFFI and therefore lie at the heart of "ethics by design": "Are there Design-Patterns or features that try to nudge users to do something they otherwise would not do? Is there a possibility to opt-in or opt-out of driving functions that are not required for driving? Is there a manual override button that can be activated by the human, for instance in case of an emergency (so-called overruling by humans)? Is there a so-called guardian angel function (i.e., the possibility that the system overrides the user's actions, e.g., an emergency brake assistant?) Does the vehicle give feedback after the human has taken control again? Does the vehicle give a reason for taking over? Are the vehicle's "decisions" transparent to the user? Are their reasons, for instance, being displayed? Are you (the developer or engineer) aware of the fact that the way how to communicate in handover situations can influence a person's trust in the system?"

---

[10]Cf. GDPR Art. 4(11), Art. 6(1) a, Art. 7, recitals 42/43.

[11]GDPR Art. 4(5).

The questions about programming also directly raise concerns about actual possibilities to make a system transparent which is one of the values we identified as being central in KoFFI and which contributes to one of the central goals of the entire KoFFI project: To raise the user's acceptance of the driver–vehicle interaction system. We ask whether open source or free software is being used, or, if not, if it would be possible to make the source code publicly available while protecting (legitimate) commercial interests. Another central question is, whether algorithms are being deployed and, if yes, if the developers can ensure that the algorithms are "fair" and that they are not reproducing existing bias, as well if they are transparent, accountable and (internally or externally) controllable. Another question is, if artificial intelligence, machine learning, or deep learning is being deployed since the ethics of AI is of particular interest in the current public debate.

Speaking of a public interest in ethics, three questions also deal directly with ethics (Ethical (accompanying) research) and the idea that ethics is a complex field that does also include professional ethics, e.g., the question if professional ethics are being incorporated.[12] It might also be the case that ethics committees already exist, or the research project or the development is being ethically monitored. Since "ethics by design" as understood in KoFFI does not come to an end, once the product is ready or the process has finished, also possible future consequences of the product are being considered. Even if human (or even corporate) responsibility in a complex world might be restricted, the question arises whether there might be aspects or results of the research that could be used for "unethical" purposes (so-called dual use, e.g., military combat drones) and if yes, if there is a possibility to contain unwanted consequences. The final question does concern the future, since ethical decision-making also means to anticipate future developments: Could undesirable consequences arise, when the development continues, e.g., when level 5, fully autonomous driving becomes marketable?

3.7.5 The 5 Ethical Imperatives

The reformulated values are the following: privacy, autonomy, trust (in technology), and transparency. To further inspire ethical reflection, we have formulated "5 ethical imperatives":

1. Do no harm!
2. Check that the system is fair!
3. Protect privacy!
4. Enable freedom of action and freedom of choice!
5. Ensure explainablity!

[12]E.g., IEEE's "Ethically Aligned Design" principles or the United Nation's Global Impact Principles.

3.8 Lessons Learned/Cooperated Driving

Research in the area of cooperated driving in KoFFI covered two main aspects: on the one hand, the project work itself ("the KoFFI project") and on the other hand, the "KoFFI system prototype."

The KoFFI project

- Evaluated ethical and legal questions relating to the highly automated driving environment (e.g., data security, decision-making sovereignty),
- Examined various takeover scenarios and offered solutions for them, and
- Developed the architecture and software for the KoFFI system prototype.

The KoFFI system prototype

- supports driving tasks, especially at SAE automation levels 1–4,
- offers a multimodal user interface with highly innovative voice dialogue technology,
- takes legal requirements into account (e.g., road traffic regulations, data logging—who was driving at what time?),
- cooperates with the driver on an equal footing, and
- is an adaptable software module.

The research results indicate that cooperative driver–vehicle interaction is able to solve traffic situations that neither people nor vehicles cannot handle alone and that it is therefore a key element for the acceptance of highly automated vehicles. Cooperative driver–vehicle interaction closes the gap between manual and automated driving. The work focused on human–machine interaction (HMI) only. Nevertheless, a huge effort is still required to develop the technical foundations for cooperative driving. This will require collaborations by interdisciplinary teams, e.g., the integration of new technologies such as 5G networks for C2X communication and highlyautomated driving. Only after integration of such new technologies in future cars the above described use cases can be experienced in real traffic environment.

We have been able to show that legal and ethical requirements need to be taken into account and how, but there is still some way to go (e.g., overriding by the vehicle).

The main challenge involves the continuous learning of new and more complex traffic situations and actions as well as integration thereof into the system. Therefore, automated systems need to learn continuously and to deal with big real-time data. This requires the use of AI with all of the associated risks and opportunities. It is therefore important to have project-related ethical and legal support, as was the case in the KoFFI project. The KoFFI Code includes an ethical self-assessment that can be used by such projects free of charge.

Acknowledgements We would like to thank the German Federal Ministry of Education and Research (BMBF Bundesministerium für Bildung und Forschung) for its financial support,

VDI/VDE Innovation + Technik GmbH for its coordination work, and our companies and institutions for their general support during the project runtime.

References

Abbink DA, Mulder M, Boer ER (2012) Haptic shared control: smoothly shifting control authority? Cogn Technol Work 14:19–28. https://doi.org/10.1007/s10111-011-0192-5

Algorithmwatch (2019) AI ethics guidelines global inventory. https://algorithmwatch.org/en/project/ai-ethics-guidelines-global-inventory/. Accessed 17 Jan 2020

Allen R, Rosenthal T, Cook M (2011) A short history of driving simulation. Handbook of driving simulation for engineering, medicine, and psychology. CRC Press. https://doi.org/10.1201/b10836-3

Amodei D, Anubhai R, Battenberg E et al (2020) Deep speech 2: end-to-end speech recognition in English and Mandarin. https://arxiv.org/pdf/1512.02595.pdf

Apple (2020) https://www.apple.com/de/siri/. Accessed 17 Jan 2020

Art. 29 Data Protection Working Party, Opinion (3/2012) on developments in biometric technologies WP 193, 2012

Autoblog (2020) Waymo self-driving cars rack up 20 million miles on public roads. https://www.autoblog.com/2020/01/07/waymo-self-driving-20-million-miles/

Awad E, Dsouza S, Kim R et al (2018) The Moral machine experiment. Nature 563:59–64. https://doi.org/10.1038/s41586-018-0637-6

Balzer T, Nugel M (2016) Das Auslesen von Fahrzeugdaten zur Unfallrekonstruktion im Zivilprozess. NJW 2016:193

Beaver I, Freeman C, Mueen A (2017) An annotated corpus of relational strategies in customer service. arXiv:1708.05449

Bengler K, Zimmermann M, Bortot D et al (2012) Interaction principles for cooperative human-machine systems interaction principles for cooperative human-machine systems. https://doi.org/10.1524/itit.2012.0680

Berndt S (2018) Das Automobil im Visier der Strafverfolgungsbehörden. NZV 2018:249

Bloomberg (2018) Who's winning the self-driving car race? https://www.bloomberg.com/news/features/2018-05-07/who-s-winning-the-self-driving-car-race

BMBF (2019) Das intelligente Auto – der beste Freund des Menschen. https://www.bmbf.de/de/das-intelligente-auto-der-beste-freund-des-menschen-10278.html

Bohus D, Rudnickey A (2003) RavenClaw: dialog management using hierarchical task decomposition and an expectation agenda. In: Eighth European conference on speech communication and technology

Boril H, Sadjadi SO, Hansen JH (2011) UTDrive: emotion and cognitive load classification for in-vehicle scenarios. In: The 5th Biennial workshop on digital signal processing for in-vehicle systems

Bosch (2019) Connected parking automated valet parking, don't get stressed, get parked. https://www.bosch.com/de/stories/automated-valet-parking/

Braunger P, Hofmann H, Werner S et al (2016) A comparative analysis of crowdsourced natural language corpora for spoken dialog systems. In: LREC

Brockmeyer H (2018) Treuhänder für Mobilitätsdaten – Zukunftsmodell für hoch- und vollautomatisierte Fahrzeuge? ZD 2018, 258

Christoffersen K, Woods DD (2002) How to make automated systems team players. Adv Hum Perform Cogn Eng Res 2:1–12. https://doi.org/10.1016/S1479-3601(02)02003-9
Cummings JJ, Bailenson JN (2016) How immersive is enough? A meta-analysis of the effect of immersive technology on user presence. Media Psychol 19(2):272–309. https://doi.org/10.1080/15213269.2015.1015740
Dekker SWA, Woods DD (2002) MABA-MABA or Abracadabra? Progress on human-automation co-ordination. Cogn Technol Work 4:240–244. https://doi.org/10.1007/s101110200022
Deng L, Liu Y (eds) (2018) Deep learning in natural language processing. Springer, Singapore
DVR (2012) German Road Safety Council (DVR) "Vision Zero". In: Schriftenreihe 16. https://www.dvr.de/download2/p3042/3042_0.pdf. Accessed 16 Jan 2020
Eliot (2020) https://www.aitrends.com/ai-insider/car-voice-commands-nlp-self-driving-cars/. Accessed 17 Jan 2020
Endsley MR (2017) From here to autonomy: lessons learned from human-automation research. Hum Factors 59:5–27. https://doi.org/10.1177/0018720816681350
Endsley MR, Kaber DB (1999) Level of automation effects on performance, situation awareness and workload in a dynamic control task
EP (2019) European Parliament "Safer roads: EU lawmakers agree on life-saving technologies for new vehicles". https://www.europarl.europa.eu/news/en/press-room/20190326IPR33205/safer-roads-eu-lawmakers-agree-on-life-saving-technologies-for-new-vehicles. Accessed 16 Jan 2020
European Data Protection Board, EDPB, Guidelines (1/2020) on processing personal data in the context of connected vehicles and mobility related applications
Ethik-Kommission Automatisiertes und Vernetztes Fahren (2017) Bericht: Juni 2017. https://www.bmvi.de/SharedDocs/DE/Publikationen/DG/bericht-der-ethik-kommission.pdf?__blob=publicationFile. Accessed 20 Jan 2020
European Union (2016) Regulation (EU) 2016/679 of the European Parliament and of the Council of 27 April 2016 on the protection of natural persons with regard to the processing of personal data and on the free movement of such data, and repealing Directive 95/46/EC: (General Data Protection Regulation) 59 (119) https://eur-lex.europa.eu/legal-content/EN/TXT/PDF/?uri=OJ:L:2016:119:FULL&from=EN. Accessed 14 Nov 2016
Fernandez R, Picard RW (2003) Modeling drivers' speech under stress. Speech Commun 40(1–2):145–159
Fischer V, Kunzmann S (2001) Bayesian information criterion based multi-style training and likelihood combination for robust hands-free speech recognition in the car. In: Proceedings of the IEEE workshop on hands-free speech communications, Kyoto, Japan
Fischer V, Ghahabi O, Kunzmann S (2018) Recent improvements to neural network based acoustic modeling in the EML realtime transcription platform. In: Proceedings of 29th conference on electronic speech signal processing. Ulm, Germany
Foot P (1967) The problem of abortion and the doctrine of the double effect. Oxford Rev 5:5–15 https://philpapers.org/rec/FOOTPO-2. Accessed 10 May 2016
Friedman B, Hendry DG (2019) Value sensitive design: shaping technology with moral imagination. MIT Press, Cambridge, Mass
German Association of the Automotive Industry (VDA)—Data protection aspects in the use of networked and non-networked vehicles (2016)
Ghahabi O, Zhou W, Fischer V (2018) A robust voice activity detection for real-time automatic speech recognition. In: Proc. of 29th Confer-ence on Electronic Speech Signal Processing, Ulm, Germany
Gleiss (2017) Lutz Automotive: Neue rechtliche Vorgaben für automatisiertes Fahren. https://www.gleisslutz.com/de/automatisiertes%20Fahren.html
Graves A, Jaitly N, Mahamed A (2013) Hybrid speech recognition with deep bidirectional LSTM. In: Proceedings of the 2013 IEEE international conference on acoustics, speech, and signal processing. Vancouver, Canada
Grimm P, Kuhnert S (2018) Funktionalität und Vertrauen: Eine interdisziplinäre Expertenrunde zum automatisierten und autonomen Fahren an der Hochschule der Medien in Stuttgart. Available via

"Wissenschaft und Forschung an der HdM". https://www.hdm-stuttgart.de/science/view_beitrag?science_beitrag_ID=452. Accessed 17 Jan 2020

Grimm P, Kuhnert S (2018) Narrative Ethik in der Forschung zum automatisierten und vernetzten Fahren. Mensch – Maschine. Franz Steiner Verlag, Stuttgart, pp 93–109

Grimm P, Kuhnert S (Upcoming 2020) Die Zusammenarbeit von Industrie, Ethik und Wissenschaft im Forschungsverbund: Kommunikation. Integration. Innovation. In: Gransche B, Manzeschke A (ed) Das geteilte Ganze: Horizonte Integrierter Forschung für künftige Mensch-Technik-Verhältnisse. Springer, Wiesbaden

Grimm P, Keber TO, Zöllner O (eds) (2019) Digitale Ethik: Leben in vernetzten Welten. Reclam, Stuttgart

Goldberg Y (2017) Neural network methods for natural language processing. Morgan & Claypool Publishers, San Rafael, CA, USA

Hannun A, Case C, Casper J et al (2014) Deep speech: scaling up end-to-end speech recognition. https://arxiv.org/pdf/1412.5567.pdf

Hansen J, Angkititrakul P, Plucienkowski J et al (2002) CU-move: analysis and corpus development for interactive in-vehicle speech systems. In: Proceedings of the 2nd annual conference on spoken language processing (INTERSPEECH). Aalborg, Denmark

Hart SG, Staveland LE (1988) Development of NASA-TLX (Task Load Index): results of empirical and theoretical research. Adv Psychol 52:139–183

Hart SG (2006) NASA-task load index (NASA_TLX): 20 years later. In: Proceedings of the human factors and ergonomics society annual meeting, vol 50, pp 904–908. Sage Publications

Hasselbalch G, Tranberg P (2016) Data ethics: the new competitive advantage. https://dataethics.eu/en/book. Accessed 20 Jan 2020

He T, Fan Y, Qian Y et al (2014) Reshaping deep neural net-work for fast decoding by node-pruning. In: Proceedings of the 2014 IEEE international conference on acoustics, speech, and signal processing. Florence, Italy

Hensch A-C, Rauh N, Schmidt et al (2020) Effects of secondary tasks and display position on glance behavior during partially automated driving. Transp Res Part F: Traffic Psychol Behav 68 23–32. https://doi.org/10.1016/j.trf.2019.11.014

Hinton G, Deng L, Yu D et al (2012) Deep neural networks for acoustic modeling in speech recognition: the shared views of four research groups. IEEE Sig Process Mag 29(6):82–97

Hoc J (2010) From human-machine interaction to human-machine cooperation, 0139. https://doi.org/10.1080/001401300409044

Hofmann H, Ehrlich U, Berton A et al (2012) Speech interaction with the Internet—a user study. In: 8th International conference on intelligent environments (IE). S. 323–326. IEEE

Hollnagelt E, Woods DD (1983) Cognitive systems engineering: new wine in new bottles, pp 583–600

Keber T, Keppeler L, Schwartmann R u.a. (Hrsg.) (2018) DS-GVO/BDSG: Datenschutz-Grundverordnung, Bundesdatenschutzgesetz, Artikel 25. Heidelberger, Kommentar

Kienle M, Damböck D, Kelsch J et al (2009) Towards an h-mode for highly automated vehicles: driving with side sticks. In: Proceedings 1st international conference on automotive user interfaces interactive vehicle applications AutomotiveUI 2009, pp 19–23. https://doi.org/10.1145/1620509.1620513

Kim B, Ryu S, Lee GG (2017) Two-stage multi-intent detection for spoken language understanding. Multimedia Tools Appl 76(9):11377–11390

Klein G, Woods DD, Bradshaw JM et al (2004) Ten challenges for making automation a "team player" in joint human-agent activity. IEEE Intell Syst 19:91–95. https://doi.org/10.1109/MIS.2004.74

Klink-Straub J, Keber T, Aktuelle Gesetzeslage zum automatisierten Fahren, NZV 2020 (forthcoming)

Klink-Straub J, Straub T, Vernetzte Fahrzeuge – portable Daten, ZD 2018, 459
Koo J, Kwac J, Ju W et al (2015) Why did my car just do that? Explaining semi-autonomous driving actions to improve driver understanding, trust, and performance. Int J Interact Des Manuf, 269–275
Kraus JM, Sturn J, Reiser JE, Baumann M (2015) Anthropomorphic agents, transparent automation and driver personality: Towards an integrative multilevel model of determinants for effective driver-vehicle cooperation in highly automated vehicles. In: Adjunct proceedings of the 7th international conference on automotive user interfaces and interactive vehicular. AutomotiveUI 2015, pp 8–13. https://doi.org/10.1145/2809730.2809738
Krausen J-M (2019) Autorecht Schaden und Beweis. ZD-Aktuell 2019:04369
Krausen J-M (2019) Unfallaufklärung 2.0, Vision Zero Datenschutz, ZD-Aktuell 2019, Dok, 06679
Landesberger J, Kornmüller D, Ehrlich U (2017) Explorative Untersuchung von Multi-Intents in Sprachdialogsystemen. Multimedia Tools Appl 76(9):11377–11390
Landesberger J, Ehrlich U (2018) Investigating strategies for resolving misunderstood utterances with multiple intents. In: Proceedings of the 22nd workshop on the semantics and pragmatics of dialogue (AixDial)
Landesberger J, Ehrlich U (2019a) Towards finding appropriate responses to multi-intents—SPM: sequential prioritisation model. In: Proceedings of the 23rd workshop on the semantics and pragmatics of dialogue (LondonLogue)
Landesberger J, Ehrlich U (2019b) Finding a metadialogue strategy for multi-intent spoken dialogue systems. In: Proceedings of the international conference on computer-human interaction research and applications (CHIRA)
Lê H, Pham TL, Meixner G (2017) A concept for a virtual reality driving simulation in combination with a real car. In: Proceedings of the 9th international conference on automotive user interfaces and interactive vehicular applications adjunct—AutomotiveUI'17, pp 77–82. https://doi.org/10.1145/3131726.3131742
Lee JD, Hoffman JD, Hayes E (2004) Collision warning design to mitigate driver distraction. In: Proceedings of the SIGCHI conference on human factors in computing systems
Lemon O (2012) Conversational interfaces. Data-driven methods for adaptive spoken dialogue systems. Springer, New York, NY, pp 1–4
Lim BY, Dey AK, Avrahami D (2009) Why and why not explanations improve the intelligibility of context-aware intelligent systems. In: Proceedings of the SIGCHI conference on human factors in computing systems. ACM, pp 2119–2128
Lucas G, Kemeny A, Paillot D, Colombet F (2020) A simulation sickness study on a driving simulator equipped with a vibration platform. Transp Res Part F: Traffic Psychol Behav 68:15–22. https://doi.org/10.1016/j.trf.2019.11.011
Lutz L (2019) Fahrzeugdaten und staatlicher Datenzugriff, DAR 2019, 125
Madsen M, Gregor S (2000) Measuring human-computer trust. In: 11th Australasian conference on information systems, vol 53, pp 6–8
Matzner T (2019) Autonome Trolleys und andere Probleme. Konfigurationen Künstlicher Intelligenz in ethischen Debatten über selbstfahrende Kraftfahrzeuge. https://doi.org/10.25969/MEDIAREP/12632
Maurer S, Rukzio E and Erbach R (2018a) Challenges for creating driver overriding mechanisms. In: Adjunct proceedings of the 9th international ACM conference on automotive user interfaces and interactive vehicular applications. Oldenburg, Germany, pp 99–103
Maurer S, Erbach R, Kraiem I, Kuhnert S, Grimm P, Rukzio E (2018b) Designing a guardian angel. In: Interaction ASIGoC-HM (ed) Proceedings of the 10th international conference on automotive user interfaces and interactive vehicular applications. ACM, pp 341–350
Mayer RC, Davis JH, Schoorman FD (1995) An integrative model of organizational trust. Acad Manag Rev 20(3):709–734

McCall R, McGee F, Meschtscherjakov A, Louveton N, Engel T (2016) Towards a taxonomy of autonomous vehicle handover situations. AutomotiveUI 2016—Proceedings of 8th international conference on automotive user interfaces and interactive vehicular applications, pp 193–200. https://doi.org/10.1145/3003715.3005456

McCormick I, Walkey F, Green D (1986) Comparative perceptions of driver ability: a confirmation and expansion. Accident analysis and prevention, vol 18, pp 205–208

McTear MF (2004) Spoken dialogue technology: toward the conversational user interface. Springer Science & Business Media

Metzger A (2019) Digitale Mobilität, Verträge über Nutzerdaten GRUR, p 129

Mikolov T, Karafiat M, Burget L et al (2010) Recurrent neural network based language model. In: Proceedings of the 11th annual conference of the international speech communication. Association (INTER-SPEECH). Makuhari, Chiba, Japan

Moreno A, Lindberg B, Draxler C et al (2000) SPEECHDAT-CAR. A large speech database for automotive environments. In: Proceedings of the 2nd international conference on language resources and evaluation (LREC). Athens, Greece

Müller M, Grimm P (2016) Narrative Medienforschung: Einführung in Methodik und Anwendung. UVK Verlagsgesellschaft mbH, Konstanz

Norman D (2013) The design of everyday things, Revised and expanded edn. Basic Books (AZ)

Nothdurft F, Ultes S, Minker W (2015) Finding appropriate interaction strategies for proactive dialogue systems—an open quest. In: Proceedings of the 2nd European and the 5th Nordic symposium on multimodal communication. Tartu, Estonia (No 110). Linköping University Electronic Press, pp 73–80

Nothdurft F, Richter F, Minker W (2014) Probabilistic human-computer trust handling. In: Proceedings of the 15th annual meeting of the special interest group on discourse and dialogue (SIGDIAL), pp 51–59

Riek LD, Howard D (2014) A code of ethics for the human-robot interaction profession. robots.law.miami.edu/2014/wp-content/uploads/2014/03/a-code-of-ethics-for-the-human-robot-interaction-profession-riek-howard.pdf. Accessed 16 Aug 2016

Rüpke G, v. Lewinski K, Eckhardt J (2018) Datenschutzrecht

SAE (2014) SAE international: taxonomy and definitions for terms related to on-road motor vehicle automated driving systems. https://saemobilus.sae.org/content/j3016_201401

Schliesky U (2019) Digitale Ethik und Recht. Neue Juristische Wochenschrift:3692–3697

Schmid W (2017) Event Data Recording für das hoch- und vollautomatisierte Kfz – eine kritische Betrachtung der neuen Regelungen im StVG. NZV 2017:357

Schwartmann R, Jacquemain T (2018) Datenschutzrechtliche Herausforderungen im Auto, RDV 2018, 247 f

Schwenke T (2018) Zulässigkeit der Nutzung von Smartcams und biometrischen Daten nach der DS-GVO. NJW 2018:823

Sheridan T, Verplank W (1978) Human and computer control of undersea teleoperators

Sørmo F, Cassens J (2004) Explanation goals in case-based reasoning. In: Proceedings of the ECCBR 2004 workshops, No 142–04, pp 165–174

Steele M, Gillespie RB (2001) Shared control between human and machine: using a haptic steering wheel to aid in land vehicle guidance. Proc Hum Factors Ergon Soc Annu Meet 45:1671–1675. https://doi.org/10.1177/154193120104502323

Steinert P, Automatisiertes Fahren, SVR 2019, 5

Tawari A, Trivedi M (2010) Speech based emotion classification framework for driver assistance system. In: 2010 IEEE intelligent vehicles symposium. IEEE, pp 174–178

Techcrunsh (2019) Completely driverless Waymo cars are on the way. https://techcrunch.com/2019/10/09/waymo-to-customers-completely-driverless-waymo-cars-are-on-the-way/

UN (1948) United Nations General Assembly, Universal Declaration of Human Rights. General Assembly resolution 217 A. https://www.un.org/en/universal-declaration-human-rights/index.html. Accessed 16 Jan 2020

UN (1949) United Nations "Convention on Road Traffic, Geneva, September 19, 1949". https://treaties.un.org/doc/Publication/MTDSG/Volume%20I/Chapter%20XI/XI-B-1.en.pdf. Accessed 16 Jan 2020

UN (1968) United Nations "Convention on road traffic, Vienna, November 8, 1968". https://treaties.un.org/doc/Treaties/1977/05/19770524%2000-13%20AM/Ch_XI_B_19.pdf. Accessed 16 Jan 2020

UNECE (2016) United Nations Economic Commission for Europe, UNECE paves the way for automated driving by updating UN international convention. https://www.unece.org/info/media/presscurrent-press-h/transport/2016/unece-paves-the-way-for-automated-driving-by-updating-un-international-convention/doc.html. Accessed 16 Jan 2020

Van den Hoven J, Miller S, Pogge T (2017) The design turn in applied ethics. In: Van den Hoven J (ed) Designing in ethics. Cambridge University Press, Cambridge, pp 11–31. https://doi.org/10.1017/9780511844317.002

Walch M, Lange K, Baumann M et al (2015) Autonomous driving: investigating the feasibility of car-driver handover assistance. In: Proceedings of the 7th international conference on automotive user interfaces and interactive vehicular applications. ACM, pp 11–18

Walch M, Baumann M, Jaksche L et al (2017a) Touch screen maneuver approval mechanisms for highly automated vehicles: a first evaluation. AutomotiveUI 2017—Adjunct proceedings of 9th international ACM conference on automotive user interfaces interactive vehicle application, pp 206–211. https://doi.org/10.1145/3131726.3131756

Walch M, Mühl K, Kraus J et al (2017b) From car-driver-handovers to cooperative interfaces: visions for driver–vehicle interaction in automated driving, pp 273–294. https://doi.org/10.1007/978-3-319-49448-7_10

Walch M, Mühl K, Baumann M, Weber M (2018) Click or hold: usability evaluation of maneuver approval techniques in highly automated driving. Conference on human factors in computing systems—Proceedings 2018-April, pp 1–6. https://doi.org/10.1145/3170427.3188614

Walch, M, Colley, M, Weber M (2019a) Driving-task-related human-machine interaction in automated driving: towards a bigger picture. In: Adjunct proceedings of the 7th international conference on automotive user interfaces and interactive vehicular. AutomotiveUI 2019, pp 427–433. https://doi.org/10.1145/3349263.3351527

Walch M, Colley M, Weber M (2019b) CooperationCaptcha: on-the-fly object labeling for highly automated vehicles. In: Conference on human factors in computing systems—Proceedings, pp 1–6. https://doi.org/10.1145/3290607.3313022

Walch M, Lehr D, Colley M, Weber M (2019c) Don't you see them? Towards gaze-based interaction adaptation for driver-vehicle cooperation. In: Adjunct proceedings—11th international ACM conference on automotive user interfaces interactive vehicle application. AutomotiveUI 2019, pp 232–237. https://doi.org/10.1145/3349263.3351338

Walch M, Woide M, Mühl K et al (2019d) Cooperative overtaking: overcoming automated vehicles' obstructed sensor range via driver help. In: Proceedings—11th international ACM conference automotive user interfaces interactive vehicle application. AutomotiveUI 2019, pp 144–155. https://doi.org/10.1145/3342197.3344531

Weng F, Angkititrakul P, Shriberg E et al (2016) Conversational in-vehicle dialog systems: the past, present, and future. IEEE Sig Process Mag 33(6):49–60

WHO (2018) World Health Organization "Global status report on road safety 2018". https://www.who.int/violence_injury_prevention/road_safety_status/2018/en/. Accessed 16 Jan 2020

WIVW (2020) Würzburger Institut für Verkehrswissenschaften. "Fahrsimulation und SILAB". https://wivw.de/de/silab. Accessed 16 Jan 2020

Woide M, Stiegemeier D, Baumann M (2019) A methodical approach to examine conflicts in context of driver—autonomous vehicle—interaction. In: Proceedings of international driving symposium on human factors in driver assessment, training, and vehicle design, pp 314–320

Young MS, Young NA, Stanton HD (2007) Driving automation: learning from aviation about design philosophies. Int J Veh Des 45(3):323–338

Yu D, Deng L (eds) (2014) Automatic speech recognition: a deep learning approach. Springer, London

Zeyer A, Doetsch P, Voigtlaender P et al (2017) A comprehensive study of deep bidirectional LSTM RNNs for acoustic modeling in speech recognition. In: Proceedings of the 2017 IEEE international conference on acoustics, speech, and signal processing. New Orleans, USA

Chapter 4
Ethical Recommendations for Cooperative Driver-Vehicle Interaction—Guidelines for Highly Automated Driving

Petra Grimm and Julia Maria Mönig

Abstract The KoFFI guidelines have been formulated during the project duration. Even though in the meantime, many guidelines have been published, we consider it as relevant, that everybody reflects about their doings and about their own ethical principles. As a means to achieve this, we offer a questionnaire for ethically reviewing a project. In a dialogue with the project partners, we have defined four values which we consider as central for the development of highly automated driving in general and the KoFFI project in particular. These values are privacy, autonomy, trust (in technology) and transparency. With the questionnaire, we suggest a tool to perform a self-check of ethical issues that might arise while doing research on highly automated driving. Slightly adapted, the questionnaire can also be used for the development of other technologies. It can be used during quality management procedures and should be reused during the ongoing of the project. Having completed the questionnaire researchers, developers and the different actors involved will probably agree that taking into account ethics can represent and advantage for the respective company.

4.1 Introduction

The societal changes caused by digitalization and especially by the not-yet foreseeable consequences of the use of artificial intelligence lead to a call that we need ethics and to a need for ethics in general. Currently, ethical guidelines seem to "have a boom". On their website, the organization Algorithmwatch lists—without claiming to be an exhaustive list—83 guidelines, principles and declarations for

P. Grimm (✉) · J. M. Mönig (✉)
Hochschule der Medien, Stuttgart, Germany
e-mail: grimm@hdm-stuttgart.de

J. M. Mönig
e-mail: moenig@hdm-stuttgart.de

G. Meixner (ed.), *Smart Automotive Mobility*,
Human–Computer Interaction Series,
https://doi.org/10.1007/978-3-030-45131-8_4

digital ethics, data ethics, AI ethics and ethical algorithms, written by NGOs, governments, consumer organizations, research institutes, companies, etc.[1] These guidelines are authored by such different organizations which show that apparently many people agree that there is a need for guidelines. We need, however, to be careful (Grimm et al. 2019).

The increasing interest in ethics should not be used as a pretext to avoid the introduction of binding regulations and laws. Self-commitments should not remain lip services. They should not be used as a method to carry out "ethics washing." In addition, "regulatory patchworks" might also lead to "ethics shopping," the phenomenon that, for instance, AI development is moved to regions of the world with "lower" ethical standards (Directorate-General for Research and Innovation (European Commission), European Group on Ethics in Science and New Technologies 2018). Wagner (2018) suggests six criteria for meaningful, sensible ethical guidelines: First, all relevant stakeholders should participate in the development process.[2] Second, a mechanism for independent oversight should be provided. Third, transparent decision-making procedures should be applied. Fourth, a list of "non-arbitrary" values should be developed, where "the selection of certain values, ethics and rights over others can be plausibly justified."[3] Fifth, it should be ensured that ethics are not "substituting" fundamental or human rights.[4] Finally, a clear statement should be made concerning the relationship between the respective statement and existing laws and regulations, especially for the case that both of them are conflicting. The ethical code at hand meets these requirements.

4.1.1 What Are Ethical Guidelines?

Ethical guidelines are standards or principles. The persons who formulate them think that the compliance with these principles seems worthwhile. Usually, they start off from a (perceived) lack of regulation or lack of compliance with existing regulations.

---

[1]The first guidelines date from 2011, however, most of them have been published in 2017 and 2018. Last Update of the list 2019-06-21, last visited 2019-07-18.

[2]This principle is well-known also in the context of value-sensitive design. To avoid—unwanted unidirectional—lobbyism, stakeholders of different interest groups should be involved. It should not be possible for companies to "soften" standards in their own interest, Metzinger (2019).

[3]The question arises, how standards can be non-arbitrary when they are freely chosen and are, after all, "only" circularly referring to existing rights or to other non-binding text, or that are being justified by themselves during the process.

[4]This criterion might be clear inside the EU; however, it must be defined, how it can be decided whether it has been tried to substitute rights, what can be done against it and how it can be avoided. This relates to criterion 6. What does it mean, when ethical commitments are in conflict and are contrasting with existing legal and regulatory frameworks? Is the law in this case unethical? Or are the ethical guidelines not "strong" enough? Are the actions and practices illegal or do they violate existing (human) rights? One example might be a company whose highest ethical goal is to maximize its own profit, without caring about its workers and employees and whether they are exploited. What does it mean, when the ethical voluntary agreements are conflicting with each other?

The ethical standards can be used as evaluation criteria. Their recommendations can go beyond existing laws. They can be valid for individuals, a certain group of people or even the whole human species. At the same time, there are only restricted ways to enforce the compliance with these rules.[5] Often they are value-based and refer to values such as human dignity that serves as a "yardstick" for acting ethically. Additionally, the individual's responsibility for one's own actions can be stressed.[6]

4.1.2 *What is Ethics?*

Ethics means discussing and reflecting moral behavior. Moral is a set of "traditional" rules and conventions and passed on possibilities how to act. As an academic discipline "ethics" is part of "practical philosophy" and provides the opportunity to reflect on morale or moral standards.

Ethics can not only provide orientation in moral issues, i.e., it does not only reflect on how the individual can act well, and what the implications of my actions are on others, but it also raises the question of how a good, successful life can look like, in current versions for the largest possible number of people. Areas in which ethics often play a role today are medicine (e.g., terminal care), environment (e.g., environmental issues, or whether apes should have human rights, technology (e.g., in robotics), but also research ethics (e.g., questions about stem cell research). Ethics has a normative, a descriptive and a volatile function. Digital ethics, for example, deals with the impact on individuals and the society. It derives normative standards from these debates, that, in turn, can set impulses for future moral actions (Grimm 2013).

4.1.3 *Implementing Ethics in Research and Development and into the New Product*

The question arises, how ethics can be implemented during the research and development process. The value-based-design or value-sensitive-design approach aims at putting ethics *by design* into practice. Since technology is never neutral, the aim is to reflect concerned values right from the beginning and to factor them in during the development of a product or a service, with the goal to make "ethical" decisions. Ethics

[5] One possibility to sanction behavior that violates guidelines would be, in case of professional ethics guidelines, to ban a person from that organization and its services.

[6] Alan Winfield calls all (industrial) standards "implicitly ethical." In contrast, a standard that is explicitly ethical could be defined as "[..] one that addresses clearly articulated ethical concerns, and seeks—through its application—to, at best remove, hopefully reduce, or at the very least highlight the potential for unethical impacts or their consequences." Winfield asks what the ethical principles are that underlie these new ethical standards Winfield (2019). Thilo Hagendorff examined the key issues of 15 ethical guidelines for AI. Privacy protection was mentioned by 14 of the examined guidelines. Further issues were fairness, transparency and security, or "hidden costs," which were only taken into account in one statement, Hagendorff (2019).

by Design can therefore mean that possibly conflicting values are taken into consideration (e.g., security against privacy), that diversity of users is taken into account, and principles for ethical action, for example, in the manufacture of products, are taken seriously. This is not a one-off matter. Rather is the development accompanied by the reflection and is not even completed after the product has reached marketability. Ethics Officers and/or Ethics Committees may be appointed for monitoring purposes, while value-based design appeals to people's moral reason. If conflicts do arise, e.g., in form of conflicting values, these institutions might act as mediators.

In order to sensitize actors for ethical questions and to illustrate ethical issues, ethical use cases are being discussed. In terms of "narrative ethics," the use cases are understood as stories that can lead to "aha-experiences," without moralizing. The persons involved are not sensitized for moral questions and ethical reasoning, and judging is encouraged.

From the beginning, therefore, there is an ethical monitoring of the research and development process, which can and should be taken up again after the completion of a process or when the development of a product is seemingly finished, because firstly, value concepts might change, even within a society, and secondly ethical conflicts might arise where none have been seen before.

The use of empirical methods from narrative research and qualitative and quantitative social research can be helpful, in order to empirically substantiate ethical research and to identify the concerns, possible fears, worries and the needs of those affected by a new technology. At the same time, research methods should be ethically screened and evaluated during the process and not only at the beginning and afterward.

4.1.4 Key Points for Ethical Guidelines

The following key points for ethical guidelines have been formulated by the ELSI research team. They relate to ethical considerations for technological developments and advances in general and highly automated driving in particular.

1. Ethics right from the start
   In the sense of Ethics by Design, ethics from the outset, accompanying the process, should be implemented. The ethical evaluation will also be carried out after the termination of the process or completion of the product.
2. Narrative ethics by design
   Narrations serve as illustrations of values and (potential) conflicts of values. The parallels between stories and examples, which in the philosophical tradition have played a major role, and use cases, that are being used, for example, in engineering disciplines and programming, can be put to productive use here.
3. Value-based ethics
   In the sense of value-based ethics, values are being considered during the whole research and development process and aftermarket launch. Values can change over time and therefore cause value conflicts that had not been anticipated beforehand.

4. Universal values
   Despite cultural differences in the weighting of values, there are—not only in highly automated driving—values that affect everyone. These values apply, without weighing them up against each other. They should apply, regardless of alleged factual constraints or cultural sensitivities. In addition, the fundamental rights provide a "visible interface" between law and ethics.
5. Involvement of all stakeholders
   In order to ensure the consideration of different interests, all research and development stakeholders, (potential) users of the technology and other representatives of interest groups are taking part in ethical discussions and the negotiation process.
6. Beware of ethics washing
   The participation of all stakeholders must not lead to unilateral lobbyism or to comprising ethical standards or basic principles. The relevant actors are identified at the beginning; it is verified during the process whether other stakeholders should be involved.
7. Sensitization and empowerment of all stakeholders
   Opportunities for ethical reasoning are created for all people. This does not only mean that professionals, such as engineers, computer scientists and programmers will be trained in ethical reflection, but that ethicists are trained in technical understanding and interdisciplinary communication alike. This requires the implementation of the corresponding contents into school, university and training curricula. Through ethical sensitization in research processes, the interest can also be awakened to undergo further ethical training. Public institutions should finance appropriate (further) educational opportunities, in order to ensure that these are not (solely) in the hands of companies and are therefore—potentially—subject to the influence of the latter.
8. Value conflicts
   In the case of conflicts of values, the discussion with all parties involved is searched. An attempt is made to find a solution that meets the social norms, which are also questioned, if necessary. At the same time, the autonomy of the individual needs to be protected. Universal rights, in particular human dignity, the right to freedom of expression and the general right of personality ("Allgemeines Persönlichkeitsrecht" in the German jurisdiction) are respected.
9. Whistle-blowing
   The public can serve as a supervisory authority. When it comes to information which should not be made public, it should be questioned what it is exactly that should be kept secret. A trade secret or security concerns differ from, for example, the deliberate deception of users and shareholders. This aspect is directly related to the value of "transparency."
10. Dual use and redlines
    Since highly automated technology and therefore highly automated driving might be used for military purposes, the question arises, which aims are envisioned. When a product is being designed or a process is being planned, it should be considered whether there are redlines that should not be crossed.

11. Ethics committees or independent ethics advisors
    To support and control the implementation of ethical standards, upon which there has been an agreement, ethics committees, ethics "officers" or independent advisory groups can be appointed. These should avoid acting as some sort of "ethics police" or "ethics nanny." They should, however, not be afraid to denounce wrongdoing. Internal enforcement should be possible, in case of, for instance, scientific or professional misconduct.
12. Future developments, e.g., moral machines
    The question, whether it is possible or makes sense to program machines that will make moral decisions, has still and continuously to be debated. Should vehicles with level-5 technology become marketable, it has to be agreed on beforehand, how to handle edge cases in which the machine has to behave in a way that includes ethical "reasoning." This is important, firstly, because programming includes a planned action in contrast to a human spontaneous and unconscious judging and a person's reaction time during an accident. Secondly, it can be problematic to train machines from the outset with parameters for (supposedly) ethically "correct" behavior, since this could mean that the machine could also be trained to behave unethically.
13. Voluntary commitments
    The companies concerned as well as other (non-)governmental actors should not only adopt ethical statements but actively commit to their realization.
14. Questionnaires and quality management
    To consider and realize potential ethical conflicts during the development of a technology, such as highly automated driving, the stakeholders can answer ethical questionnaires, as a self-assessment, e.g., as part of their foreseen quality management process.

4.1.5 Guidelines for Highly Automated Driving

According to *Nature News* "the only governmental guidance on self-driving cars" has been published in Germany in 2017 (Maxmen 2018). In the research project at hand, we are dealing with the challenges that come with automation level 3–4. It focuses on the handover between human and machine and the cooperation between human and vehicle in highly automated driving.

4.1.6 What to Do? Practical Examples and Questions for Self-Assessment

Even if ethics and acting ethical cannot be reduced to a checklist, question lists can be helpful for self-assessment and thus initiate reflection processes.

1. **Check for discrimination**

1. *Fair algorithms and data*: Different attempts have been made to program "fair" algorithms from the outset. The programmers have to become aware of their bias, prejudices and "social imprint," the data basis must be tested for inherent bias, systematic errors and distortions should be avoided from the beginning, also with technical means.[7]
2. *Gender*: Should the speech assistant have a male or a female voice? A well-known language assistant does not only have a female voice, but even a female name as an activation word, even if other words can be used and the device itself has a different name. Obviously, the problem has been observed that robots are treated like "slaves," and that devotion is expected of them. At the same time, it is being discussed that we might transfer our behavior toward robots to our contact with humans; for worse (cf. Kant's brutalization argument with regard to the cruelty to animals) as well as for better [cf. the argument that also compassion is practiced on animals, (Hagendorff 2019)].[8] In relation to gender, there is the proposal to use a modulated voice whose frequency is 153 Hz, which is perceived as gender-neutral.[9]

2. **Data protection and privacy**: To facilitate compliance with the legal obligations set by the General Data Protection Regulation (GDPR), such as implementing "privacy by design" and the principle of data minimization, the following questions can help to identify possible privacy violations: Will you carry out a data protection impact assessment (GDPR Art. 35)?

   1. *Connected vehicles*: Even if highly automated—and autonomous—vehicles could work offline and without being connected to each other and the street's infrastructure, currently connected cars are being developed. Regardless of the advantages, we have to ask what happens to the data, to whom the data are being transmitted and to whom the data can and should be transmitted. 1. Who might be interested in using the data in general and what does this mean for the product in question (law enforcement authorities, insurance companies, criminals, betrayed spouses, car manufacturers, repair shops, …)? 2. Which consequences might (a) the use and (b) the processing of data has for the car passengers?
   2. *Always listening*: In the—narrative—interviews conducted within the project, the interviewees thought of a car that is always listening to ongoing conversations as awkward. Instead of an always-listening language assistant software, the activation through a hardware button is possible. Another possibility to

[7]However, we have to be careful that the discussion about the technical realization of fair algorithms doesn't obscure bias's root causes, lead to people being exposed to additional harms or detract from "bigger" problems Powles (2018).

[8]That people treat machines/robots if not as humans but at least as living beings can be shown by the fact that some people tend to name their cars—as for instance one of the interviewees mentions—and their Roombas, Biever (2014).

[9]https://www.genderlessvoice.com/However, to make its use mandatory, even if it might seem like a good idea from an ethical, emancipatory point of view, might be perceived of as paternalistic.

strengthen data protection would be the local, on-board storage of data of the collected (voice) data, instead of storing it in a cloud.[10]

3. *Privacy self-protection* ("Selbstdatenschutz") in the connected vehicle: Can users individually control the collected and stored data (e.g., with an information tool, or the possibility to opt-in/opt-out?) If yes, does the limited data processing or collection limit the scope of use of the product?
    1. Which possibilities are foreseen to withdraw consent from/object the storing, use and possible sharing of the collected data?
    2. Which data are really necessary for the operation of the vehicle?
    3. Can people who only take the car temporarily decide what happens to their data?
    4. Is the possibility to object the use of the collected data made transparent? Can people with different levels of technical knowledge make use of this objection?
    5. What happens to the data of accompanying children (a) the children of the car owner and (b) friends whom one give's occasionally a lift (e.g., the friends of one's own children)? Can their parents simply object to the data processing? Is this solution practicable?
4. *Profiling* To avoid the possibility to combine data with unique identifiers, is it possible to separate data and information that are not directly related to each other or for which it is not immediately necessary to connect them or to collect them in a combined manner?

3. **Values concerned**
Which values are concerned when we talk about value-sensitive design in highly automated driving? The answer to this question can vary. Within the value-sensitive design literature, it is being discussed whether it does make sense at all to predefine a list of values (Simon 2016; Spiekermann 2019). From a liberal-democratic perspective, the societal preconditions have to exist, to be able to "realize" all ethical values. Furthermore, all humans concerned have to be empowered and enabled to realize and enforce the values that concern them. In (eudaimonist) virtue ethics, the overarching value is "eudaimonia," from ancient Greek translatable as "happiness, well-being or flourishing." In our view, this includes a just social order. The list of values is meant as an universal and global list, even if the values are weighted differently in different societies.[11]

    (a) Privacy
    Privacy is more than data protection. Even though the protection of personal data and the reason why these should be protected are part of human privacy, the phenomenon that has to be protected is more than this. To grasp the

---

[10]This would be one example for the relationship between law and ethics, in which the ethical requirements do exceed the legal ones. The cloud-based storage of data is legal, as long as it is "secure."

[11]Cf. Also the results of Awad et al.: When deciding whom the "moral machine" should kill, clear cultural differences can be seen, however, universally valid values can be seen, too Awad et al. (2018).

broad meaning of "privacy," one common classification in privacy research is three dimensions: local, informational and decisional privacy (Rössler 2001). Questions about data protection are considered as "informational privacy." However, the dimensions are often overlapping. Local privacy, for instance, includes the respect for one's home (Art. 8 ECHR). Many people have a clear idea about the question who is allowed to enter their home at what point. This is relevant in the context of highly automated cars, since the car sometimes is considered as a sort of a "second living-room."[12] The phrase "who is allowed" also includes the decisional dimension of privacy. A person in a liberal-democratic system has to have the right and the possibility to decide about herself, her life, and therefore the access to information about herself at any point.

(b) Autonomy
Closely connected to the protection of privacy is the discussion about a person's autonomy. In a liberal-democratic society, we need citizens who make their life decisions independently and critically and act accordingly. The moral autonomy of a person has to be guaranteed, to be able to judge morally. A person's autonomy should not be influenced by "total surveillance" (German Federal Ministry of Transport and Digital Infrastructure 2017).

(c) Trust (in technology)
According to Niklas Luhman, trust is a mechanism to reduce complexity (Luhmann 2014). The complex phenomenon of trust in technology is a central problem when talking about self-driving cars. Do I trust a machine? Am I trusting its ability to drive me safely to my destination? Central aspects, why humans are trusting a machine, are the transparency of the machine's "decisions" and the assumption that the machine does not make mistakes. Trusting machines and robots and trustworthiness of machines and robots is different from trust relations between human beings (Coeckelbergh 2012).

(d) Transparency
Technology should be transparent and explainable for its users. Programming code should be made public, and how algorithms function should be explainable. Also, in relation to current debates about whistle-blowing, transparency is a principle to make things visible. People concerned can and should ask themselves why a certain action or certain information should not be made public (e.g., the willful deceit of shareholders and clients about emission standard violations).

[12] The narrative interviews, conducted within the project, showed that the interviewees conceive of the(ir) car as a private room in which they can relax and in which they can have private conversations, sometimes there might even be no other (literal and metaphorical) room, time and space to do so. Some of them are, however, aware that, e.g., through navigational systems, data are being or can be recorded and that connected driving will raise data protection issues in the (near) future.

4. **Research methods**
   Applied research methods, e.g., dealing with test persons when using wizard-of-Oz-technology, should be screened ethically, at the beginning of, after the completion of, as well as during the research process.[13]
5. **Use cases**
   Thought experiments such as the so-called trolley-problem, in which direction a driverless car should swerve, i.e., which person should be overrun by an autonomous vehicle, have to be discussed, because they can make visible (potential) discrimination (Liu 2018). The discussion of so-called edge cases that seem absurd or improbable is "best practice" in software engineering (Lin 2017). Considering *ethical* use cases would mean taking extreme moral cases into consideration in order to anticipate and possibly eliminate them. At the same time, it should be factored in that use cases are not covering discrimination, etc. (Ammicht Quinn 2014).
6. **After completing the development process/aftermarket launch**
   To take into consideration ethical concerns after the market launch, it is possible to design an online feedback form. This way, users have the possibility to express concerns, critique, remarks, complaints or suggestions for improvement. The remarks are being reviewed by a person responsible and discussed by the ethics commission. If necessary, an expert might be consulted, or steps of the Ethics by Design process are repeated (e.g., answering the questionnaire below). The person that has submitted the comments gets feedback to show that concerns are treated seriously. If the person likes, they can be given credit.[14]
7. **Ethical questionnaire**
   Below you will find a questionnaire, which is meant as a suggestion for ethical reflection during the development of highly automated vehicles. Following the theoretical reflections of the above-described ethics code, the 44 questions comprise general questions about the project, questions about dealing with test persons during the research process, questions about the product design, questions about programming, about ethics and about possible consequences of the technology.

---

[13]Riek and Howard (2014) When dealing with test persons, researchers should be aware of the so-called turing-deception. To avoid shame, deception and false expectations, it is being discussed, whether and to what extent test persons should be informed that it is not an autonomous car (Riek and Howard 2014, 3). The company Daimler, for instance, simulated a situation in which the driver was wearing a suit looking like the car seat and in which the car had a makeshift sensor on top of the roof, to create the illusion that the car was driving autonomously https://blog.daimler.com/2018/11/26/fussgaenger-autonom-fahrzeug-immendingen-sicherheit-zukunft-test/2019-05-08. In this case the test persons were not operating the vehicle but interacting with the car as pedestrians.

[14]Cf. The "Bosch Product Security Incident Response Team," which the company Bosch uses in product security. "Security researchers, partner or clients/customers" can communicate via an encrypted connection identified potential security flaws or incidents in Bosch products or on the Bosch website or can ask "data protection requests" https://psirt.bosch.com/de/ "Data Protection Requests" can be asked as well this way. A similar system could be set up for ethical questions and concerns.

4.1.7 Questionnaire for Ethical Self-assessment

Every new technology raises new ethical, moral and, not least, legal questions. Technology is never (value-)neutral, because world views and facts are always being inscribed into it. Therefore, those, who are developing it, are assuming moral responsibility. Ethical research and development are deeply rooted in the principle of human dignity and its protection (Art. 1, para. 1) German Constitution "Grundgesetz" (GG), Art.1 Charter of Fundamental Rights of the European Union (CFR)). In terms of highly automated driving other values are concerned, too. Due to "connected driving," especially questions about privacy and data protection are relevant (Art. 2, para. 1 GG, Art 7, 11 para 1 CFR in conjunction with GDPR). In terms of "ethics by design" the ethical reflection process monitors the development process of new technologies from the beginning and after completion of the product or the process. It is very important to communicate the guidelines and to find the right moment to do so. Within the scope of the ongoing digitalization and especially the—supposedly—ongoing "AI-race," taking into account ethical questions and viewing people as human beings—and not only as (potential) users or clients—and reflecting one's own actions in the social context may offer competitive advantages.

The questionnaire at hand aims at helping enterprises, companies, programmers, developers, data scientists, research teams and other parties and persons involved in the research and development of highly automated driving, to identify ethical questions during the development process. By addressing potential ethical conflicts early, the implementation of Ethics by Design and Privacy by Design (Art. 25 GDPR) can be facilitated. The questionnaire might be used for instance during foreseen processes of quality management or can be instated as new part of the quality insurance process. The process should be documented in order to be able to discuss emerging questions with experts and to evaluate whether ethical questions have been continuously considered. Furthermore, this way questions that arise during the product development can be addressed. Not least the questionnaire can serve as an inspiration to reflect one's own approach to product development.

The 44 questions cover the following topics:

1. About the project
2. Dealing with involved persons
3. Design
4. Programming
5. Ethical evaluation
6. Possible future consequences of the invention.

1. About the project

1.1 Name of the project:

1.2 Project partners:

1.3 Please describe the goal of your project in a few words:

1.4 Which stakeholders are involved?

1.5 Are the needs of potential users taken into account?

☐ yes ☐ no

1.6 Who could be affected by possible design choices?

1.7 Whom does the technology serve? Who benefits from it?

1.7.1 What is the purpose of the technology?

1.7.2 Does the technology serve members of „vulnerable groups“ (children, elderly people, people with disabilities,….)?

☐ yes ☐ no

1.7.3 Does the technology improve people's quality of life?

☐ yes ☐ no

2. **Dealing with involved persons**

2.1 Are test persons involved into the development of the product?

☐ yes ☐ no

2.2 Are the needs of different user groups assessed?

☐ yes ☐ no

2.3 Do the test persons have the possibility to consent to the collection and processing of personal data (GDPR Art. 4 (11), art. 6 (1) a, art. 7, recitals 42/43)?

☐ yes ☐ no

2.4 Is it possible to collect the data anonymously?

☐ yes ☐ no

2.4.1 If no, is it possible to render anonymous the collected data (i.e. no possibility to (subsequently) attribute them to a specific data subject) or at least pseudonymous (GDPR Art. 4 (5)) ?

☐ yes ☐ no

2.5 Is a Wizard-of-Oz vehicle being used?

☐ yes ☐ no

2.5.1 If yes, do the test persons know that they are not driving an autonomous vehicle?

☐ yes ☐ no

2.5.2 When are the test persons told that they are not driving fully automated?

☐before the ride ☐ after the ride ☐during the ride

2.5.3 If after the ride, are precautions being taken to avoid so-called Turing-Deceptions (feeling deceived because of not having been informed beforehand)?

☐ yes ☐ no

2.6 Are „Use Cases“ being discussed during the development?

☐ yes ☐ no

2.6.1 If yes, are the different (moral) perspectives of the various stakeholder represented?

☐ yes ☐ no

2.6.2 Are also unlikely edge cases being discussed?

☐ yes ☐ no

3. **Design**

3.1 Are there Design-Patterns or features that try to nudge users to do something they otherwise would not do?

☐ yes ☐ no

3.2 Is there a possibility to opt-in or opt-out of driving functions that are not required for driving?

☐ yes ☐ no

3.3 Is there a manual override button that can be activated by the human, for instance in case of emergency (so-called “overruling” by humans)?

☐ yes ☐ no

3.4 Is there a so-called “guardian angel” function (i.e. the possibility that the system overrides the user’s actions, e.g. an emergency brake assistant?)

☐ yes ☐ no

3.5 Does the vehicle give feedback after the human has taken control again?

☐ yes ☐ no

3.6 Does the vehicle give a reason for taking over?

☐ yes ☐ no

3.7 Are the vehicle's "decisions" transparent to the user? Are their reasons, for instance, being displayed?

☐ yes ☐ no

3.8 Are you aware of the fact the way how to communicate in handover situations can influence a person's trust in the system?

☐ yes ☐ no

4. Programming

4.1 Is Open Source- or Free Software being use?

☐ yes ☐ no

4.2 If no, would it be possible to make the source code publicly available while protecting (legitimate) commercial interests?

☐ yes ☐ no

4.3 Are algorithms being deployed?

☐ yes ☐ no

4.3.1 If yes, can you guarantee / ensure that the algorithms are "faire" and that they are not reproducing existing bias?

☐ yes ☐ no

4.3.2 Are the algorithms transparent, accountable, controllable (internally or externally)?

☐ yes ☐ no

4.4 Is Artificial Intelligence / Machine Learning / Deep Learning being deployed?

☐ yes ☐ no

5. **Ethical (accompanying) research**

5.1 Are professional ethics being incorporated (e.g. IEEE's "Ethically Aligned Design" principles or the United Nation's Global Impact Principles)?

☐ yes ☐ no

5.2 Is the research project or the development being monitored ethically?

☐ yes ☐ no

5.3 Is there an organizational unit or body responsible for ethical questions / issues?

☐ yes ☐ no

6. **Possible future consequences of the product**

6.1 Are there aspects or results of the research that could be used for "unethical" purposes (so-called dual use, e.g. military combat drones)?

☐ yes ☐ no

6.1.1 If yes, is there a possibility to contain unwanted consequences?

☐ yes ☐ no

6.1.2 Could undesirable consequences arise, when the development continues (e.g. level 5 fully autonomous driving?)?

☐ yes ☐ no

Translated from German by author with the aid of www.DeepL.com/Translator.

References

Ammicht Quinn R (ed) (2014) Sicherheitsethik. Springer, Berlin

Awad E, Dsouza S, Kim R, Schulz J, Henrich J, Shariff A, Bonnefon J-F, Rahwan I (2018) The moral machine experiment. Nature 563:59–64

Biever C (2014) Roomba creator: robot doubles need more charisma. New Sci (2961). https://www.newscientist.com/article/dn25253-roomba-creator-robot-doubles-need-more-charisma/. Accessed 10 Sept 2019

Coeckelbergh M (2012) Can we trust robots? Ethics Inf Technol 14:53–60. https://doi.org/10.1007/s10676-011-9279-1

Directorate-General for Research and Innovation (European Commission), European Group on Ethics in Science and New Technologies (2018) Statement on artificial intelligence, robotics and 'autonomous' systems. https://publications.europa.eu/en/publication-detail/-/publication/dfebe62e-4ce9-11e8-be1d-01aa75ed71a1/language-en/format-PDF/source-84689254. Accessed 23 July 2019

Grimm P, Keber T, Zöllner O (eds) (2019) Digitale Ethik Reclam Stuttgart

Grimm P (2013) Werte- und Normenaspekte der Online- Medien – Positionsbeschreibung einer digitalen Ethik. In: Karmasin M, Rath M, Thomaß B (eds) Normativität in der Kommunikationswissenschaft, vol 08. Springer Fachmedien Wiesbaden, Wiesbaden, s.l., pp 371–395

Hagendorff T (2019) The ethics of AI ethics: an evaluation of guidelines. https://arxiv.org/abs/1903.03425. Accessed 3 May 2019

Lin P (2017) Robot cars and fake ethical dilemmas. https://www.forbes.com/sites/patricklin/2017/04/03/robot-cars-and-fake-ethical-dilemmas/#64c250013a26. Accessed 12 Sept 2019

Liu H-Y (2018) Three types of structural discrimination introduced by autonomous vehicles. UC Davis Law Rev Online 51:149–180

Luhmann N (2014) Vertrauen: Ein Mechanismus der Reduktion sozialer Komplexität, 5th edn. UVK Verlagsgesellschaft mbH

Maxmen A (2018) Self-driving car dilemmas reveal that moral choices are not universal. https://www.nature.com/articles/d41586-018-07135-0. Accessed 9 May 2019

Metzinger T (2019) Ethik-Waschmaschinen made in Europe. https://background.tagesspiegel.de/ethik-waschmaschinen-made-in-europe. Accessed 26 April 2019

Powles J (2018) The seductive diversion of 'solving' bias in artificial intelligence. Coauthored with Helen Nissenbaum. https://medium.com/s/story/the-seductivediversion-of-solving-bias-in-artificial-intelligence-890df5e5ef53. Accessed 1 May 2019

Riek LD, Howard D (2014) A code of ethics for the human-robot interaction profession. robots.law.miami.edu/2014/wp-content/uploads/2014/03/a-code-of-ethics-for-the-human-robot-interaction-profession-riek-howard.pdf. Accessed 16 Aug 2016

Rössler B (2001) Der Wert des Privaten. Suhrkamp, Frankfurt am Main

Simon J (2016) Values in design. In: Heesen J (ed) Handbuch Medien- und Informationsethik. J. B. Metzler

Spiekermann S (2019) Digitale Ethik: Ein Wertesystem für das 21. Jahrhundert

Wagner B (2018) Ethics as an escape from regulation: from "Ethics-Washing" to Ethics-Shopping? In: Baraliuc I, Bayamlıoğlu İE, Hildebrandt M, Janssens L (eds) Being profiled: Cogitas ergo sum: 10 years of 'profiling the European citizen'. Amsterdam University Press, Amsterdam

Winfield A (2019) Ethical standards in robotics and AI. Nat Electron 2:46–48

Chapter 5
Vorreiter: Manoeuvre-Based Steering Gestures for Partially and Highly Automated Driving

Frank Flemisch, Frederik Diederichs, Ronald Meyer, Nicolas Herzberger, Ralph Baier, Eugen Altendorf, Julia Spies, Marcel Usai, Vera Kaim, Bernhard Doentgen, Anja Valeria Bopp-Bertenbreiter, Harald Widlroither, Simone Ruth-Schumacher, Clemens Arzt, Evin Bozbayir, Sven Bischoff, Daniel Diers, Reto Wechner, Anna Sommer, Emre Aydin, Verena Kaschub, Tobias Kiefer, Katharina Hottelart, Patrice Reilhac, Gina Weßel, and Frank Kaiser

Abstract Automated driving is dramatically changing the way we perceive and control cars and lorries. In partially automated driving, the driver is still responsible for the vehicle control, but is strongly supported. In highly automated driving, the driver can give control and responsibility to the automation for a certain time and can get control back, e.g. when the automation encounters limits. However, an intuitive way to interact with these automated modes is largely missing. Vorreiter is addressing this by using the inspiration of a rider and a horse to provide intuitive steering gestures on the steering wheel or an alternative device, which initiate manoeuvres to be executed by the automation and to be supervised, influenced or interrupted by the driver. The gestures are built up in a design-for-all which helps all drivers, including beginners and drivers with disabilities. A consortium of RWTH Aachen University, Fraunhofer IAO, University of Stuttgart, Valeo and Hochschule für Wirtschaft und

---

F. Flemisch · R. Meyer · N. Herzberger (✉) · R. Baier · E. Altendorf · J. Spies · M. Usai · V. Kaim · G. Weßel
Institute of Industrial Engineering and Ergonomics IAW, RWTH Aachen University, Aachen, Germany
e-mail: n.herzberger@iaw.rwth-aachen.de

F. Diederichs · H. Widlroither · V. Kaschub
Fraunhofer Institute for Industrial Engineering IAO, Stuttgart, Germany

B. Doentgen · E. Bozbayir · T. Kiefer · K. Hottelart · P. Reilhac · F. Kaiser
Valeo Schalter und Sensoren GmbH, Bietigheim-Bissingen, Germany

A. V. Bopp-Bertenbreiter · S. Bischoff · D. Diers · R. Wechner · A. Sommer · E. Aydin
Institute of Human Factors and Technology Management IAT, Stuttgart, Germany

S. Ruth-Schumacher · C. Arzt
Berlin School of Economics and Law (HWR Berlin), Berlin Institute for Safety and Security Research (FÖPS Berlin), Berlin, Germany

G. Meixner (ed.), *Smart Automotive Mobility*,
Human–Computer Interaction Series,
https://doi.org/10.1007/978-3-030-45131-8_5

Recht Berlin defined a concept, built up prototypes and investigated its impact in workshops and simulator experiments.

5.1 Biologically Inspired Manoeuvre Gestures for Partially and Highly Automated Driving: Introduction and Overview

5.1.1 Introduction: Humans, Automation and the Quest for Intuitive Interaction and Cooperation

Einstein is quoted having said, that 'There are only two ways to live our lives. One is as though nothing is a miracle, the other is as though everything is a miracle.'[1] Before we focus on technical and ergonomic aspects of steering gestures again in the rest of the chapter – assuming they do not arise by miracles alone, but in proper combination with scientific and engineering insights, art, knowledge and know-how – let us allow ourselves to stretch our minds a little and first take a look through the more miraculous of Einstein's perspectives on life.

What makes us Homo sapiens unique compared to other species in the biosphere, and what is our "miracle"? Besides many communalities with most other species like the ability to eat, drink, digest and reproduce—often in competition with other individuals and species—and communalities with many other species like the ability to communicate, cooperate, play and enjoy life, some features differ significantly. One of them, shared with only a minority of species, is the use of artefacts, which extend our own limits. Early examples for that are the use of stones as tools as weapons, ca. 1.2 million years ago (e.g. Harcourt and de Waal 1992), or the use of wooden spears, ca. 270.000 years ago shown by the artefacts found in Schoeningen (Thieme 1997). Tomasello (2014) describes how the use of these tools also shaped our way of thinking and minds. Marean (2015) describes that it was not only new tools but a unique combination of the new tools and the ability to interact and to cooperate that allowed Homo sapiens to leave Africa, to spread out over almost the entire planet and have a severe impact on the biosphere (Fig. 5.1).

Today, several hundred thousand years after the first tools extended our physical limits, we use physical extensions like bicycles, cars, trains, etc., and we increasingly invent tools like books or computers to also extend our cognitive limits. In the form of automation, computers can increasingly take over cognitive processes and steer physical processes previously assigned to humans. Examples are autopilot and flight management systems in aircraft, process automations in manufacturing, or increasing assistance and automation in cars and lorries. Is it again only the new cognitive tools, or is it once more the combination of the tools with the ability to interact and

[1] https://www.goodreads.com/author/quotes/9810.Albert_Einstein.

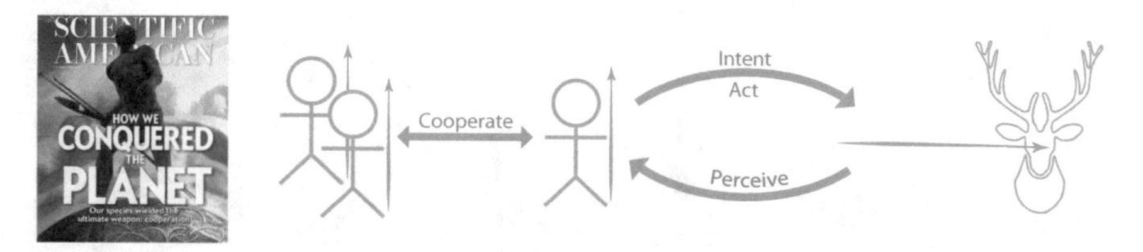

Fig. 5.1 How we conquered the planet (Marean/Scientific American 2015) (left). Tools and cooperation as key enablers for Homo sapiens (right)

cooperate between humans and humans, and between humans and computers, which might again change our lives and with that influence the whole biosphere?

5.1.2 Vehicle Automation and Autonomous Functions

Since the early days of vehicle assistance and automation, two main motives and streams were prevailing, sometimes seen as completely different efforts, but always connected in the network of science and engineering: one stream focused on increasing the autonomy of machines, i.e. the ability to act in the environment within a certain operational domain without a human. An example for that is the Sperry autopilot, which automated the stabilization of a course of an aircraft and which later led to flight management systems able to fly an aircraft fully automated from take-off to landing. Another example is fully automated vehicles starting, e.g. with Tsugawa et al. (2000), Dickmanns et al. (1987), handling both longitudinal and lateral controls, which caused a strong movement towards so-called autonomous vehicles in the first two decades of the 3rd millennium (e.g. Thrun et al. 2006). The second stream focused on the assistance of the human to perform his or her task. Examples are radio beacon navigation in aviation, assistance systems for planes (e.g. Onken and Flemisch 1998; Onken and Schulte 2010) or assistance systems in cars and lorries like lane-keeping assistance system (LKAS).

In 1997ff, these two streams were willingly combined to describe how autonomous functions can be formed into a powerful co-system, which could also act autonomously if necessary, but usually cooperates with the human in a way that the human has a choice of a few, intuitively understandable modes of assistance and automation. The scientific background was formulated as human–machine cooperation (e.g. Hoc and Lemoine 1998; Flemisch and Onken 1998; Flemisch et al. 2003; Flemisch et al. 2015; Pacaux-Lemoine and Flemisch 2019) and shared/cooperative control (e.g. Griffiths and Gillespie 2004; Abbink 2006; Flemisch et al. 2010; Flemisch et al. 2019; Schneemann and Diederichs, 2019). To make a complex issue easier to understand, a design metaphor was crafted (see Fig. 5.2) which describes that human and machine work together like a rider and a horse (or a driver and a horse cart) where both cooperate to different extents on the control of the vehicle, and interact with control devices comparable to a tight rein, loose rein or secured

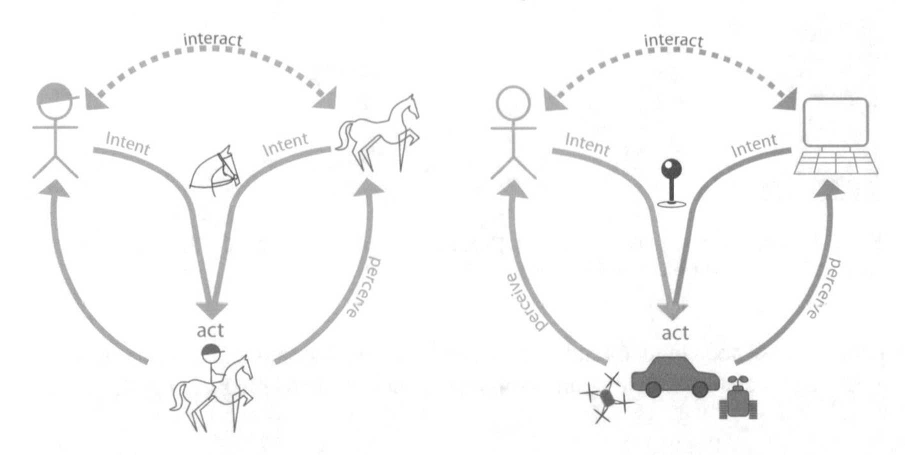

Fig. 5.2 Rider and horse as a cooperative system with shared and cooperative control (left). Human and machine as a cooperative system with shared and cooperative control (right) (extended from Flemisch et al. 2003ff.; see also Baltzer 2020)

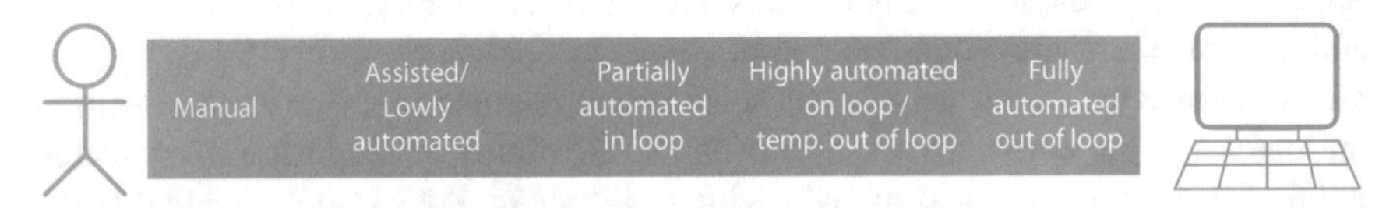

Fig. 5.3 Control distribution between human and machine with levels of assistance and automation (based on Flemisch et al. 2003; Flemisch et al 2008; 2013ff)

rein (Flemisch 1997, 2000; Flemisch et al. 2003). Based on this h(orse)-metaphor, a haptic-multimodal way of interaction with a highly automated vehicle was developed and tested (e.g. Flemisch et al. 2005; Altendorf et al. 2015a,b). This interaction is realized through a human–machine mediator, which separates the machine into an automation and a mediator part.

After the formulation as a metaphor, this spectrum or scale of assistance and automation (Flemisch et al. 2008) was translated into levels of vehicle automation by BASt (Gasser et al. 2012) and SAE (2016), and into prototypes and serial products (e.g. highly automated driving). Examples are assisted, partially automated, highly automated and fully automated driving (Fig. 5.3).

5.1.3 *Manoeuvre-Based Driving and Gestures*

Besides the fluid transitions between different levels of assistance and automation, another essential feature of the H-mode, already indicated by the H-metaphor in Flemisch et al. (2003) was the easy initiation of manoeuvres by steering gestures (e.g. Loeper et al. 2008). In a historic coincidence, Winner et al. (2006) and Flemisch et al. (2006) were presenting on manoeuvre-based driving in the same track at the same conference (FAS, Löwenstein) and discovered many connecting aspects and increasingly cooperated on human–vehicle cooperation (e.g. Flemisch et al. 2014), including manoeuvre-based driving. An example for this is the work of Benjamin Franz et al. on Pie-Drive, a manoeuvre-based touch interface (Franz et al. 2012).

5.1.4 Gestures for Manoeuvre-Based Driving in the BMBF Project "Vorreiter"

While the projects described above were mainly basic research projects, the goal of the Vorreiter project (Logo; see Fig. 5.4.) was the applied, product-oriented research and development of a core set of manoeuvre gestures for partially and highly automated driving. These gestures were developed in a "design-for-all" approach to make this complex technology accessible to all drivers—i.e. the mass market as well as drivers currently limited by disabilities or higher age. The inspiration is again the H-metaphor, a biological metaphor of rider and horse (Flemisch et al. 2003).

Over three years, the partners Institute of Industrial Engineering and Ergonomics of the RWTH Aachen University (IAW), Institute of Human Factors and Technology Management of the University of Stuttgart (IAT), Fraunhofer Institute for Industrial Engineering (IAO), Berlin School of Economics and Law (HWR), Valeo and (in the first two years) Paravan—a medium size company specializing on vehicles for people with handicap and on drive-by-wire systems—were researching the use, design and evaluation space of manoeuvre gestures in an iterative, explorative approach (see Sect. 5.2). The consortium was designing alternative gestures, developing a new steering wheel, testing them in driving simulators in Stuttgart and Aachen, always checking for legal plausibility (see Sect. 5.7), performance, usability and safety. The basic principle is shown in Fig. 5.5: based on manoeuvres proposed by the automation, the driver initiates one manoeuvre, which is then executed by the automation and can still be influenced or interrupted by the driver (see Sect. 5.8).

The starting point of the consortium was pushing and twisting gestures on the steering wheel. During the explorations of the first years with potential users, a new kind of "swipe" gesture was discovered at two locations in parallel: swipe gestures

Fig. 5.4 Logo of the Vorreiter project

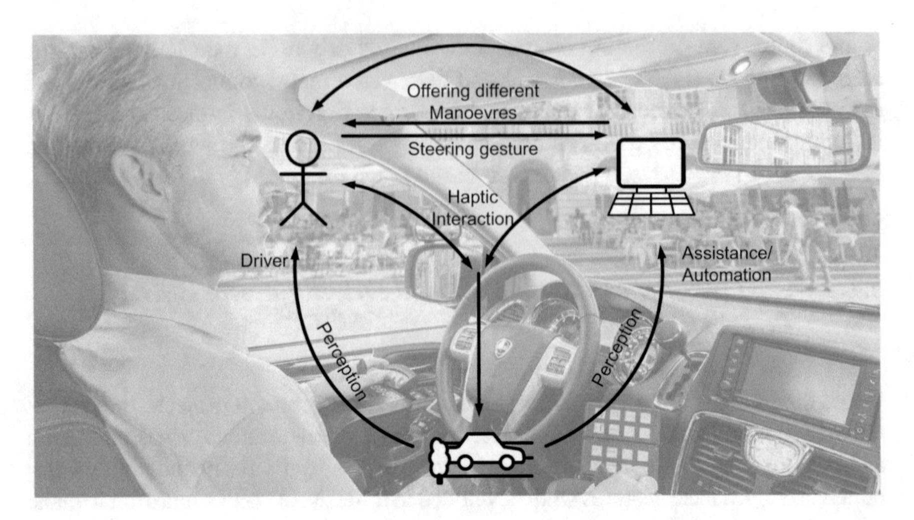

Fig. 5.5 Gestures for partially and highly automated manoeuvre-based driving

with a hand at a steering wheel at IAO, described by Sommer et al. (2019), and with a thumb in an exploration workshop of Gina Wessel at Paravan, described by Spies et al., in this issue. The swipe gestures were developed together and tested successfully. Whether pushing, twisting or swiping, the gestures are based on the same principles described in a gesture catalogue (see Sect. 5.8), and the haptic interaction is complemented with a visual interaction and, where necessary, an auditive interaction. The implementation of the gestures into a Wizard-of-Oz (Green and Wei-Haas 1985) vehicle was prepared, but due to a dropout of Paravan—originally responsible for setting up the vehicle—the tests were moved to the simulators (see Sects. 5.10 and 5.11). In addition to the steering wheel and based on the fruitful experience with sidesticks and highly automated vehicles, first implementations of alternative steering devices were explored, including a drivestick for disabled drivers provided by Paravan (Sommer et al. 2019), an alternative sidestick from Valeo and a virtually represented Holowheel from Fraunhofer IAO (see Sect. 5.10). The successful application of various input devices shows the great potential of generalizing the concept of cooperative control and steering gestures for manoeuvre-based driving in automated vehicles.

5.2 Design and Use Space of Manoeuvre Gestures

5.2.1 Introduction

The participatory design approach of the project Vorreiter is based on a dynamically balanced development process, which combines user-centred design phases

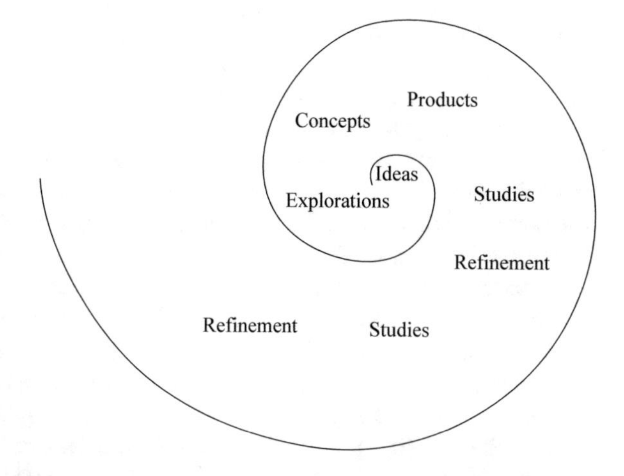

Fig. 5.6 Iterative design approach utilized in the project Vorreiter and originally used in EU project HAVEit (Flemisch and Schieben 2010)

with technology and organization-centred perspectives. This approach is utilized as a systematic investigative method for the development of human–machine systems and implies the integration of user groups in the design process (Altendorf et al. 2015a,b; Flemisch et al. 2017). The design process runs iteratively through the corresponding stages which are illustrated in Fig. 5.6. The process begins with the generation of ideas for the system design. Already in this phase, these ideas are structured by applying a system theoretical methodology by defining the known final goal of Vorreiter as a work system with certain objectives and user groups. This allows a structured definition of the system itself and its placement and causes and effects in its metasystem; i.e. the surrounding and affecting systems are considered as influencing entities (Haberfellner et al. 2015). The decisive definition of the work system and the metasystem directly influenced the formation of the consortium for the Vorreiter project, where the development of the theoretical idea of manoeuvre gestures and their conversion into concrete concepts and finally into working prototypes was the main goal.

An essential element of the concept and prototyping development method which has been used in the Vorreiter project was the realization of explorative design workshops. Exploration is a process paradigm on a similar level like "experiment", which is—in contrast to experiment—not focused on critical testing of hypothesis, but on getting an understanding of the design space and constructive generating design solutions, which can then serve again as hypothesis for critical experiments (Flemisch et al. 2019a, b).

Such exploration workshops are conducted in a simulated, controlled but yet open environment which is related to the target domain of the design goal and can be engaged with common, i.e. naïve, users or with experts from the domain. Over the years, we developed a concept of exploroscopes, specially prepared laboratories with interactive entities from the design space, and users and designers can explore new methods and varieties in defined use spaces—and thus practice exploratory design sessions.

Explorative design workshops can basically be conducted at every stage of the design process, even if most of the final design does not yet exist. Thus, design elements from the target design or domain can be used and be interconnected with rapid prototyping elements such as paper-based prototypes and the application of the theatre method but may also reach to the application of complex technical implementations to fine-tune visual, auditory or haptic feedback with explorative tools. The combination with the theatre method (Flemisch et al. 2005, 2017) is an extension of the "Wizard-of-Oz" method, where designers, developers, stakeholders and future users can play through use cases of new systems and develop new interaction methods in an agile environment. By applying the theatre method, new technology can be emulated by a human (Flemisch et al. 2012). In this setting, mental models of the design team can be synchronized in an early stage of the development process (ibid.), which is especially valuable among an interdisciplinary consortium as congregated in the Vorreiter project. Furthermore, the designers may use the setting as a participatory design tool (ibid.) within the user-oriented balanced design process, which has been applied in the Vorreiter project. In order to specify the requirement space preceding the exploration phases, the Vorreiter consortium conducted workshops to create a model for the evaluation of human–machine systems. The result was a procedure to specify concrete system qualities, which were allocated to specific requirements. The approach is briefly depicted in the following section.

5.2.2 Requirement Analysis and System Qualities

Sneed's (1987) Devil's Square and further relevant literature form the theoretical basis of the model. The result of this workshop is the Expanded Devil's Square (Flemisch et al. 2019a, b), which is illustrated in Fig. 5.7.

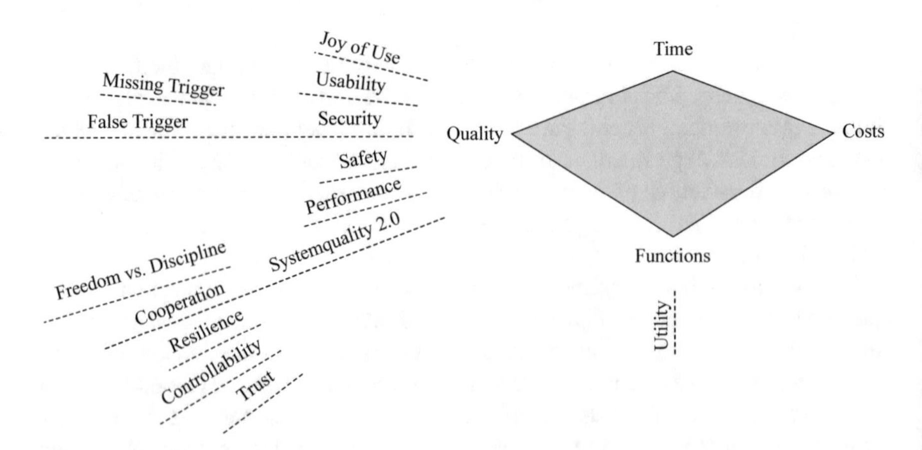

Fig. 5.7 Extended Devil's Square (Flemisch et al. 2019a, b, based on Sneed 1987)

Sneed's Devil's Square (1987) has its origin in the project management theory. It describes the antagonistic interaction of the project resources time, quality, cost and scope; e.g. if a project is to be completed in less time and at lower cost, the scope of services and quality is also reduced.

The Devil's Square can also be used to evaluate a human–machine system. Thus, when developing a human–machine system, it is necessary to take into account that increasing functionalities (scope) are associated with development costs that are passed on to the customer. In addition, the time for the development of a system is limited. According to Flemisch et al. (2019a, b), the system requirements have to be considered at every step of system development and evaluated iteratively to guarantee that development is moving towards the target concept. To ensure that the costs are spent on the target, the quality of the system has to be assessed from various aspects (see Fig. 5.6). To achieve this, several tests and processes are required to define standardized metrics and methods for evaluation and comparison (Flemisch et al. 2019a, b).

In the case of highly automated and partially automated driving systems, it has to be considered that with increasing level of automation the driver has fewer tasks to perform than in manual driving. Consequently, this development is changing the role of drivers from active to supervisory (Flemisch et al. 2017; Pöhler et al. 2016). For this reason, users should be integrated in the early stages of evaluating automated driving systems in order to test and evaluate the system; i.e. the application of a participatory design approach is recommended.

Thus, the evaluation of the quality of the system does extend not only to the quantitative measurement of the system performance, the missing triggers or the false triggers during different driving manoeuvres, but also to the qualitative dimensions. These dimensions are joy of use, which describes the enjoyment of using the system (Burmester et al. 2002), usability, which can be defined as the ease of use and acceptability of a system (Holzinger 2004), and the evaluation of the security against a threat of intentional attacks or safety of data stored by the machine (Makrushin et al. 2008). These dimensions are added to the extended Devil's Square.

In addition, the second-order system qualities, which describe complex qualities of the system itself, are also qualitative dimensions of the extended Devil's Square. With regard to human–machine systems, the dimension freedom vs. discipline, cooperation and controllability aim on the way on how the human shall interact with a machine (Flemisch et al. 2012). For example, in the case of automated driving systems, driving tasks can be distributed between driver and automation (Altendorf et al. 2015a, b; Flemisch et al. 2015). This has an impact on the way a driver and machine cooperate and how authority (freedom vs. discipline) and control are distributed. Furthermore, trust is another dimension of the system qualities 2.0. Trust in the context of human–machine interaction is described by Lee and See (2004) as "the belief that a system helps the individual to achieve certain goals in a situation of uncertainty and vulnerability". This uncertainty is particularly evident in automated driving systems due to the changing role of humans. Resilience is also part of the system qualities 2.0, describing the robustness and agility of the system against disturbances (Wieland and Wallenburg 2012).

5.2.3 *Structures and Dimensions of the System: Design and Use Space*

To cover every meaningful interaction between humans and machine system, Flemisch et al. recommend getting an overview about the interaction possibilities. This can be achieved by forming the design space and the use space to a reusable structure. The main idea of structuring the use and design space into dimensions originates to Rasmussen's "space of possibilities", where Flemisch et al. (2013) indicate that it is crucial to give the many subjective decisions in the exploration process an objective structure and to document the whole exploration process. In the Vorreiter project, so-called exploration maps were used, which reflect the design and use space. Thus, the designer group and/or experts prepare the use space and the design space in an abstract form, which allows to rearrange and further document the explored interaction during the exploration process. Flemisch et al. (ibid.) state that every result of usability assessment can be documented in an exploration map, though a beneficial exploratory design session postulates a well-structured design space and use space. The structure of both is described in the next section and used to illustrate the procedure as it was conducted in the project Vorreiter.

5.2.4 *Iterative Design: Structuring of Design Space and Use Space*

The final design of an interaction and cooperation concept in the Vorreiter project emerges from the selection of options based on specifications and criteria, which were explored in the design explorations and in the requirement analysis. The design space itself is enclosed by various boundaries, which are, for example, from the areas of work that are not economically viable or too heavy delimited. It is therefore itself a selection of suitable interaction possibilities.

In the Vorreiter project, expert milestone workshops were conducted on a regular basis over the whole project phase of three years, where all experts of the consortium from Vorreiter's interdisciplinary domains participated and combined their knowledge and experience to jointly organize and design the planned work packages and supported the iterative and participatory process. The results of the iteratively conducted expert milestone workshops are briefly described in the following paragraph.

5.2.4.1 Use Space: Use Cases and Scenarios

To describe the use cases and their specific scenarios derived from the manoeuvres, an ontology proposed by Geyer et al. (2014) was used. Table 5.1 illustrates the identified use cases, which are objectives of investigation in the Vorreiter project and

Table 5.1 Relevant use cases and common basis for exploration workshops

Start moving	Move faster
Stop moving	Move slower
Change lane right	Change lane left
Turn left	Turn right
Overtake	U-turn
Emergency brake	

are implemented during the iterative design process. The table is a brief summary of a use case catalogue, which was formed into a gesture design catalogue over multiple iterative steps. The result of this iterative process is a catalogue of gestures for specific manoeuvres, further described in Sect. 5.8.

5.2.4.2 Structure and Exploration of Design Space

The structure of the design space was conducted for the interactive elements that are part of the Vorreiter design.

An important part of the design space is the interplay of interaction resources of the human. Wickens et al. (2008) describe how visual and auditory resources can be used simultaneously, if they do not use the same verbal or spatial encoding. Flemisch et al. (2014) (Lecture series bHSI) extend this model with the haptic channel, which can be used to communicate more directly. In Altendorf et al. (2015a,b), they describe how the haptic channel shows what should be done in the presence, while the visual channel communicates how the situation develops into the future. Flemisch et al. (2019a, b) fuse this model with a model of layers of cooperation and describe how the stages of interaction described by Wickens can be used on a tactical, operational, strategic and co-operational layer (Fig. 5.8).

Another part of the design space is the coupling between human and machine with different coupling schemes and patterns. An example is the direct coupling of human and machine on the steering wheel, which could also be decoupled, e.g. with a drive-by-wire steering or a virtual steering wheel described in Sect. 5.10. Figure 5.9 shows a sketch of the design space, where different parts of the coupling can be haptic and/or complemented with other multimodal interaction.

All necessary resources are documented in an exploration map—where all design dimensions are congregated and pre-structured. An example of an exploration map used in Vorreiter is illustrated in Fig. 5.8: it shows an advanced design space that was used to explore the steering gestures. It covers possible design elements, which are the classification of the system qualities, the drivers or vehicles' intention, the level of automation and a possible transition, different interaction modalities such as haptic, visual or acoustic feedback and/or input, all available components in the design space, the drivers' attention state and extended charts and diagrams that may be related to the drivers' state. These pre-structured elements were investigated during the explorations in meaningful combinations—i.e. for every manoeuvre and use case. Explorations were conducted iteratively at different development stages of the project, and results were carried over into the next development phase.

Figure 5.9 shows a running exploration with two experts documenting the whole exploration process and one expert user interacting with a confederate. An explorative

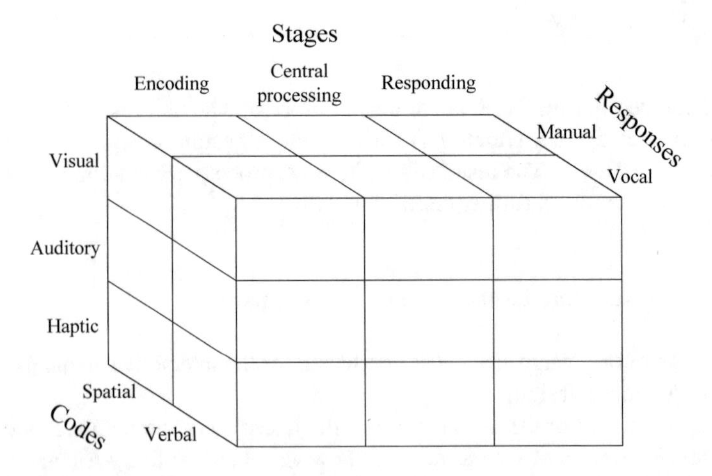

Fig. 5.8 Extended multiple resource model (Flemisch et al. 2014, based on Wickens 2008)

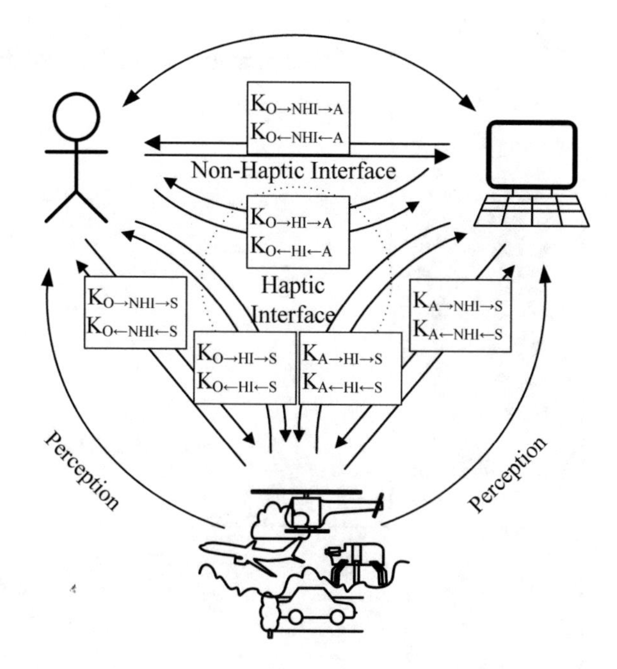

Fig. 5.9 Haptic-multimodal coupling between humans and automation as part of the design space of human–machine cooperation (Flemisch et al. 2010)

pilot study for a haptic interface design using force feedback sidesticks was also conducted by Meyer et al. (2018). Figure 5.10 shows an exemplary exploration map, which has been used to explore possible steering gestures for the shown use cases "lane change left" and "lane change right", and the ongoing exploration process with two design experts, an expert user and a confederate, is shown in Fig. 5.11. The empty space of the exploration map is used for documentation while an ongoing exploration with possible ideas and solutions that are elaborated among the exploration session.

5.2.5 *Further Development of Manoeuvres in Their Use Space and User Groups*

The derived manoeuvres were further analysed in the context of their specific user groups by applying the dynamically balanced design process. Elaborated design spaces were discussed with experts and users, e.g. with engineers of the Paravan GmbH (see Sect. 5.4) and pilots and experts from the Vorreiter consortium. Details about the typical Vorreiter user group are given in the next section.

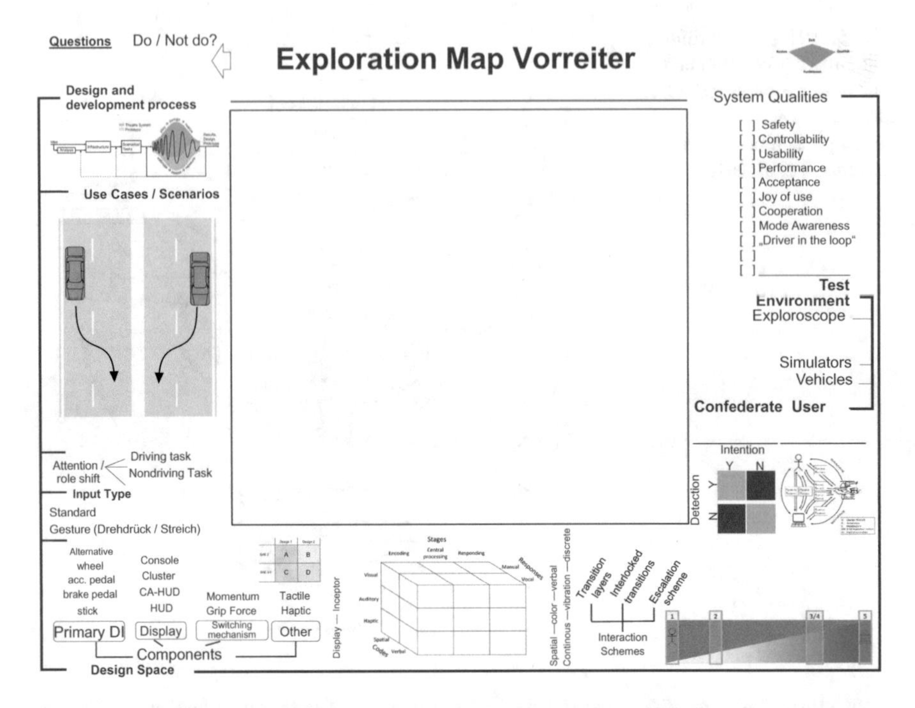

Fig. 5.10 Exemplary exploration map with design and use space for designing the manoeuvre gestures used in project Vorreiter

Fig. 5.11 Snapshot of an exploration with two design experts, expert user and confederate (e.g. RWTH Aachen University, IAW, Exploroscope)

5.3 Vorreiter User Groups and Their Requirements Based on Empirical User Research

5.3.1 Introduction

Based on the theoretical foundations described in Sect. 5.2, empirical user research was conducted to assess the characteristics and requirements of the user groups as well as situations where users need support by a driving automation. This included the explorative process that will be described in Sect. 5.5 as well as focus groups with geriatric nurses and persons using electric wheelchairs daily. Besides the mass market, three other user groups are likely to benefit from the cooperation concept: novice drivers and people currently limited by age or physical disabilities. For the USA alone, it is estimated that the annual mileage might increase by 14% due to people using automated vehicles which are currently restricted by medical conditions or age (Harper et al. 2016). Future cooperation and HMI concepts should include their needs, as access to transportation is an important factor to increase inclusion (Maunder et al. 2004).

5.3.2 Mass Market

The term mass market is "used to describe a product that is intended to be sold to as many people as possible […]" (Cambridge University Press 2020). "Mass-market" users were defined as members of the general population driving automobiles of the SAE levels 0–2. Mass-market users stated the following requirements towards automated vehicles during the explorative design and development described in Sect. 5.5: the car needs to collect real-time data through Car2x and is able to recognize/interpret the current driving context. A driver (state) recognition to prevent drunk drivers and children from driving is implemented, and the driving automation can be switched on and off at the user's discretion. Regarding the HMI, the users expect hints on whether or not the driving automation is able/going to intervene. They also want the automated vehicle to prompt takeover requests early enough for the user to create sufficient situation awareness. The users also wish for unambiguous information about the current responsibility: Is the human or the machine in control?

The gestures used to input wishes to the vehicle need to be customizable, simple, familiar (use existing operating elements) and comfortable for both hands. The users appreciate direct feedback on whether a gesture was successfully recognized or not. Lastly, the users expect the machine to differentiate between random touches and intentional gestures.

On the motorway, the users of the mass market need support during congestion, construction sides, entering and leaving the motorway. On a rural road, they appreciate support during overtaking and when startled by animal crossing. In urban traffic, support is needed during the search for a parking space, in 30 km/h zones and play streets. In general, users of the mass market want support during encounters with tailgaters and dawdlers, lane keeping with headwind, during lane changes and when their right of way is ignored. They also need support when uncertain about the priority in traffic, under poor weather and visibility conditions and when distracted and would like to drive with adaptable lane keeping.

5.3.3 Novice and Younger Drivers

Novice drivers under 25 are the group with the highest risk for traffic accidents by far in Germany (Statistisches Bundesamt (Destatis) 2019). Reasons include distraction by the use of electronic devices, faulty situational awareness and driving too fast for the prevailing conditions (Twisk 1993). According to our findings, novice drivers still have difficulties controlling their vehicle in critical situations a long time after obtaining their driving licence and assess the consequences of their actions in a less differentiated manner, and their perception routines in relation to vehicle control are not yet fully developed. Therefore, young drivers could benefit from an automated driving function that supports them in difficult situations and reduces complexity by enabling manoeuvre-based driving.

5.3.4 Older Drivers Who Are no Longer Fully Fit to Drive

Individuals are seen as “elderly” when reaching the age of 65 years (World Health Organization 2002). Their driving is affected by the onset of age-related shortcomings, such as musculoskeletal symptoms, loss of visual/auditory abilities, decrease in physical strength and slower reaction times (Shaheen und Niemeier 2001). This might explain why older drivers often feel stressed in traffic and, consequently, cease driving—especially women, although still healthy (Siren et al. 2004)—or continue to drive, but do not drive safely (Siren und Haustein 2016). Cessation of driving seems to affect seniors’ health negatively (Chihuri et al. 2015). The literature falls in line with the results of our focus group with geriatric nurses ($N = 8$): the Vorreiter concept could increase older drivers’ subjective sense of security and the objective safety of other road users, and could enable them to participate in daily life independently for longer. The design and development process described in Sect. 5.5 elicited the following requirements: senior drivers want the automation to adapt to a certain driving style, to take control during emergencies, to keep the vehicle in the lane and to react proactively, but never without confirmation from the driver. The execution of a manoeuvre has to be abortable via breaking or counter-steering, and gestures should be as simple as a normal, manual action/intervention. The users are open-minded towards voice control.

Beyond the use cases of the mass market, older drivers want to be supported when overtaking on extremely winding roads, doing shoulder checks, entering congested crossroads, entering and leaving roundabouts, and encountering T-junctions or obstacles. The following more general use scenarios were stated by the geriatric nurses: visits to family and friends, visits to the doctor and pharmacy, going shopping and going on holiday.

5.3.5 People with Disabilities

In 2020, people with physical disabilities of all ages often still rely on the assistance of others to travel, according to the focus group we conducted with people using an electric wheelchair daily ($N = 4$). All interviewees of the focus group agreed on the need to deal with the skills and requirements of each person with disability individually, so the only general requirements were technical feasibility and easy access to the vehicle with wheelchairs.

The people with disabilities would use the Vorreiter concept for going on holiday, to visit friends and family, on long trips on the motorway and rural road, and to get to work.

5.4 Interviews at Paravan: Exploration of Driving Gestures for People with Disabilities

5.4.1 Introduction

As described in the introduction, one of the goals of the Vorreiter project is a design-for-all set of manoeuvre gestures, which also works for people with disabilities. One goal is to give people with disabilities, with the help of automation, their mobility back and thus improve their quality of life. Due to the necessary, highly individual solution spaces, which are primarily dependent on the respective health impairment of the driver, it is essential to design a multimodal input interface that can be adapted to the individual condition. This often requires solutions which, due to the lowest possible force potential as well as severe movement restrictions of the persons, make use of not only X-by-wire but also of an adaptable user interface.

In order to specify particular requirements of users with disabilities for the design-for-all, IAW carried out open narrative explorative interviews with five experts for driving with disabilities: experts of Paravan, a company for vehicle modifications for disabled persons in Germany. In the first part, the experts were interviewed about which disabilities they encounter most frequently in their everyday work. Moreover, they were asked which previous solutions already exist to enable people with physical disabilities to drive. In the second part, they were asked to explore possible driving gestures for people with disabilities by using different control devices. Each interview lasted 1–1.5 h and was documented with video and sound recordings.

5.4.2 Clinical Pictures and Previous Driving Solutions for Physical Disabled People

The experts reported that the customers of Paravan are suffering from paraplegia (paralysis from the waist down, which results in restricted leg movement), tetraplegia (restriction of leg and arm movement) and missing limbs. In the case of paraplegia and tetraplegia, affected people are restricted in their degree of movement, which depends on the degree of paralysis. Depending on the level of paralysis, some solutions already exist. There are already existing control devices that allow the affected people to remain mobile, for example hand-controlling aids like joysticks for fine motor skills, miniature steering wheels or pushing and braking aids for non-fine motor skills. In the case of affected people who are missing, there are also already existing solutions. Plates in the footwell allow the affected people to steer the car with one foot. There are also steering aids like joysticks available, which can be moved with the toes, legs or chin. Even though there are already possibilities to enable people with disability to drive with diverse assistive aids, there is still a range of people who have not yet been able to drive, for example, affected people who have insufficient fine motor skills and too much spasticity, or who are completely paralysed and can only speak.

Gesture-based movement control of vehicles could be a possible solution in order to enable people with physical disabilities to drive. In order to identify possible gestures for gesture-based movement, the IAW and the experts of Paravan explored possible gestures for different control devices. For structuring the results, a model of the multiple resource model according to Wickens (1992) and extended by Flemisch et al. (2010) is used. The model states that the resources are visually, verbally and haptically essential for structuring dimensions of interpersonal interactions, but also between humans and technology.

5.4.3 Exploration Results

In the explorative part of the interview, the participants were shown four pictures for manoeuvres used daily in road traffic: turning manoeuvres, lane change manoeuvres, avoidance and overtaking manoeuvres and speed control. For these manoeuvres, the participants should consider different gestures that are possible for people with physical disabilities (see Figs. 5.12 and 5.13). For the exploration, Paravan provided various devices (joysticks, mini steering wheels, accelerator and brake sliders). Apart from that, the participants were free in their choice of input modalities. In the following, the exploration results are summarized regarding the input modalities, camera-based input (visual), speech input (acoustic) and haptic input (haptic).

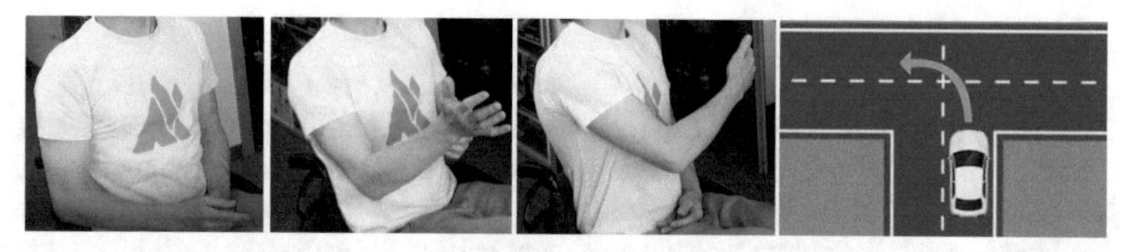

Fig. 5.12 Proposed hand gesture to turn left at a T-junction

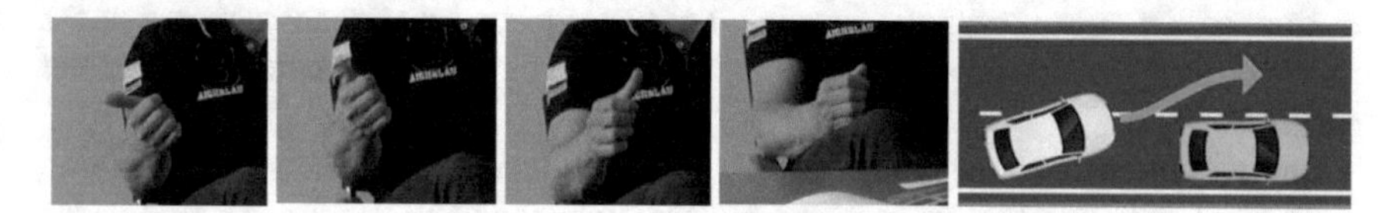

Fig. 5.13 Proposed "swipe gesture" for overtaking manoeuvres

5.4.3.1 Proposed Camera-Based Manoeuvre Gestures

For the initiation of the manoeuvres, the participants could imagine different hand and head movements, which could be detected by different cameras in the vehicle. For example, a lane change manoeuvre or a turning manoeuvre could be initiated with an unambiguous head movement or by pointing in the appropriate direction with a hand gesture (see Fig. 5.12). The main problem that the participants reported was the difficulty in distinguishing between the gestures for lane change manoeuvres or turning manoeuvres. The participants saw a solution to the problem in making the gesture twice, the first time for changing lanes and the second time for initiating a turn. Another solution was to combine different input modalities for a better distinction. A specific requirement identified by the participants is the fact that disabled persons are often restricted in their head and hand movements. Therefore, it was determined that the gestures should be able to be executed with as few degrees of freedom as possible.

5.4.3.2 Proposed Speech Input

The participants had the idea of initiating a turning manoeuvre or lane change manoeuvre by means of voice input. For example, the participants could imagine initiating the manoeuvres with the help of short voice commands, such as "next left", "sharp right", "straight ahead" and "lane change right" or "lane change left". The participants saw the difficulty in the fact that such input could have a high error detection rate. However, the participants also mentioned that in a combination with another input device speech input could be an option in order to distinguish between the different manoeuvres.

5.4.3.3 Proposed Haptic Input

The participants had explored different gestures by using joysticks, mini steering wheels, accelerator and brake sliders. In order to initiate a turning manoeuvre, it has been suggested to tilt the joystick right or left to turn. The participants also used the joystick to explore a clear movement pattern for avoidance and overtaking manoeuvres. In the clear sequence of an avoidance gesture (lane change to the left) and two overtaking gestures (driving past the object and shearing off to the right in front of the object) with the joystick, the participants saw the advantage that the manoeuvring gestures are unambiguous and do not have to be separated from other gestures.

In order to initiate a lane change manoeuvre, the participants explored a swiping movement with the thumb over the steering wheel in the desired direction (see Fig. 5.13). The participants also explored gestures for speed control. In this case, the participants used the joystick and the accelerator and brake sliders. In order to increase the speed, the control device is pushed away, and to slow down it is pulled towards the driver. Here, the participants saw the difficulty of determining the degree of breaking intensity. For example, it could be difficult to initiate a slow roll-out or an emergency stop with a single gesture. The participants saw a solution in the fact that the degree of braking intensity could be set in several steps. In addition, in higher automation levels the driver could adjust with only one impulse. The adaptation of the speed to the situation would then be handled by the automation.

5.4.4 Discussion of the Gestures Proposed by People with Disabilities

In summary, it must be noted that the different clinical pictures of people with disabilities lead to the demand of a high flexibility of a gesture concept. Depending on the nature and degree of the limitations, an inclusive design concept can be a starting point, which has to be significantly adapted. Moreover, the manoeuvres that are presented are specific to the interviewed participants. Nevertheless, a few conclusions can be drawn from the interview, which are crucial for a design-for-all concept. The results of the explorative part of the interview have shown that the gestures should be very simple, as unambiguous as possible, and able to be executed with as few degrees of freedom as possible, so that the gestures can also be performed by people with a higher degree of physical limitation. However, it should be noted that a lower number of degrees of freedom can also result in an increase in the error rate, whereas the exploration has shown that a combination of multiple, redundant interaction resources has a good chance to reduce the error rate. The difference between turning and lane change manoeuvres might induce an unnecessary complexity, which might be solved simpler, as described later.

5.5 Explorative Design and Development of Swipe Gestures for Interaction with Highly Automated Cars

5.5.1 Introduction

Based on the general introduction to manoeuvre gestures described in Sect. 5.1 in this issue, this section describes the explorative design process and iterative development of the swipe gestures on the steering wheel. The process that led finally to the invention of swipe gestures included diary studies, future workshops, the exploration of the concepts with extreme users, evaluation by means of a questionnaire study and several future workshops with different user groups (see Sect. 5.3). They accumulated in a driving simulator study to compare the swipe gestures with voice input and a joystick provided by Paravan (Sect. 5.5) and in a virtual reality study to compare the swipe gestures with a joystick provided by Valeo. This also included a pure holographic representation of a steering wheel (Sect. 5.10) as an explorative approach. Further, a driving simulator study to compare swipe gestures with push/twist gestures was performed (Sect. 5.11).

5.5.2 Diary Studies and Future Workshops

As a starting point to identify interaction gestures with automated vehicles, we conducted a diary study ($N = 5$) and future workshops ($N = 12$ users, *mean age* = 53.58 years, *range* = 24–79) based on the method from Vavoula and Sharples (2007) (Kaschub et al. 2018).

Based on the Vorreiter Use Case Catalogue (Sect. 5.2.4.1) and from user scenarios in the diary studies and future workshops, we derived manoeuvres where users wished for support by a driving automation. $N = 10$ participants saw videos in a static driving simulator with these manoeuvres and were asked to execute a gesture input that they thought would initiate the presented manoeuvres. We visualized the results in a first gesture catalogue (Kaschub et al. 2018). A catalogue which fusions and condenses over all the explorations a preferred set of gestures is presented in Sect. 5.8.2.

5.5.3 Exploration with Extreme Users

In order to increase the scope of possible solutions, we also conducted interviews with extreme users: a Eurofighter pilot, a pantomime and a sign language specialist. We chose these extreme users, because the pilot is used to interact with automated (flying) vehicles, and the pantomime and sign language specialists are experts for intuitive and consistent gesture languages. They were presented videos of different manoeuvres (pre-recorded in the driving simulator) and were asked to identify what gestures would be suitable for this manoeuvre. Afterwards, the participants evaluated the gestures we developed earlier in the project using the card sorting method (e.g. Nurmuliani et al. 2004) and to add their own suggestions for gestures where necessary.

As a result, the extreme users agreed on the idea of the use cases as self-contained units of action, so that a user can always initiate even more complex manoeuvres (e.g. overtaking another vehicle) with a single gesture. At the same time, the driving automation shall only execute a manoeuvre if the manoeuvre as a whole does not violate the traffic regulations. The basic elements of a manoeuvre are described in Sect. 5.2.4. Further results of this explorative development are the requirement of performing gestures with one hand only and optionally with two hands. This is important to perform manoeuvres also with one occupied hand. Another important aspect is a very low physical demand, especially for elderly and many disabled persons (Kaschub et al. 2018; Sommer et al. 2019), which replicates the findings of Spies et al. (see Sect. 5.5). A unified general concept of the gesture set, as described in Sect. 5.8, may also be even more important than individual intuitiveness of gestures.

5.5.4 Future Workshops and Swipe Gesture Design

We conducted three future workshops with $n = 32$ participants (*mean age* = 26 years, *range* 21–29). In each workshop, we further divided the participants into small groups of 5–6 persons and asked some of them to wear an age simulation suit to assess the concepts from the point of view of the elderly. First, we presented the H-metaphor and H-mode (Flemisch et al. 2014), manoeuvre-based driving as a general concept and the cooperation concepts (gestures and voice control) based on the questionnaire study above. We then handed visualized concepts of the gestures and voice control commands to the participants who tried to carry out these concepts on the first-generation steering wheel prototype from Valeo. The participants used the thinking aloud method (e.g. discussed in Boren and Ramey 2000) to evaluate the concepts during this phase. By doing so, the participants developed a self-contained concept for the driving manoeuvres and later on used the paper prototyping method (e.g. described in Rettig 1994) to develop a holistic interface design for the cooperation concept in their groups. At the end of the future workshops, the participants discussed the cooperation concepts they just experienced as well as their interface designs.

The results of the future workshops showed that swiping over the steering wheel surface was an appealing interaction concept. Participants wearing an age simulation suit preferred smaller gestures on the steering wheel airbag cover, while participants without such a suit liked the more familiar gestures on the steering wheel rim better. Therefore, physical demand seems like an important factor that one should keep in mind when designing cooperation concepts for the elderly. Interestingly, all workshop groups agreed on the same consistent concept for voice control independently and wished for a visual feedback for both touch and gesture recognition via the LEDs in the steering wheel rim. Participants want the driving machine to display possible manoeuvres as well as the trajectory chosen by the driver in the head-up display

(HUD) of the vehicle. As the different groups of participants agreed on these cooperation and HMI requirements without consulting each other, the concepts seem rather intuitive (Sommer et al. 2019).

5.5.5 *Joint Application Development with Usability Experts*

We used the Joint Application Development method (e.g. Carmel et al. 1993) to discuss issues and possible solutions in a workshop with five usability experts in order to define a consistent cooperation concept. Our aim was for this concept to meet the users' demands, suffice legal and technical requirements and consider psychological, cognitive and ergonomic aspects as well. Therefore, we consolidated the results of both the future workshops and the earlier results, and the usability experts furthermore experienced the gesture concepts on the first-generation steering wheel prototype, integrated in the Fraunhofer driving simulator in automated driving scenarios. As a result, suitable swipe gestures on the steering wheel, suitable voice commands and assigned manoeuvres were identified for a driving simulator study (Sommer et al. 2019) (Fig. 5.14).

5.5.6 *Evaluation in the Driving Simulator (Wizard-of-Oz)*

We evaluated and compared the aforementioned swipe gesture concept on the steering wheel and the voice command concept. In the last drive, the participants interacted with the concept joystick (the input element was provided by Paravan and integrated into the Fraunhofer IAO driving simulator).

The participants did not have to drive themselves; instead, they were driven by a driving automation and had to initiate manoeuvres using one of the three interaction concepts. This allowed us to evaluate the concepts and its strengths and weaknesses from the users' point of view and to assess further user requirements with a more

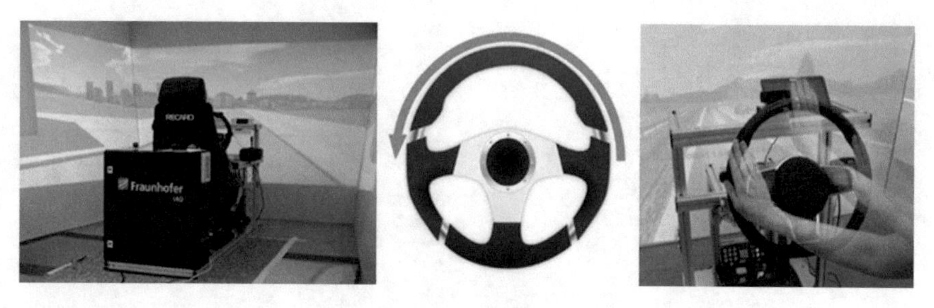

Fig. 5.14 Driving simulator with integrated Valeo steering wheel and swipe gestures

realistic prototype than in the paper prototyping workshops. We used the Wizard-of-Oz method to do so, which means that a member of the team simulated the parts of the system that were not yet implemented (Green and Wei-Haas 1985). The input of each manoeuvre was confirmed by a sound played on the stereo speakers.

The simulator study took place in a 180° projection area—one screen in front and two on the left and right of the participant—with a mock-up including a driver seat, stereo speakers and the moveable capacitive steering wheel prototype provided by Valeo. The steering wheel moved synchronously with the movements of the vehicle in the simulation, and the necessary steering angle was derived from the simulation software SILAB (2019) using a CANopen interface. We used the system usability scale (SUS; Brooke 1996) and questions assessing the constructs "joy of use", "safety", "performance", "freedom vs. discipline", "cooperation", "controllability" and "trust". The evaluation of the human–machine interface was based on the norm regarding "dialogue principles" (DIN EN ISO 9241-110:2008-09 2008).

Overall, $N = 46$ valid users (30 males, *mean age* = 43 years, *range* 21–74) participated in the study as 4 participants had to be excluded from analysis due to simulator sickness (assessed continuously during the drive). Seven older drivers participated (*mean age* = 65 years, *range* 62–74), 10 participants wore an age simulation suit, and furthermore, 29 users from the mass market (*mean age* = 36 years, *range* 21–59) participated. After a familiarization drive, the participants listened to a presentation explaining the concept of the gestures on the steering wheel for different manoeuvres. They then drove through the whole course, first using the voice control concept, filled in the first questionnaire, drove the whole course using the gestures on the steering wheel, then answered the corresponding questionnaire, then did the same with the joystick concept, answered the questionnaire and then completed a half-structured interview about the gestures on the steering wheel.

The gestures on the steering wheel achieved $M = 75.27$ in the SUS, and the voice control commands scored $M = 76.90$ on the SUS, which means that both concepts reached an above average usability. The participants did not seem to clearly prefer either of the concepts at hand, as the questions regarding the different constructs (joy of use, safety, performance, freedom vs. discipline, cooperation, controllability and trust) did not differ significantly as well.

Important results from the half-structured interview include:

Capacitive zones on the steering wheel rim need a better visualization where to swipe.
Direct feedback on whether a gesture was accepted on the steering wheel.
Gestures need to work anywhere on the steering wheel, not only on the upper parts in case the steering wheel turns.
Gestures should be less physically demanding, e.g. should work with a shorter "swipe distance".
A haptic emergency stop button should serve as a backup for the voice control concept, but is not necessary for the gesture concept.
Most participants would favour an interaction concept that combines voice control and gestures on the steering wheel.
A gesture to abort manoeuvres is not necessarily needed as participants rather wish to overwrite previously entered manoeuvres with new ones.

We consolidated all of the results and used them to generate the concept for the next generation of capacitive steering wheels (implemented by Valeo) in a design-thinking workshop (see Sect. 5.9; Sommer et al. 2019). Hence, the user requirements from the explorative development were implemented in the second generation of the Valeo steering wheel (Sect. 5.9) using the catalogue of steering gestures as described in Sect. 5.8.2, which was evaluated in the user studies described in Sects. 5.10 and 5.11.

5.6 Wizard-of-Oz Vehicle for the Test of Manoeuvre Gestures

5.6.1 Introduction

Based on the general idea and overview of manoeuvre gestures presented in Sect. 5.1, this section describes the concept and build-up of a test vehicle based on the Wizard-of-Oz (WOz) and the theatre methods.

The goal was to design a test vehicle for gesture-based steering via manoeuvres that could be operated in real road traffic. In this environment, the vehicle serves as a platform to carry a test subject, a WOz operator and a safety driver while providing the required equipment to test manoeuvre-based driving.

5.6.2 History of the Wizard-of-Oz Method Applied to Vehicles and the General Test Set-Up

As described in Sects. 5.1 and 5.2, it is the test subject's task to drive the vehicle a certain course by using a set of gestures. To realize this experiment from a legal perspective (see Sect. 5.7) and to guarantee safety, a WOz experiment approach was chosen.

In a WOz experiment, the illusion of a real automation is created for the test subject by another person playing the automation covertly. For a detailed description of this method, please refer to Kelley (1984). As a further methodology and notwithstanding the foregoing, Flemisch developed the idea of the theatre method in 2002, which was first described in depth by Schieben et al. (2009a, b). While in the WOz method, the (metaphorical) curtain is closed and the user does not know that the technology is only emulated, in the theatre method, the curtain is open. So, the participant and a confederate (equivalent to test subject and WOz) playing the technology act together and can think and drive through different variations of the technology.

Both the WOz and the theatre methods were refined and applied in the course of the H-mode project by Flemisch et al. (2003, 2005). The idea to apply the WOz method to a real vehicle was born in a discussion between Volkswagen and DLR in 2003 on this basis. Müller et al. (2019) provide an overview of WOz vehicle concepts.

The WOz set-up for the Vorreiter vehicle was derived in close discussions with experts from Paravan GmbH,[2] especially Georg Kotrotsios and Madeleine Höschle.

In the Vorreiter test vehicle, the test subject sits in the left seat of a right-hand drive car. Via the pedals and a force feedback steering wheel, the test subject instructs the vehicle by using the predefined gestures described in Sect. 5.8. The WOz operator (or theatre confederate) who actually controls the vehicle during the experiment sits in a seat in the back that replaces the rear bench. A third person who serves as a safety measure sits on the right side. Through the original steering wheel and pedals, the safety driver is capable of taking over the vehicle controls anytime.

5.6.3 The Vorreiter Test Vehicle Set-Up

The Vorreiter test vehicle is based on a right-hand drive Volkswagen Caddy Maxi Trendline and was converted by Paravan GmbH. Therefore, all existing vehicle control systems are laid out for the safety driver to have primary control over the vehicle. Specifically, this means the control takeover capabilities are designed in a way that the safety driver is able to take over and keep full control over the vehicle at any time. Furthermore, the WOz operator can take over control from the test subject, but not vice versa.

---

[2]https://www.paravan.de/.

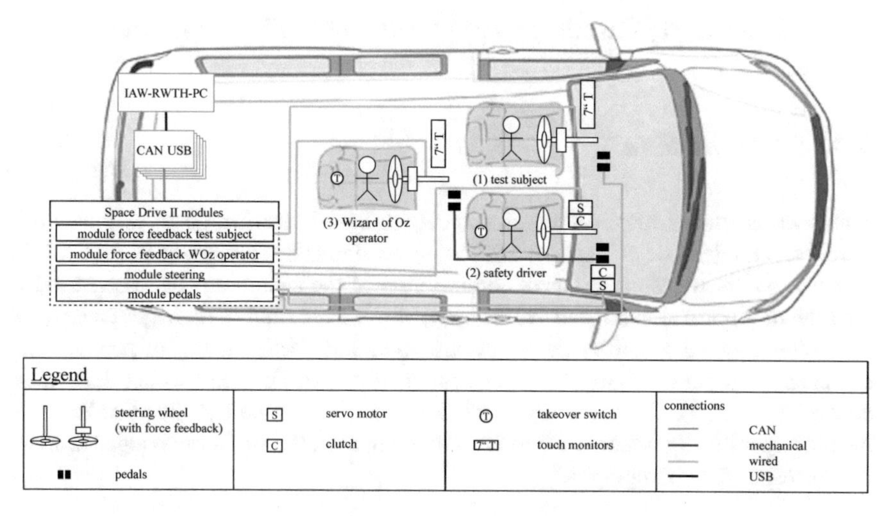

Fig. 5.15 Schematic layout of the Vorreiter Wizard-of-Oz test vehicle

As shown in Fig. 5.15, all input devices are directly connected to the respective Space Drive® II modules. These process their input signals and control the vehicle by means of the existing steering and pedals via servo motors. Each module is connected to the IAW-RWTH-PC via a CAN-USB interface.

The test subject can use a force feedback steering wheel and pedals like in a normal car. Both are drive-by-wire input devices that connect to the Space Drive system. Their signals are either used to drive the vehicle directly over the drive-by-wire system or are processed in the IAW-RWTH-PC to recognize gestures according to the selected gesture type. In addition, there is a touch-sensitive monitor for secondary driving tasks.

The WOz operator has an identical steering and touch-sensitive monitor to the test subject, whereas the pedals are mechanically connected to the stock ones. In order to get the instruction, the WOz operator has a secondary monitor where the recognized gestures are displayed. To start the experimental ride, there is a button for taking over control from the test subject.

As a safety measure, the safety driver has the highest authority regarding the vehicle controls: in case of an emergency, the safety driver can use the original vehicle control devices. In order to gain full control, there is a button that disengages the clutches between the steering and the pedals and their servo motors to separate the others from the vehicle controls.

Due to the dropout of the project partner Paravan GmbH in the second year of the project, the test vehicle could not be used in an experiment. At the end of the project Vorreiter, the vehicle was handed over to Fraunhofer IAO for further use in research.

5.7 Manoeuvre-Based Steering Gestures—The Legal Point of View

5.7.1 Introduction

This section grasps legal issues of manoeuvre-based steering gestures under the Vorreiter paradigm, described in the following section and summarized in Sect. 5.7.2. The section discusses questions of admissibility of the technology as referred to by the Law of Approval and Road Traffic Law, of the regulation of damages caused by the technology, i.e. Liability Law, and of a potentially better access of people with handicap to attain a driving licence as provided for in Driving Licence Law (see Sects. 5.7.3–5.7.5). The summary (Sect. 5.7.7) will stress the legal results. We will mention specifically whether we refer to international, EU or German legislation in the corresponding paragraphs.

5.7.2 The Technology and Its Objectives

Manoeuvre gestures developed in the Vorreiter project can be applied to all kinds of physical interfaces, but at the current state use the steering wheel for gesture input. This input can be performed by a push/twist gesture or a swipe gesture with the hand over the surface of the steering wheel rim, without moving the steering wheel into any direction. In any case in Vorreiter, such gesture itself is not sufficient to exercise the driving task in all its aspects, e.g. in all its operational, tactical and strategic aspects (cf. SAE 2016, No. 3.8 definition of "Dynamic Driving Task", ECE/Global Forum for Road Traffic Safety 2018: definition of "dynamic control"). Instead, an automation system is necessary in order to recognize the gesture, process it along with other signals and implement the steering command contained, i.e. to take over the operational and tactical aspects of the driving task. Input of a gesture is assigned to the strategic or tactical level; moreover, it does not directly change the vehicle's situation in road traffic. A gesture can only be used when the vehicle is driven in the automated mode.

Combined with a partially automated driving system—SAE level 2 (SAE 2016, No. 4)—the gesture can be used for a driving command in the automated mode while the driver is monitoring the traffic situation, as outlined by the technical description of this level. The driver has to decide whether a gesture is sufficient that is to say whether automated driving is possible, or whether personal steering is needed, thus ending the automated mode. On the other hand, a combination of manoeuvre gestures with highly automated driving—SAE level 4 (SAE 2016, No. 4)—enables the driver to gesture input without monitoring the traffic situation because the automation system is technically supposed to handle all traffic situations while driving on its own. Hence, this combination allows for realization of the steering command at the first opportunity.

Apart from the concept of steering by gesture as an automation tool in the mass market, the very idea of the Vorreiter project was to enable a broader variety of people to apply for a driving licence focusing especially on physically handicapped people.

5.7.3 Admissibility of (Use of) Technology

Whether manoeuvre gestures can be used for cooperative driving depends on the one hand on the legal possibility to approve the technology (Law of Approval) and on the other hand on the legal admissibility to use the technology in traffic (Law of Road Traffic Conduct). Although this is not uncontested, both branches of law are considered to be related to each other in legal literature (von Bodungen and Hoffmann 2016; Lutz 2014): technology which cannot be approved cannot be used, while technology which cannot be used in traffic according to road traffic rules cannot be approved (theory of unity of laws). Moreover, both branches of law are widely shaped by international law which therefore has to be taken into account for judgement of manoeuvre-based gestures.

5.7.3.1 Aspects of Law of Approval: Safety Regulations

The Law of Approval establishes safety requirements for motor vehicles. The requirements mainly rest upon European law, whereas national law is of minor importance. Directive 2007/46/EC establishing a framework for the approval of motor vehicles (…) (Framework Directive) is transformed into national law by EC Vehicle Approval Ordinance (*EG-Fahrzeuggenehmigungsverordnung—EG-FGV*). The Directive will be repealed by Regulation (EU) 2018/858 on the approval and market surveillance of motor vehicles (…) from 1 September 2020 (cf. Article 88 Regulation). Having direct effect in national law (cf. Article 288 § 2 Treaty on the Functioning of the EU), it will render transformation unnecessary. EU legislation refers broadly to UNECE Regulations (cf. Article 8, Annex IV of the Directive, Article 5 § 1, Annex II of the Regulation) established in the framework of the Agreement concerning the Adoption of Harmonized Technical United Nations Regulations for Wheeled Vehicles, Equipment and Parts which can be Fitted and/or be Used on Wheeled Vehicles (…) (1958 Geneva Agreement).

Special provisions for technical equipment capable to perform manoeuvre-based gestures do not exist. Since such equipment is based on vehicle automation systems, the corresponding safety regulations have to be applied first. This field is determined by high dynamics of legislative activity. UN Regulation 79 concerning steering equipment has been amended as to provide rules for some functions of Advanced Driver Assistance Steering Systems among which are different categories of Automatically Commanded Steering Functions (cf. ECE 2018: definitions in no. 2.4.4 and constructive provisions in no. 5.6). These rules address automation systems of SAE levels 1 and 2, i.e. up to partially automated driving, but not yet higher degrees. Furthermore, a framework document for SAE level 3 and higher (which is not binding) has been developed containing general safety requirements defined as "not causing any non-tolerable risk" and "free of unreasonable risk", respectively (ECE/World Forum for Harmonization of Vehicle Regulations 2019, No. 7 and 9.a.).

As long as binding construction rules for automation systems have not yet been established, EU legislation allows for exemptions from existing rules for new technology and concepts (cf. Article 20 § 1 Directive 2007/46/EC respective Article 39 § 1 Regulation (EU) 2018/858), thus offering a possibility for ongoing technical progress.

5.7.3.2 Aspects of Law of Conduct: Road Traffic Rules

Following the theory of unity of laws, a vehicle with advanced technical capacities cannot be approved—even if it fulfils the requirements of technical safety regulations—if road traffic rules prohibit the use of those capacities. As for manoeuvre-based gestures, their use depends on the admissibility of the use of automated systems with which they are indivisibly connected, at least under the Vorreiter approach.

In international law, the Vienna Convention on Road Traffic provides guidance. According to Article 8 § 5bis, vehicle systems influencing the way vehicles are driven are in accordance with the requirement of vehicle control in Article 8 § 1 and Article 13 § 1 when either (i) these systems are in conformity with conditions of construction, fitting and utilization according to international legal instruments concerning wheeled vehicles like the 1958 Geneva Agreement or (ii) there is a possibility for the driver to override or switch off the system. Internationally, this is interpreted as encompassing assistance and automated systems, except driverless systems, thus admitting SAE levels 1–3 because a driver is needed as fall back, challenging SAE level 4 since the system performs without a driver in its operational design domain, and refusing SAE level 5 where no driver is necessary at all (ECE/Global Forum for Road Traffic Safety 2017, No. 14 and 23; cf. also Lutz 2016; contested by von Bodungen and Hoffmann 2017).

On the national level, the 2017 amendment of the German Road Traffic Act *(Straßenverkehrsgesetz - StVG)* in § 1a Section 1 permits highly and fully automated systems to be used under certain circumstances established in § 1a Section 3. Among these circumstances, there is the approval according to international safety regulations or an exemption in accordance with Article 20 of the Directive 2007/46/EC (cf. 7.3.1). Highly and fully automated systems according to § 1a *StVG* correspond to SAE levels 3 and 4 (Deutscher Bundestag 2017). Partially automated systems may be used in accordance with the Road Traffic Regulations (*Straßenverkehrsordnung—StVO*). Especially, § 1 provides for constant care and mutual respect of human drivers, thus requiring permanent observance and a constant ability of the driver to override the system.

Having in mind this legal setting, the use of manoeuvre-based gestures on the basis of an automation system is allowed taking into consideration that

- legal uncertainties with respect to highly automated vehicles under the Vienna Convention exist,
- performing a gesture is not equivalent to overriding an automated system as established by the Vienna Convention Article 8 § 5 bis or by § 1a Section 2 No. 3 *StVG* since in both provisions, overriding refers to retaking control of the vehicle by driving it personally, i.e. without automation system, whereas the use of gesture requires an activated automation system,
- the driver has to differentiate between the possibility of a manoeuvre gesture and the necessity to override the system (cf. § 1b Section 2 *StVG*: driver obliged to be sufficiently attentive to take over driving),
- input of a manoeuvre gesture is allowed without paying attention to the traffic situation in accordance with § 1b Section 1 *StVG* when using a highly automated system but not when using a partially automated system since § 1 *StVO* in that case requires permanent observance,
- using a manoeuvre gesture to overcome specific limitations of the automation system, e.g. automatic stop at streets with equal priority, is possible if driver decides paying attention to the traffic situation (as part of the automation system's determined use according to § 1a *StVG* with regard to highly automated vehicles or as part of the obligation to constant care according to § 1 *StVO* regarding partially automated systems).

Table 5.2 shows the legal admissibility of manoeuvre-based gestures for different levels of automation as developed in the Vorreiter project.

5.7.4 Liability Law: Settlement of Damages

Damages caused by an accident of the automated vehicle will first of all be settled under Road Traffic Liability Law. Secondly, damages can raise questions of Product Liability Law.

Table 5.2 Legal admissibility of steering by gestures for different levels of automation based on gesture design of RWTH Aachen (see Sect. 5.8.2)

	Manual driving	Push/twist in SAE level 2	Swipe in SAE level 2	Push/twist in SAE level 4	Swipe in SAE level 4
Vienna Convention on Road Traffic Safety					
German Road Traffic Act					

Green light: admissible; yellow light: probably admissible in future; red light: inadmissible

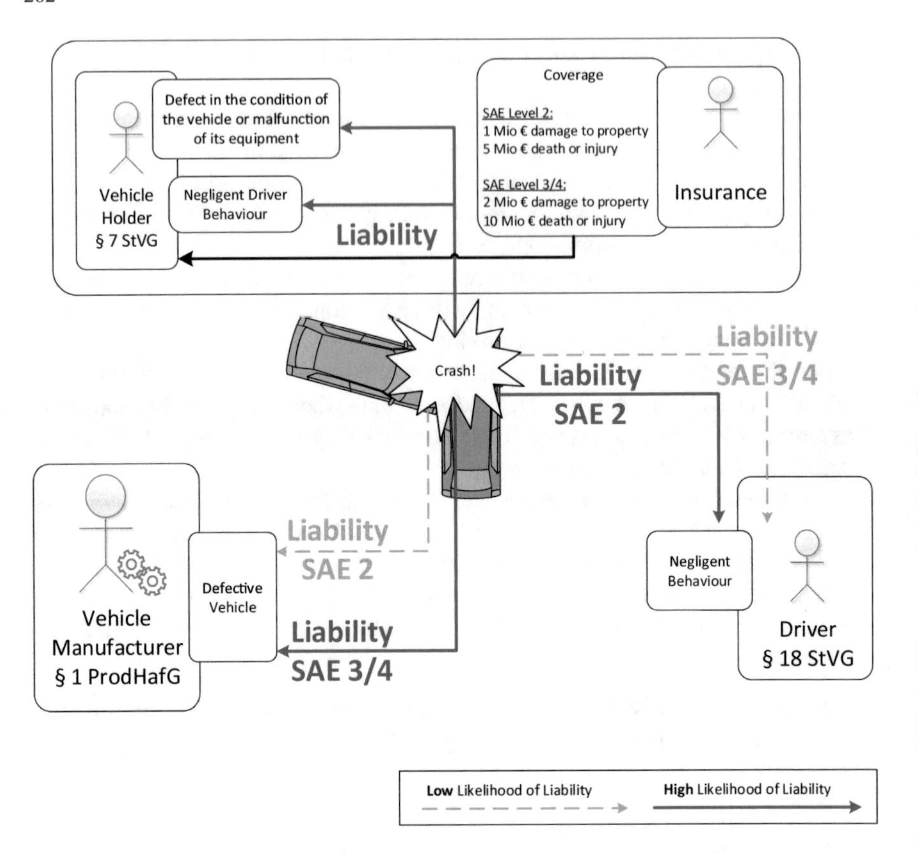

Fig. 5.16 Liabilities of the stakeholders and shift of liability in case of an accident with partially and highly automated vehicle involved

Figure 5.16 shows who is held liable on which legal reason in case a partially or highly automated vehicle causes an accident. outline will be clarified in Sects. 5.7.4.1 and 5.7.4.2.

5.7.4.1 Road Traffic Liability Law: Liability of Vehicle Owner and Driver

According to the *StVG*, holder and driver can be held liable if their power-driven vehicle has caused an accident in public traffic. Whereas the liability of the holder is based on risk (strict liability), liability of the driver rests on negligence. The 2017 *StVG* reform has kept this two-tiered system unaltered. Solely, the maximum amount of compensation has been doubled up to 10 million Euro for damages of injured persons and up to 2 million for damages to property regarding accidents involving highly or fully automated vehicles (cf. § 12 *StVG*). Subject to § 1 German Mandatory

Insurance Act (*Pflichtversicherungsgesetz—PflVG),* the holder has the obligation to take out insurance covering these damages.

According to § 7 Section 1 *StVG,* the holder of a vehicle is liable if the risk of operating the vehicle has materialized in damages to a third party. This also applies to vehicles with automation systems and use of gestures. Such materializing of risk can result from technical failure (cf. § 17 Section 3 *StVG*) including failure of the automation system, or from negligent behaviour of the driver.

Additionally, the driver is held liable according to § 18 *StVG* for negligent behaviour. While using gestures driving a partially automated system, negligent behaviour can regularly be established since the driver is obliged to constantly monitor the environment and to take over driving if necessary. In particular, the driver has to carefully decide whether input of a gesture is (still) possible or whether his personal intervention is needed. Though, if the vehicle operates in an automated mode higher than partial automation, negligent behaviour is regularly excluded (Gasser et al. 2015, 2013; questioned by Berndt 2017). This is also true in case the driver enters a gesture while the automation system drives the vehicle, because the gesture is only one of the signals processed. Nonetheless, the driver has to fulfil his duties according to § *1b StVG,* especially to be sufficiently attentive to take over.

Apart from the degree of automation, two situations have been identified in the Vorreiter project which may entail driver liability: first, push/twist gestures on the steering wheel, if they cause a trace offset that the driver has to take into consideration, and second, if the gesture is used to overcome limits of the automation system, e.g. manoeuvring decisions by the driver at crossroads of equal priority.

5.7.4.2 Product Liability Law: Liability for Defective Products

According to § 1 Product Liability Act (*Produkthaftungsgesetz—ProdHaftG*), the manufacturer is held liable if his defective product injured persons or damaged property. German product liability is based on Council Directive on the approximation of the laws, regulations and administrative provisions of the Member States concerning liability for defective products (85/374/EEC). It will not be applicable if the accident is caused by failure of the driver but supplements the vehicle holders liability in case the accident is caused by a defect in the condition of the vehicle or a malfunction of its equipment (cf. § 17 Section 3 *StVG*). Regarding this, it should be expected that liability will shift from vehicle holders to vehicle manufacturers with ongoing vehicle automation (Lutz 2015), in particular by regress of the mandatory insurance of holders.

According to § 3 Section 1 *ProdHaftG*, the product is defective if it does not provide the safety standards users are entitled to expect at the time when this product is put into circulation. Safety expectations are not determined by the expectations of any individual user but rather by an impartial user horizon including that one of an innocent bystander (Oechsler 2018, No. 15–17; Federal Court of Justice 2009a, b). Such expectations can especially be focused on the construction of the product itself or the instructions about its use. Constructive measures to prevent risks and damages take priority over instructive ones. Thus, the manufacturer is not able to avoid safety-friendly construction by substituting constructive means by detailed and comprehensive instruction of use (cf. Oechsler 2018, No. 46). Constructive measure must be inspired by the state of science and technology as provided for in technical guidelines and, much more rarely, in law (cf. Oechsler 2018, No. 20a; Foerste 2012, No. 19–23 and 47–48; Federal Court of Justice 2009b). These principles have to be adapted for new technologies, where constructive standards are still in development, as it is the case in automation systems allowing for manoeuvre gestures. One can very generally request that no additional safety risks must be caused by such a new technology compared to personal driving. This encompasses at least that

- the driver must always be able to safely recognize whether he/she is in the automated or manual mode, since gestures are only possible during automated driving,
- the safe transition from automated to manual mode is secured and
- the safe recognition of manoeuvre gestures is guaranteed especially when overriding system limitations.

Instructive measures complement the necessary constructive measures. Instructions should cover information about functions of the automated system and its limits on the one hand as well as information about the admissible and possible use of manoeuvre gestures on the other hand.

5.7.5 Driving Licence Law: Rules on Entitlement to Application

Manoeuvre-based steering gestures under the Vorreiter approach offer the possibility to drive a vehicle in cooperation with a vehicle automation system. Although they do not compensate for all physical handicaps, they largely minimize the physical effort to steer a vehicle. Hence, there might be a potential to raise the number of people entitled to gain a driving licence.

The driving licence is a prerequisite for driving a motor vehicle in public traffic according to § 2 Section 1 *StVG* and § 4 Section 1 Driving Licence Regulation (*Fahrerlaubnisverordnung—FeV*). Notwithstanding the fact that this prerequisite is based in national law, the conditions for a successful application for a driving licence are widely harmonized by European law, i.e. the Directive 2006/126/EC of the European Parliament and of the Council as amended Commission Directive 2009/113/EC. This European legislation is transformed into German law particularly by the *FeV* (Dauer 2019, No. 2).

Both legislative acts provide for certain physical abilities to be fulfilled in order to gain a driving licence. According to Article 7 Section 1 of the Directive, the applicant has to meet a certain medical standard which is especially detailed in Annex III (cf. requirements for persons with a locomotor disability who could mainly profit from steering by gesture). Accordingly, § 11 Section 1 *FeV* stipulates for the necessary physical requirements by the applicant. Despite the potential of steering gestures to allow more people to gain a driving licence, this promise cannot be kept on a general basis as far as technology and law are contemporarily (February 2020) developed. Vehicle automation systems as the bedrock of steering gestures are not sufficiently far developed as to allow possible door-to-door journeys. There will still be parts of the journey where the driver has to drive personally. According to legislative requirements (see Sect. 5.7.4.2), any driver is required to take over driving at the limits of the vehicle automation system. Systems without the possibility for personal intervention, i.e. direct exercise of the entire driving task by the driver, are not allowed to be used in public traffic at all. Thus, besides progress in technology there is also a need for legislative amendments, in a first instance concerning the Law of Conduct. This would lay a basis for participation of persons who are now excluded from applying for a driving licence: Driving Licence Law namely provides for the restriction of driving licences for persons with special needs to specially equipped vehicles in § 23 Section 2 *FeV* (cf. also Annex III Directive 2006/126/EC concerning persons with a locomotor disability). In the near future, specially equipped vehicles can be imagined as automated vehicles cooperating with the driver by manoeuvre gestures as they are designed in the Vorreiter project.

5.7.6 Summary

The results of the legal assessment can be summarized as follows:

The use of manoeuvre-based gestures in combination with automated driving is admissible as far as

- the technical equipment can be approved under Law of Approval (which still requires exemptions for new technology and concepts) and
- the use of automated systems is allowed according to the Law of Conduct.

Damages caused while using an automation system allowing for input of gestures are covered by German Liability Law according to the following principles:

- The vehicle holder is held responsible in any case, which is covered by the mandatory insurance (the insurance coverage has been doubled to 10 million Euros for injured persons).
- The driver is regularly held responsible while driving partially automated (SAE level 2).
- With respect to highly automated driving (SAE level 4), the driver is regularly not held responsible, except when he/she entered a gesture at the system limit and when push/twist gestures change the situation of the vehicle in traffic (trace offset).
- The manufacturer is held responsible if the technology used involves new safety risks.

Manoeuvre-based gestures do not extend the number of persons entitled to attain a driving licence, based upon the actual status of technology, the Law of Conduct and Driving Licence Law.

5.8 Towards a Catalogue of Manoeuvre Gestures for Partially and Highly Automated Driving: Results from Vorreiter

5.8.1 Introduction

As described in Sect. 5.1, driving with manoeuvre gestures consists of an action performed by the driver which triggers the machine to execute the appropriate manoeuvre. The fundamental difference to manual driving is that the human driver is decoupled from operational tasks and may only give manoeuvre or even only navigational input (c.f. Flemisch et al. 2019a, b). In the Vorreiter project, a manoeuvre-based driving was implemented for partially automated driving (similar to SAE level 2) and highly automated driving (similar to SAE level 4). The main difference between both levels of automation is that when driving partially automated, the driver is still responsible for the operational input to the vehicle according to Sect. 5.7 and should therefore have a direct control over the steering or acceleration of the vehicle, whereas in highly automated driving mode, the automation and therefore the manufacturer would be responsible for the driving. The driver might even be entirely decoupled from this operational layer (c.f. Flemisch et al. 2019a, b), and only the vehicle is in charge of it. Regarding the input of manoeuvres, automation and human may give conflicting input on which manoeuvre to execute, which should be resolved by a third entity that mediates clear decisions between automation and human. This splits the "automation system" into two parts: the automation and a human–machine mediator. Respecting legal aspects as well as those of the design of the automation levels, mediation depends on the level of automation.

From a driver's perspective, when the driver intervenes in the driving manoeuvre decision, he or she first has to decide on a manoeuvre and then communicate it to the machine for the machine to execute it. To communicate the will of the driver to the machine, it is sufficient to insinuate the manoeuvre decision to the machine instead of explicitly formulating the necessary operational steps. They may be expressed haptically, verbally or visually.

In the course of the Vorreiter project, push/twist and swipe gestures have been formulated, designed and investigated. This section deals with our final design of driving with manoeuvre gestures in partially and highly automated driving in the form of a gesture catalogue. As a next step, we describe the mediator in general and focus on the process of the mediator understanding the human gestures and manoeuvre arbitration based on human input and current driving situation. We close with some remarks on the final design.

5.8.2 First Gesture Catalogue for Manoeuvre-Based Driving

To narrow down the number of use cases that the machine has to differentiate, we first collected situations in which we expect users to want to intervene with the driving situation (see Sect. 5.2). In a second step, we generalized all situations into a set of use cases (Sect. 5.2). Based on the set of use cases, we developed a first gesture catalogue. The catalogue is structured as follows. Based on the use cases formulated in Sect. 5.2.4.1, we present our developed gesture set in an abstract way as well as two implementations of this gesture set, push/twist and swipe designs on the steering wheel and possible other devices. The gesture design covers the communication part of manoeuvre-based driving.

All use cases can be referenced to the current state of the vehicle and aim to change it into a given direction (see Fig. 5.17 left). To each basic use case, a direction in respect of the current driving situation may be associated. This results in four directions to distinguish, namely “to front”, “to back”, “to left” and “to right”, which cover the abstract set of gestures. These gestures were derived from the design space described in Sect. 5.2, and the explorations and design workshop described in Sects. 5.3, 5.4 and 5.5. They are designed to be easily understood and memorized, and their symmetry is meant to support intuitiveness and spontaneous change of mind. The abstract gestures are defined as pars pro toto movements or actions in the desired direction of change of the vehicle state (see Fig. 5.17 centre). Our push/twist and swipe designs are developed on the base of this abstract set of gestures. That is, for all four abstract gestures, there is a dedicated push/twist and swipe gesture (see Fig. 5.17 right).

The purpose of the gesture catalogue is to show the connection between the basic use cases described in Sect. 5.2.4.1 and the corresponding gesture to trigger a manoeuvre as described in the use case for both implemented gesture sets.

In this project, two different sets of gestures based on the general gesture set were implemented. The first design contains push and twist gestures, which can be

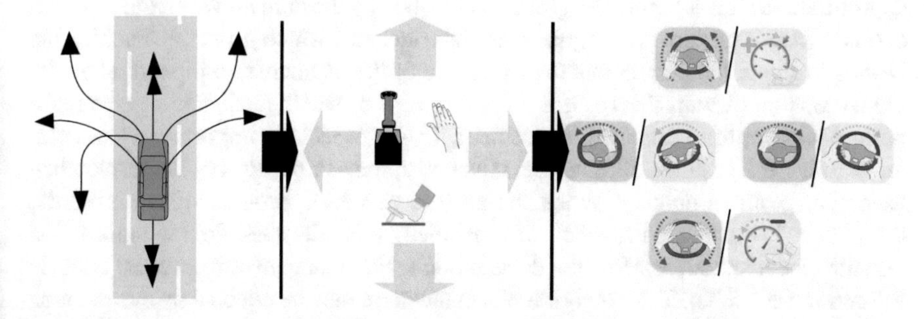

Fig. 5.17 Stages of development of the gesture catalogue. Desired change in direction of the vehicle based on a use case (left). Derived abstracted movement in direction of the desired change of, e.g., hand or foot or any device, which forms an abstract gesture (centre). Development of a specific set of gestures based on the abstract gesture set, e.g. push and twist or swipe gestures (right)

implemented, e.g., on a steering wheel, pedals or a sidestick. They are designed as a set of gestures for which the human needs full haptic coupling to the input device to execute. The other design includes swipe gestures on the steering wheel. In this design, the human only has to touch the input device lightly by swiping with a hand or finger along a touch-sensitive surface.

Both sets of gestures are described for each use case and corresponding abstract gestures in Table 5.3, in which, however, use cases that are associated with the same direction are grouped. All non-basic use cases were established to investigate the concatenation of simple manoeuvre gestures to initiate more complex manoeuvres.

5.8.3 Details of the Gesture Recognition

A good recognition of manoeuvre gestures has to bridge two tension fields, first between (1a) being intuitive to the driver and (1b) being logical and implementable by the machine, and second regarding the different gestures between (2a) having a common logical behaviour and (2b) allowing specifics of gesture implementation also to be formulated and implemented. To reach both, a logic core design of gestures and their potential interruption has been designed in the form of a state machine, where the first logical steps are independent from the implementation as push/twist or swipe gesture, while the follow-on steps instantiate the specific interaction, e.g. for push/twist or for swipe.

All gestures are given two properties: direction and progress. Depending on which input device is used, progress may be continuous (twist of steering wheel) or discrete (number of touched zones on a steering wheel). This progress property is meant for the machine to detect gestures early and give appropriate feedback to the human already while the gesture is still being executed. When the progress of a gesture exceeds 50%, a gesture is recognized as "armed"; i.e. the machine recognizes that progress towards a given gesture is made. This is crucial for the machine to give appropriate feedback before the gesture is finished by the human. When the progress exceeds 75%, a gesture is recognized as "interlocked", which gives the machine the chance to signal the human that the execution of the manoeuvre will start when the human continues with the gesture. In addition, it provides feedback on whether a gesture was not fully executed by accident. This is especially important when users are learning the gestures. Skilled users, however, may skip stages of feedback when executing gestures quickly. When the gesture execution progress exceeds 100%, the gesture is considered finished. If a manoeuvre is still possible, the automation executes the manoeuvre. How the decision on which manoeuvre to execute is made follows in Sect. 5.8.4. The execution of a manoeuvre may be rejected by the machine if there is no manoeuvre associated with the gesture's direction that is executable at that moment. The rejection of a gesture does not require the gesture to be finalized; it may be decided as soon as a gesture is armed or when the situation changes while the gesture is still made by the driver.

Table 5.3 Gesture catalogue derived from the use case catalogue developed in Sect. 5.2.4.1

Use case(s) (gesture type)	Push/twist gesture		Swipe gesture	
Start moving Move faster (To front)		On pedal: Give a small impulse on gas pedal or push pedal for a longer time to increase speed until target value is reached On steering wheel: give a small push-impulse forward on steering wheel		On steering wheel: Swipe with both hands from bottom to top of steering wheel; hold at the end to gradually increase speed until target value is reached
Move slower, stop moving (to back)		On pedal: give a small impulse on brake pedal or push pedal for a longer time to increase speed until target value is reached On steering wheel: give a small pull-impulse on steering wheel		On steering wheel: swipe with both hands from top to bottom of steering wheel; hold at the end to gradually increase speed until target value is reached
Change lane left/right, turn left/right (to left/right)		On steering wheel: small steering impulse in left/right direction		On steering wheel: swipe on the steering wheel in left/right direction
U-turn (combined)	Combine two "to left" gestures			
Emergency brake (special)	On pedal: full hit of brake pedal (triggers transition to manual) On steering wheel: not applicable		Not applicable	
Overtake (combined)	Combine "to left" and "to front" gestures			

5.8.4 Mediation of Driving Manoeuvres

In the general architecture derived for cooperative automation, the machine consists of two parts: an automation and a mediator (Altendorf et al. 2015a, b). The automation detects and proposes all possible manoeuvres, ranking them based on their likeliness to be executed. It gives control inputs to execute manoeuvres that result from the mediation process. The mediator arbitrates between automation and human input (Baltzer et al. 2014) on an operational, tactical (i.e. manoeuvre and trajectory) and strategic layer, as they are described in Flemisch et al. (2019a, b).

The following section describes the machine's understanding of gestures and the mediator parts that combine gesture and automation input and a driving situation to a driving manoeuvre. From the driver's perspective, the mediator connects gestures to manoeuvres.

The design of the mediation of gesture input is based on an interaction pattern between human and machine. A generalized pattern is displayed in Fig. 5.18. While the machine performs a manoeuvre M1 and another manoeuvre is possible, this is constantly, but unobtrusively proposed to the human through any human–machine interface (HMI). The human may input a gesture at any time, which is then evaluated by the machine (Fig. 5.18, Phase 1) and either accepted (because the gesture could be connected to a proposed manoeuvre; Fig. 5.18, Phase 2.1) or rejected, if the gesture was either not complete or cannot be connected to a proposed manoeuvre (Fig. 5.18, Phase 2.2). The human is given feedback on any action of the machine. While the machine executes a manoeuvre based on a gesture input of the human, the human may abort the execution with a gesture in the opposite direction of the gesture that triggered the execution (see Fig. 5.18, Phase 3). After executing a manoeuvre, the automation returns to keeping its lane and another manoeuvre may be triggered (Fig. 5.18, Phase 4).

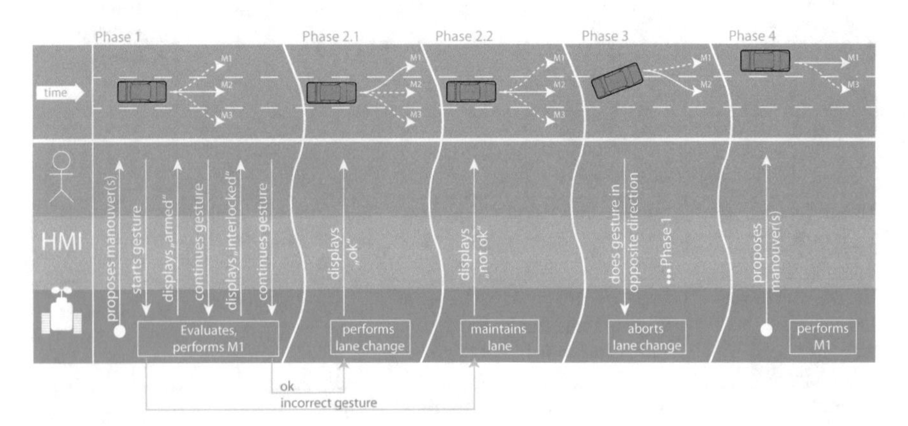

Fig. 5.18 General design of the pattern for manoeuvre-based driving through gestures. Human and machine communicate through any HMI

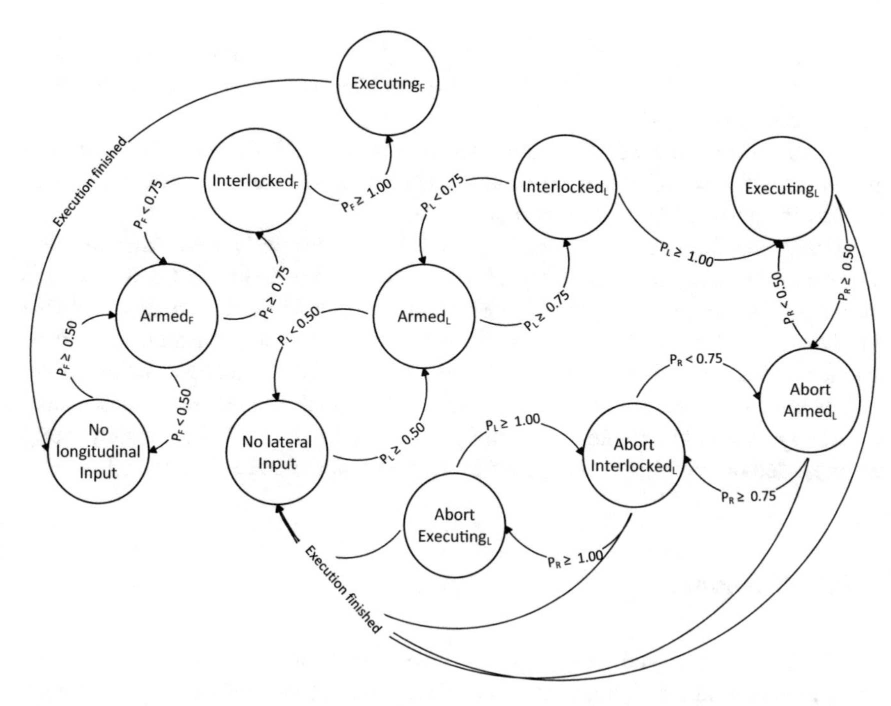

Fig. 5.19 State machine of gesture states and mediator processes for gesture input "to left" and "to front". States for gesture input "to back" and "to right" are ordered, respectively

Based on this, general as well as specialized patterns for each basic use case, the design of the mediation process and its connection to the automation were created.

Figure 5.19 depicts how the mediator processes gesture input of the human for the cases of "to front" and "to left". States for gesture input "to back" and "to right" are ordered, respectively. In general, the mediator receives gesture and automation input, gives feedback and checks which manoeuvre to connect to the gesture and whether it is possible for the automation to execute (in partially automated driving mode) or whether it is possible and legal (including safe) for the automation to execute (in highly automated driving mode). When no gesture input is registered and no manoeuvre is executed, the mediator idles in the "no input" state. When a gesture progress rises above $P \geq 0.5$, the respective gesture is considered armed. Therefore, the mediator switches to "armed" state and so on. When a gesture progress decreases before a manoeuvre is executed, the mediator switches back. Lateral manoeuvres are considered more complex than longitudinal ones. When a lateral manoeuvre is executed, it can be aborted by inputting the opposite gesture. Whenever the mediator is in an "executing" or "abort" state and the current manoeuvre is completed, the mediator switches to "no input" state regardless of human input. When the human inputs a gesture at that exact time, it triggers a new "armed" process. In each state, the mediator gives feedback to the human through the human–machine interface. Note that the mediator does not necessarily reside for a given time in any state; i.e.

it may happen that the mediator switches from "no input" to any "executing" state faster than the human can notice any feedback whenever gesture progress increases (or decreases) very fast.

In the process of mediation, feedback is important. Which is why the user is presented with a set of possible manoeuvres as well as the current execution state of the driver's gesture(s) and current manoeuvre.

At any time, the user may input a gesture, which is then considered by the mediator. The user gets feedback on any progress of the gesture, i.e. feedback when gesture is armed, interlocked, in execution or rejected. In the execution state, the most probable manoeuvre is forwarded to automation. While executing, any manoeuvre may be aborted by the user. This is achieved by inputting a gesture in the opposite direction.

For the basic manoeuvre set, lateral and longitudinal gestures may be mediated separately, as it was implemented for this project (see Fig. 5.19). When implementing an expanded or combined set of gestures, however, both need to be merged.

5.8.5 *Summary*

A basic gesture catalogue with a generic core of logic applied to two different gesture sets (push/twist and swipe gestures) was developed. These gestures cover the need for an easy-to-use and slim design, at least from the designer's perspective. To be able to satisfy legal and technical needs if and which manoeuvre the vehicle should execute based on a gesture input, a human–machine mediation was designed.

Both gesture sets and the human–machine mediation were implemented and evaluated in both partially and highly automated driving modes (see Sects. 5.9 and 5.10).

For the implementation of swipe gestures, a new touch-sensitive steering wheel was developed by Valeo, which is described in Sect. 5.9.

5.9 The Valeo Steering Wheel as an Innovative HMI for Steering Gestures

5.9.1 *Introduction*

Sections 5.1–5.5 explained how the steering gesture concept was scientifically derived based on exploratory workshops and user tests. This section explains Valeo's technical implementation of a human–machine interface for the concept of manoeuvre-based driving. There are three major needs the interaction concept targets and challenges the technical implementation:

1. Offer a self-determined driving experience for users by providing an interface that allows users to input steering gestures in a cooperative way, avoiding a complete deactivation of the automation.
2. Gain and calibrate users' trust in assisted and automated driving functions to ensure a high frequent use of these systems.
3. Easy and lean integration to ensure the path to industrialization.

5.9.2 Interface for Steering Gestures

Besides the three general needs mentioned above, two major functions are set as requirements for the development of an interaction concept for cooperative manoeuvre-based driving:

1. Smooth transition from manual to automated driving and vice versa.
2. Detection of steering gestures to perform manoeuvre-based driving while avoiding unintended use.

For the purpose of detecting steering gestures, it is required to recognize a moving or swiping hand from one certain position of the steering wheel to another. Furthermore, the aim is to not limit the area of sensitivity, which means that it should not make any difference, whether a motion takes place more on the outer or inner ring of the steering wheel or even tendentially at its backside. Besides this, the airbag area is used for detections.

All such detections could theoretically be done by a camera, but it is difficult to manage a clear line of sight from the camera to all inclined parts of a steering wheel.

Even if few potential suitable positions in the vehicle were identified, a camera with depth perception would be needed to have precise detections. Another difficulty of using a camera-based system is the light condition. In order to be independent from external light conditions such as daylight, an active illumination would have been needed to be installed, which would come along with high system costs. For this purpose, camera-based systems have been deselected in the concept development and sensors directly installed into the steering wheel were preferred.

In terms of directly installed sensors in the steering wheel, there are two different basic technologies fulfilling the project requirements mentioned above: force-sensitive and capacitive. Capacitive sensing supports a short reaction time and a high granularity in the sensor segmentation. This goes directly hand in hand with resolution. A high resolution supports shorter gestures, shorter travel-ways and a more accurate detection. Regarding failure compensation algorithms, small sensing segments bring along advantages.

The capacitive system architecture and signal processing can be kept rather simple. This is valid for force-sensitive detection as well, but the fact that a small but recognizable force has to be applied during the whole gesture motion was assessed as unfavourable.

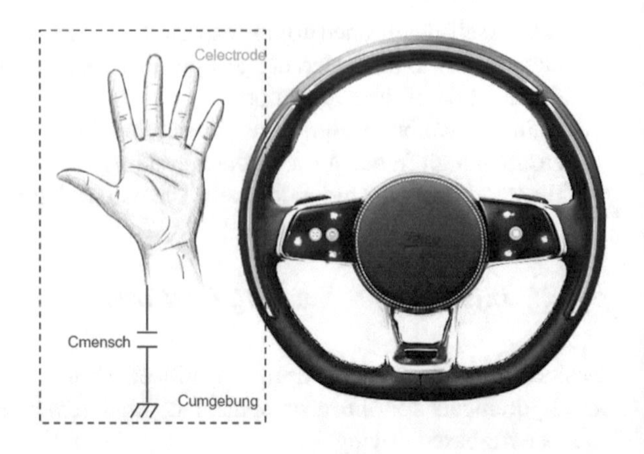

Fig. 5.20 Principle of capacitive touch detection

Furthermore, there are various influencing factors which need to be considered by the detection design for swipe gestures on the steering wheel. The following three factors are perceived as crucial to detect a hand or touch on the steering wheel:

1. Human—size of hand and fingers.
2. Vehicle—steering wheel size and rotation of wheel during driving.
3. Environmental factors—humidity and temperature.

Considering these influencing factors, the preferred technology to detect swipe gestures on the steering wheel is capacitive touch as this offers sensing of the haptic and/or tactile interaction between hand and steering wheel. As a cost-sensitive solution, it is beneficial for industrialization purposes. Figure 5.20 shows the principle of capacitive touch detection.

In general, the capacitive touch electronic control unit (ECU) is able to measure a change in capacitance. When there is no hand next to the sensor electrode, the ECU only measures the capacitance of the electrode to the environment ($C_{\text{environment}} = C_{\text{electrode}}$). By bringing the hand closer to the electrode, the capacitance of the environment increases ($C_{\text{environment}} = C_{\text{electrode}} + C_{\text{human}}$). This shift can be detected and brought into relation of two fingers up to two hands touching the wheel.

Considering the defined swipe gestures in Sect. 5.5, to reliably detect these gestures, it is needed to differentiate intended and unintended gestures. This avoids a mis-triggering on the one side (gesture was not recognized) or a false triggering of a driving manoeuvre (gesture was recognized, but not intended by the driver). To improve the prototype and overcome the challenge of false- and mis-triggering, a design-thinking workshop (see Fig. 5.21 left) was organized by Valeo and the project partners. Based on the results of the workshop, final decisions on the number of touch-sensitive segments as well as the layout of the segmentation were made (see Fig. 5.21 right).

This configuration allows to gain more detailed information on the position and movement of the driver's hands. The touch detection is activating as soon as the driver swipes over these segments. This way you can interpret related swipe gestures and avoid false- or mis-triggering.

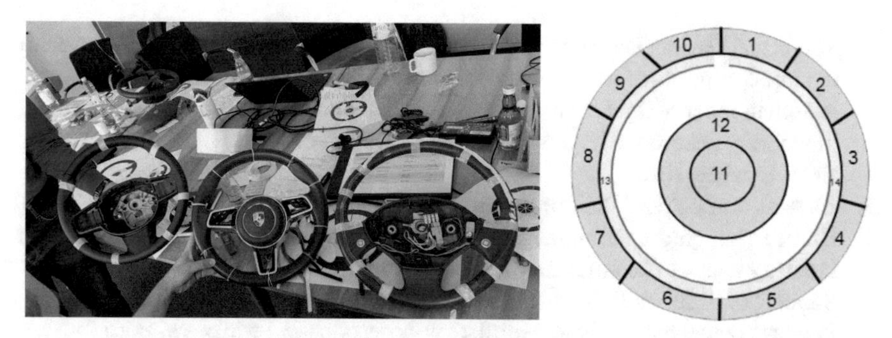

Fig. 5.21 Exploration of touch-sensitive zones on a steering wheel in a design-thinking workshop and final zone layout

Ten touch-sensitive segments are integrated around the steering wheel rim in an equidistant manner. This ensures a reliable detection even with or during rotation of the steering wheel.

With two additional segments on the backside of the steering wheel rim, a full grasp of the hand can be detected used to recognize the availability of the driver to take back control (from automated to manual driving).

By using the same size of touch-sensitive segments and a symmetrical location to the axis of rotation, the swipe gestures can be detected independently from the rotation of the steering wheel.

5.9.3 *Gain and Increase Trust to Assisted and Automated Driving Functions to Ensure a High Frequent Use of These*

To gain trust in assistance and automated driving functions, the driver needs to be fully aware about the (in-)active and assistance functions as well as the currently performed actions by the vehicle and the environmental conditions (e.g. other road users, pedestrians, traffic signs and lights).

For the development of the interaction prototype (swipe gestures on the steering wheel), the following requirements are considered.

Clear indication which driving mode is active (manual or automated driving).

The indication shall be at the eye level of the driver—potential manoeuvres shall be shown.

Include the related steering gesture.

Request to take back control.

From a technical perspective, the following requirements were derived based on the research and user studies executed with a first-generation steering wheel prototype:

1. The visual feedback area on the steering wheel rim shall be extended to support the input of gestures with a visual feedback.
2. A high number of light sources shall be used to incrementally feed back the input of steering gestures with a consecutive activation of single light sources in the steering wheel rim.
3. At least four colours shall be represented (ideally eight).
4. Due to changing environmental light conditions (day, night, dawn), the light intensity needs to be adjustable (at least 3, optimal 8).
5. Support of different light effects (permanent light, blinking, pulsating).
6. Increase perceived brightness of display-based steering wheel switches.

5.9.3.1 Implementation of Requirements

Lightband

A lightband is integrated into the steering wheel rim, which is supported by 103 RGB LEDs and which can be controlled individually. This is realized by using a new and innovative LED technology.

Various light patterns, colours and intensity of brightness can be controlled with these LEDs.

Display-Based Steering Wheel Switch

The display-based steering wheel switches offer the possibility to freely apply pictures to each of the switches. Potential driving manoeuvres and the related steering gestures can be illustrated.

To enable a smooth appearance in the vehicle interior, a black panel effect was realized, meaning if the switch is not needed, a seamless black surface is visible (the display is invisible).

By using an active haptic feedback on the display-based switch, the driver knows exactly when a function is activated.

Figure 5.22 shows the Vorreiter steering wheel with integrated lightband and the display-based steering wheel switches.

5.9.4 Easy and Lean Integration to Ensure the Path to Industrialization

To ensure that the technologies developed with Vorreiter project find their way to industrialization, it was taken care that major components are already now based on automotive standards.

Fig. 5.22 Vorreiter steering wheel prototype—second generation

Important to ensure industrialization is the flexibility and applicability to various types of steering wheels. To ensure this, the electronic computing unit is integrated in the display-based switches to control the features of the steering wheel (hands-on/hands-off detection and switches). The states of the hands-on/hands-off detection and the display-based steering wheel switches including the lightband can be controlled via controller area network (CAN), which is a standard automotive communication interface and which ensures a high-speed communication between the steering wheel and the rest of the vehicle.

Figure 5.23 shows the bidirectional communication interface of the steering wheel.

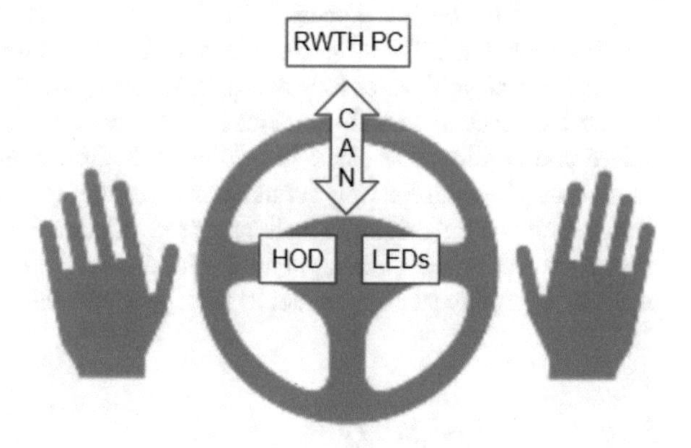

Fig. 5.23 Communication interface—Vorreiter steering wheel

5.10 Experimental Results for Swipe Gestures Versus Joystick in a Virtual Reality User Study

5.10.1 Introduction

Based on the general direction of the Vorreiter project described in Sect. 5.1 and the generic gesture catalogue in Sect. 5.8, we compared a capacitive steering wheel for swipe gestures (see Sect. 5.1) and a four-way sidestick ("joystick") for push gestures from Valeo, supported with the gesture recognition algorithm provided by RWTH-IAW described in Sect. 5.11—and explored the futuristic possibility of a vehicle with a holographic steering wheel ("holowheel") in two user studies in VR. We used driving scenarios representing different motivations to change lanes. In the first study, 17 participants evaluated the devices with single- and double-lane changes. In the second study, 16 participants (6 with age simulation suit) executed single-lane changes. The joystick was perceived as significantly more usable and intuitive and physically less demanding in both studies, but the participants clearly favoured the visual feedback of the two steering wheels over the joysticks.

5.10.2 Hypotheses

We expected a difference between steering wheel and joystick regarding intuitive use and regarding usability. The selection of those input devices follows a hypothetical technical development path of automated vehicles' interior design and incorporates the results of the design process in the Vorreiter project described in Sects. 5.4 and 5.5 (Sommer et al. 2019). Steering wheels are and will remain the predominant input devices for road vehicles as long as manual driving is required for a noticeable amount of time. However, on long distance automated drives a joystick is ergonomically better placed and requires less space in the car packaging. It might become a second and later on maybe even the only input device for driving. Physical input devices might disappear if manual driving is no longer predominant. For those cars, our holographic steering wheel is a possible input device for drivers who still want to influence the car as the holographic steering wheel as metaphor provides well-known input gestures.

5.10.3 Method Study 1

We tested the following experimental set-up and trained the Wizard-of-Oz technique (Green and Wei-Haas 1985) for the holowheel during a pilot study. We used a with-in design and presented the steering wheel and the joystick in counterbalanced order.

Fig. 5.24 VR Lab: physical elements (two seating bucks, HMD and the prototypes)

5.10.3.1 Development and Implementation of the VR Lab

In a workshop with experts for user research and software and hardware implementation, we agreed on using a VR environment that would incorporate the physical prototypes and enable the presentation of a holographic steering wheel. Manoeuvres with strong accelerations and operations with small angles are likely to induce motion sickness (Mourant et al. 2007). Hence, we decided to use the lane change manoeuvre on a straight road and to implement different scenarios that make a lane change probable in our user studies (Fig. 10.24).

To enable four combinations of the devices (steering wheel, joystick, holowheel and steering wheel + joystick), the setting incorporated a head-mounted display (HMD), two car seats and the two prototypes provided by Valeo. The hand tracking software of a Leap Motion allowed us to use virtual models of the participants' hands. As shown in Fig. 5.25, the prototypes are connected to the computers running the simulation software and the gesture recognition via CAN bus. The simulation software reads the touch inputs on the steering wheel and the input commands on the joystick and controls their LEDs. A second CAN bus is used to synchronize the rotations of the virtual and the physical steering wheel.

5.10.3.2 Implementation of the VR Scenarios

Every lane change scenario represented a different motivation to change lanes: long-term use, to avoid pollution, comfort and sense of security. For the long-term use, e.g. to drive on vacation, we created a scenario based on the normed lane change

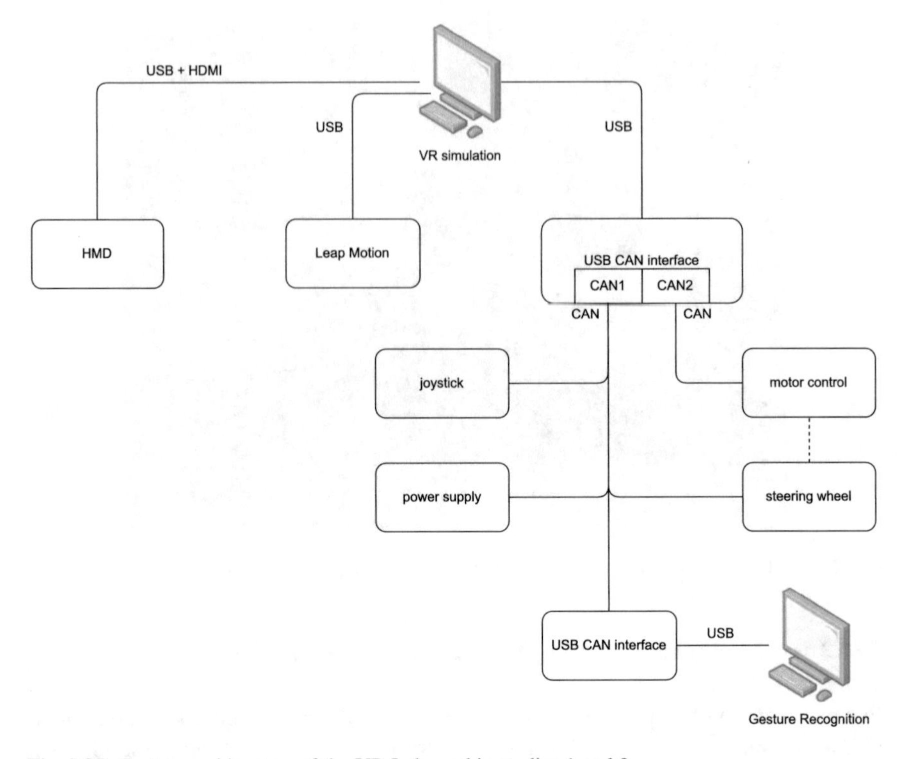

Fig. 5.25 System architecture of the VR Lab used in studies 1 and 2

task (ISO 26022:2010(en) 2010) where the participants had to comply with the signs on both sides of the road. We added a construction site scenario where participants changed lanes to avoid being soiled by the excavator; see Fig. 5.26. For the dimension comfort, we created a scenario where the ego-vehicle followed a truck with excessive

Fig. 5.26 Three input devices in VR: steering wheel and joystick (left), and holowheel (right). The construction scenario (left) and the lane change test (right)

exhausts. In the last scenario, the participants encountered an accident situation with two cars and a police vehicle.

If the participants did not change lanes, the vehicle continued its way in the same lane, but the obstacle felt uncomfortably close. In the comparison drive, the long-term scenario had 30 lane changes so the participants could test both physical input devices alternatingly.

Regarding technical implementation, each lane change during the scenarios consisted of a trial separated into three phases: Phase 1 started when the obstacle or sign became visible to the participant. The participant then decided how to respond to the obstacle presented (change lanes or not) and to input their decision. Phase 1 ended when the virtual car reached a "decision point", and Phase 2 started, no further inputs possible. During Phase 2, the response chosen during Phase 1 was carried out. Phase 2 ended when the car completely passed the obstacle, and Phase 3 started. Phase 3 ended when the participant reached the first phase of the next trial. The current trial phase was determined by the distance of the virtual car to the current obstacle.

5.10.3.3 Human–Machine Interface

We used the Valeo capacitive steering wheel described in Sect. 5.9 and the RWTH gesture recognition software for swipe gestures along the steering wheel rim (see Sect. 5.11.2.4). The steering wheel and its virtual twin contained a band of LEDs, and the joysticks contained LEDs as well. The LEDs lit up in magenta in the section that was touched. When the software accepted a gesture, all the LEDs along the steering wheel lit up once in chartreuse (lemon yellow). When a gesture failed (e.g. when it was fragmentary or too slow), the LEDs lit up in subtle red. We used the Wizard-of-Oz technique (Green and Wei-Haas 1985) for the holowheel implemented by Fraunhofer IAO. The holowheel was represented as a light blue and almost transparent steering wheel.

5.10.3.4 Sample and Procedure

$N = 17$ drivers (9 male) participated in the first study, their mean age was 34.0 years ($SD = 11.79$, $min = 18$, $max = 55$), and the precondition was a valid driving licence. The participants were recruited using the Fraunhofer IAO database and received financial compensation. The participants were instructed to evaluate three different input devices for a fully automated vehicle.

The procedure is shown in Fig. 5.27. The questionnaire consisted of two standardized questionnaires: Questionnaire for Intuitive Use (QUESI; Naumann and Hurtienne 2010) and system usability scale (SUS; Brooke 1996). Furthermore, questions regarding different dimensions of driving (comfort, long-term use and visual feedback) were implemented. We also asked the participants to choose their favourite input device after the comparison drive (steering wheel vs. joystick). Overall, one test run lasted about one hour.

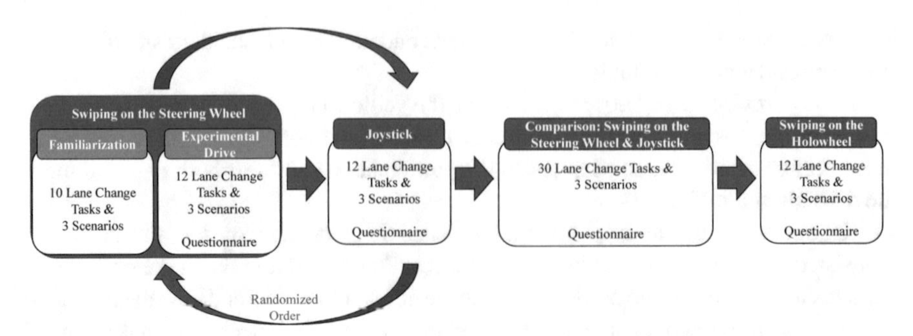

Fig. 5.27 Procedure of the user studies for the comparison of swipe gestures and joystick and exploration of the holowheel

5.10.4 Results of VR Study 1

All values are converted for better interpretability; higher values are always more positive. Normality was tested for the difference between steering wheel and joystick for QUESI and SUS scores with Shapiro–Wilk tests in order to calculate paired-sample *t*-tests.

Intuitive Use. On average, using the joystick ($M = 3.67$, SD = 0.91) is perceived as more intuitive than using the steering wheel gestures ($M = 2.60$, SD = 0.81). This difference was significant $t(16) = -3.80$, $p = 0.002$ (2-tailed) and represents a large effect, $r = 0.69$. In addition, the holowheel was perceived as slightly positive ($M = 3.33$, SD = 0.87).

Usability. On average, the joystick ($M = 70.74$, SD = 23.71) is perceived as more usable than the steering wheel gestures ($M = 51.91$, SD = 25.12). This difference was significant $t(16) = -3.15$, $p = 0.006$ (2-tailed) and represents a large effect, $r = 0.62$. The holowheel lays between the other input devices ($M = 56.47$, SD = 21.78).

Comfort, Long-term use, Visual feedback. The Participants rated the steering wheel ($M = 2.81$, SD = 0.91) as rather neutral, the joystick as more positive ($M = 3.76$, SD = 0.89) and the holowheel as neutral ($M = 3.04$, SD = 1.08) regarding comfort. Participants favoured the joystick for long-term use ($M = 3.63$, SD = 1.08), the steering wheel ($M = 3.00$, SD = 1.11) received neutral and the holowheel ($M = 2.63$, SD = 1.13) rather negative results. The visual feedback of the steering wheel ($M = 3.51$, SD = 0.95) is favoured over the joysticks ($M = 2.86$, SD = 0.92). One participant did not notice visual feedback for the joystick and was therefore excluded from the analysis. The holowheel ($M = 4.10$, SD = 0.88) received especially good ratings for visual feedback.

Direct comparison steering wheel versus joystick and qualitative data. The participants alternately tried out both input devices up until the 17th lane change, and then they also tried out both options, but used the joystick more often. Afterwards, 12 out of 17 participants chose the joystick as favourite input device, 4 the steering wheel and 1 person the "other" option. Those in favour of the steering wheel reported "it reacted more precisely" was "more familiar" (×2), "because of the visual feedback".

Fig. 5.28 Usage of input device during the comparison drive of study 1, $N = 17$. Participants tried out both input devices alternately up until lane change 17, when a trend towards using the joystick seemed to emerge

Reasons for choosing the joystick include ease of use (×6)—e.g. "simple and clear operation", less physical demand (×5) and better feedback (×3) (Fig. 5.28).

5.10.5 Discussion of Study 1

The participants assessed the swipe gestures as below average regarding intuitive use and usability. This might be due to the physical demand of the gesture, as it was rather high compared to the joystick, the recognition software allowed little variations in swiping velocity and time for input, and some participants did not correctly interpret the visual feedback to indicate acceptance or refusal of a gesture. The same pattern was found for the dimensions' comfort and long-term use, as the swipe gestures for both steering wheel and holowheel might have been too physically demanding over time.

5.10.6 Method of VR Study 2

We changed the long-term scenario so that participants did not execute double-lane changes anymore to ensure they evaluated just the interaction and added a paragraph to the instruction to communicate the prototypical nature of the devices more clearly. We used an age simulation suit to simulate elder users.

5.10.6.1 Sample

We had to exclude one participant due to technical issues, resulting in a valid total of $N = 16$ drivers (8 male) for the second study. We asked six younger participants to wear the age simulation suit GERT (Wolfgang 2020) during the test drives to assess the devices for the elderly (as elder participants are prone to motion sickness; Park et al. 2006). The remaining participants' mean age was 37.8 years (SD = 11.42, min = 22, max = 62).

5.10.7 Results of VR Study 2

Intuitive Use. On average, using the joystick ($M = 4.48$, SD = 0.60) is perceived as more intuitive than using the steering wheel gestures ($M = 3.43$, SD = 0.76). This difference was significant $t(15) = -4.37$, $p = 0.001$ (2-tailed) and represents a large effect, $r = 0.75$. The holowheel lays between the ratings of the physical devices ($M = 3.89$, SD = 0.92).

Usability. On average, using the joystick ($M = 84.06$, SD = 12.37) is perceived as significantly more usable than using the steering wheel gestures ($M = 64.38$, SD = 19.75). This difference was significant $t(15) = -3.75$, $p = 0.001$ (2-tailed) and represents a large effect, $r = 0.70$. The holowheel scored better than the steering wheel ($M = 66.72$, SD = 19.78).

Comfort, Long-term use, Visual feedback. Participants rated the steering wheel as neutral ($M = 2.91$, SD = 0.73), the joystick as positive ($M = 4.17$, SD = 0.75) and the holowheel as rather neutral ($M = 3.11$, SD = 0.80) regarding comfort. In the category long-term use, the steering wheel ($M = 3.08$, SD = 1.12) received neutral results, the joystick is seen as positive ($M = 4.10$, SD = 0.87), while the holowheel was assessed as slightly negative ($M = 2.79$, SD = 1.05). The visual feedback of the steering wheel ($M = 4.19$, SD = 0.82) and of the holowheel ($M = 4.19$, SD = 0.70) is received as positive. The joystick ($M = 3.47$, SD = 0.85) is perceived as neutral to positive regarding visual feedback. Six participants did not notice any kind of visual feedback for the joystick at all and were therefore excluded from analysis.

Direct comparison steering wheel versus joystick and qualitative data. The participants alternately tried out both steering wheel and joystick up until the 15th lane change, and then more participants preferred the joystick. Afterwards, 14 participants chose the joystick as their favourite input device, none the steering wheel, and 2 persons would like to use a combination. The 14 participants preferred the joystick due to the ease of use/fun/comfort (×10)—e.g. "is effortless and is more fun", lower physical demand (×9), faster operation (×6), less mental demand (×3), more reliable (×3), clearer how to use (×2) (Fig. 5.29).

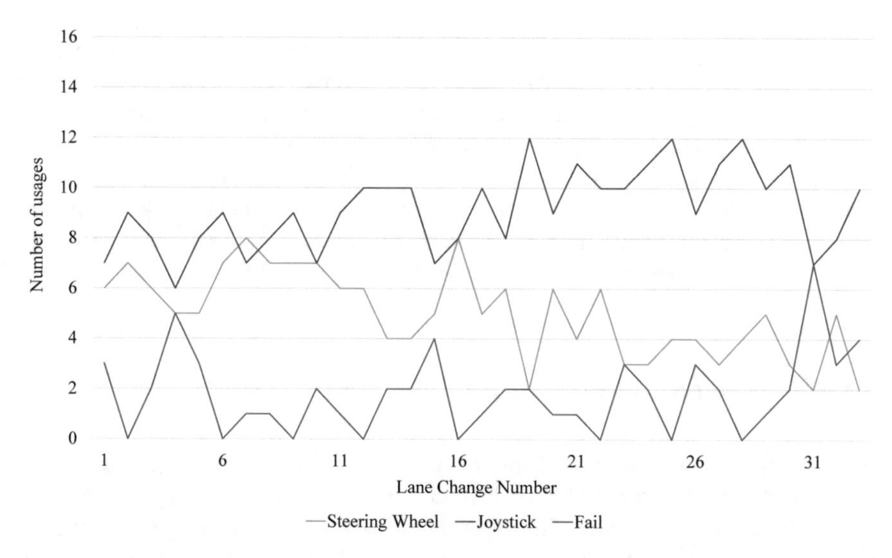

Fig. 5.29 Usage of input device during the comparison of study 2, $N = 16$. Participants tried out both input devices alternately up until lane change 15, and then most used the joystick

5.10.8 Discussion of VR Study 2

To sum up, all input devices scored slightly better than in study 1, probably because the participants could focus particularly on the differences between the input devices. First results were presented at the "Intuitive Partially and Highly Automated Driving" symposium in Aachen (Diederichs et al. 2019a, b).

5.10.9 General Discussion

The participants perceived the gestures on the steering wheel as less intuitive than the joystick, although the steering wheel should be a more common metaphor. Intuitive use is defined as "the subconscious application of prior knowledge (…)" (Naumann and Hurtienne 2010). The participants' prior knowledge of the steering wheel may have interfered with the gestures as they are used to moving the whole steering wheel. The perception of the gestures may change as automated driving becomes more common, because the users may be less used to actively steer with the steering wheel all the time. Participants also expressed uncertainty about how long and in what time they had to swipe to get a gesture accepted. They learned how to execute a valid gesture during the familiarization drive, but the gesture recognition software should accept a broader range of inputs. This is probably the reason why the participants perceived the holowheel as more intuitive than the steering wheel as the Wizard accepted a higher variability. The answers in the questionnaire and the qualitative data suggest that we need to decrease the gestures' physical demand and that we need clearer, more salient way to communicate whether the machine accepted the swipe gesture, especially for the joystick: visual feedback on gesture acceptance to the head-up display (HUD) as described in the following section or earcons or hybricons (Diederichs et al. 2010) could help. Overall, the participants in these studies preferred the joystick to communicate manoeuvre wishes to the driving automation, although a steering wheel might be a more familiar input element. More research is needed to compare different means of communicating the wish for a different manoeuvre.

5.10.10 Qualitative Trial with Expert User

Additionally, we conducted a qualitative test run with a user that uses an electric wheelchair daily and currently uses human assistants or public transport, but plans on obtaining a driving licence. The expert reported that he liked the joystick better as it was less exhausting and very similar to the control element of his wheelchair. The expert estimated that 70–80% of persons with the same physical disability as him could profit from the Vorreiter concept and would gain more autonomy, freedom and flexibility in daily life.

5.11 Experimental Results for Push/Twist Versus Swipe Gestures on a Steering Wheel in Partially and Highly Automated Driving

5.11.1 Introduction

While the last section compared conventional with unconventional HMIs for manoeuvre gestures, it can be assumed that the steering wheel will continue to be the most widely used input device in the coming years (see Sect. 5.1). The following study should therefore compare the two families of manoeuvre gestures (push/twist versus swipe) applied to the two levels of automation SAE level 2 and SAE level 3/4 (partially and highly automated) on a steering wheel and pedal in a static driving simulator. In contrast to the study described in Sect. 5.10, it addresses a wider range for traffic situations, e.g. with crossroads, motorway and rural roads. A special focus was on the implementation of the gesture controls and the automated system.

5.11.2 *Method*

5.11.2.1 Driving Simulator and Scenarios

The study was carried out in the static driving simulator in the Exploroscope of the Institute of Industrial Engineering and Ergonomics (IAW) at the RWTH Aachen University (see Fig. 5.30).

The driving simulator software used was SILAB 6.0 developed by Würzburger Institut für Verkehrswissenschaften GmbH (WIVW). The capacitive steering wheel by Valeo (see Sect. 5.9) and a gesture recognition software by IAW, RWTH (see Sect. 5.11.2.4), were used for vehicle guidance.

Eight driving scenarios were selected, which addressed the use case catalogue described in Sect. 5.2 and enabled the selected gestures (see Sect. 5.2.4.1). The used driving scenarios were: city centre, country road with X-junction, motorway entrance, motorway, road work scenario, motorway exit, country road with T-junction and finally a city centre with a parking bay again. This route was driven with both gesture controls (push/twist and swipe; see Sect. 5.8) in both levels of automation (SAE 2 and 3/4) and in the baseline (manual driving). The SAE level 2 system was designed to brake automatically and keep the lane. Every other driving task such as accelerating, lane changing and turning manoeuvres had to be performed by the driver. It was also necessary to keep at least one hand on the steering wheel. If the driver took his or her hand from the steering wheel for too long, a hands-on request was given. Furthermore, the driver was always fully responsible for the driving task and had to be able to take over the vehicle guidance at any time. The highly automated driving system corresponded to a SAE level 3/4 and had all the functions of the SAE level 2 system, plus the ability to change lanes and take a turn on its own, and

Fig. 5.30 Static driving simulator in the Exploroscope of the IAW

to adhere to the road traffic regulations, such as slowing down at a speed limit or adhering to the keep-right regulation. In the higher automation levels, the driver was not obligated to monitor the driving task.

5.11.2.2 Human–Machine Interface (HMI)

In manual driving (SAE level 0), the driver receives instant feedback and has a direct impact on the vehicle and the environment. However, when using steer-by-wire steering gestures, this direct feedback is not automatically available: depending on the level of automation and in combination with gesture controls, there is no longer a direct coupling of the steering wheel with the steering. Consequently, if non-executable gestures are entered by the driver, these do not result in a direct reaction of the vehicle. For a good driving experience, however, seamless feedback on every action of the driver is essential. Therefore, in the study carried out, special emphasis was placed on feedback and the HMI. This was achieved by using the Valeo capacitive steering wheel with LED strips (see Section "Lightband") and the head-up display (HUD).

In this study, the feedback from the HMI depends on the gesture type and the level of automation and was designed according to the gesture catalogue described in Sect. 5.8.2. A main aspect of the feedback was the trajectories shown in the HUD (see Fig. 5.31). Here, the current trajectory is highlighted in the colour of the current level of automation (blue for SAE level 2 and magenta for SAE level 3/4) and indicates the currently selected future direction of movement. Other detected and possible trajectories are shown in grey. There are transition trajectories, which, e.g., connect adjacent lanes (Fig. 5.31 right) or help to decide on turning manoeuvres at crossroads (Fig. 5.31 left).

The left part of Fig. 5.31 depicts a crossroad situation in partially automated driving (SAE level 2), where the driver entered a swipe gesture to turn right at the upcoming crossroad. As feedback to the driver's input, the selected trajectory leading to the right is highlighted in blue. On the right is a motorway situation in highly automated driving mode, in which the vehicle changes from the middle to

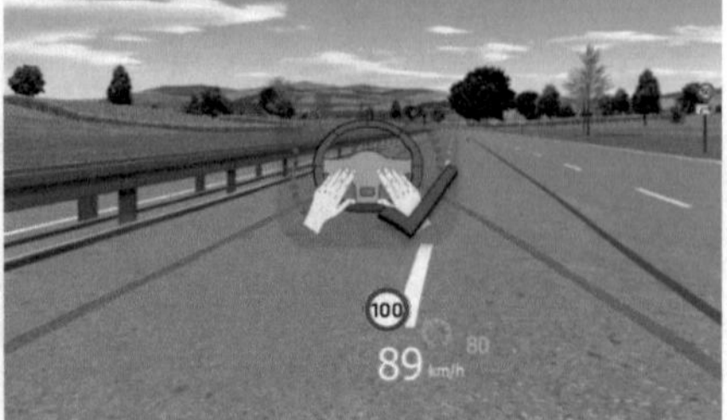

Fig. 5.31 Visual feedback in partially automated driving while turning right (left) and in highly automated driving while changing lanes and decelerating (right)

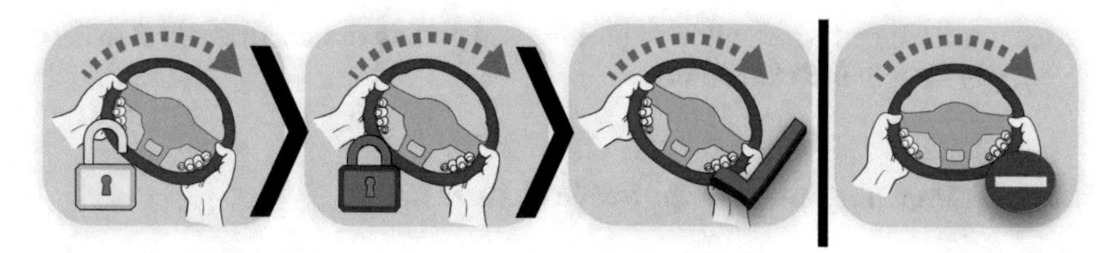

Fig. 5.32 Example set of twist gesture input feedback in the HUD

the left lane. The feedback in the HUD differs from the feedback in the partially automated driving mode by the colour of the trajectory (magenta).

In addition to highlighting the trajectories, a feedback symbol is displayed when a gesture is made, e.g. "to right". The selection feedback varies in four different steps. When the gesture is armed, this is indicated by a grey open lock. If the gesture is continued, it reaches an interlocked state, represented by a coloured lock. When the gesture is accepted and a manoeuvre is executed, a coloured check mark is displayed. However, if a gesture is rejected because, for example, a lane change is not possible, a no entry symbol is displayed (Fig. 5.32, from left to right).

The feedback symbols are similar for all gesture concepts. Depending on the level of automation, they are displayed either in blue (SAE level 2) or in magenta (SAE level 4). In addition to the feedback icons, the HUD displays the current speed in white, the set target speed in cruise control in green and the speed limit as a traffic sign.

5.11.2.3 Implementation of the Push/Twist Gesture Set

The idea of the push/twist gestures is to steer the machine in the desired direction in the same way as a driver would initiate the manoeuvre with a conventional steering wheel. Thus, the input of the "to left" or "to right" gestures is done by a slight rotating impulse on the steering wheel in the desired direction. The push gestures on the braking or accelerator pedal ("to back" or "to front" gestures) were implemented as described in Sect. 5.8.

Both devices were parameterized by a threshold value at which the activation of the gesture is defined as executed (which also implicitly defined the thresholds for being armed and interlocked).

The implementation of the gestures at the steering wheel followed the approach used for the pedals. The steering wheel was also parameterized by a threshold value that defined when the gesture "to right" was triggered due to the steering wheel being turned clockwise and when the gesture "to left" was triggered by the steering wheel being turned counterclockwise. The progression of the gesture was calculated as a fraction of the current steering wheel deflection to the gesture execution threshold. For the execution of longitudinal gestures on the steering wheel, a new pushing device was designed which allowed pushing and pulling the steering wheel to detect gestures in longitudinal direction.

5.11.2.4 Implementation of the Swipe Gesture Set

The implemented swipe gestures allowed the driver to enter the desired driving direction by swiping over the steering wheel rim without actually turning the steering wheel. The swipe gestures designed in Sect. 5.5 and represented in the gesture catalogue of Sect. 5.8 were implemented on a capacitive steering wheel developed by Valeo Schalter und Sensoren GmbH (see Sect. 5.9). The swipe gestures are recognized as a pattern of touched areas around the steering wheel in a clockwise or counterclockwise direction to detect movements in the right or left direction. Increasing the number of zones that can be detected results in a finer detection resolution. However, the smallest number of zones available is the number of zones that can be touched with one hand at a time, plus three more to detect the three feedback levels of a gesture input (armed, interlocked and execution). These patterns cover the "to right" and "to left" directions.

The directions "to front" and "to back" can be covered by introducing two-handed gestures. If the driver swipes upwards with both hands on the outside of the steering wheel, a "to front" gesture is initiated. Swiping down in the opposite direction initiates the "to back" gesture. Although all manoeuvres could be performed using the swipe gestures on the steering wheel, other input devices can also be used. For example, as with the push/twist gestures, the pedals can be used as input devices alongside the Valeo steering wheel. In this case, the device registers the gesture input regardless of the device from which it originates.

5.11.2.5 Non-driving Related Task (NDRT)

In this study, the surrogate reference task (SuRT) was used as a non-driving related task (NDRT). The SuRT is a visual–manual task in which the participants have to identify a single larger circle from a large number of circles of equal size. The selection was made using the arrow keys on a number pad, which was mounted in front of the centre armrest. During the test drives, the participants had the opportunity to carry out the NDRT. In SAE automation levels 0 and 2, they were instructed that they always were fully responsible and should act as they would in real road traffic. Carrying out the NRDT during these rides represented a misuse. In SAE automation level 3/4, the participants were also offered the NDRT but were instructed that the automated driving system would be in charge during an automated ride.

5.11.2.6 Sample and Procedure

$N = 26$ subjects participated in the study (26.9% female and 73.1% male). A valid driving licence and no uncorrected visual impairment were required to participate. The age of the test persons was between 19 and 64 years ($M = 28.96$ years, SD = 13.25 years). The participants were recruited using the database of the IAW, RWTH, and received financial compensation.

After a short welcome, the participants signed a declaration of consent and a confidentiality agreement. The participants then completed a socio-demographic questionnaire and took a vision test. Afterwards, the participants entered the simulator and received a short general introduction. The test procedure was also explained to them at this point (Fig. 5.33).

Each test run started with an introductory run (SAE level 0) to familiarize the participants with the simulator and the driving characteristics. The test subjects were also told that they would complete five test cycles, which always start with a naive run, followed by a short questionnaire, an explanation, a trained run and an associated questionnaire. The naive runs were done without prior instruction in order to assess the intuitive comprehensibility and the learnability. Afterwards, the participants adjusted their seating position and started the introductory run after the eye-tracker (Smart Eye Pro 3D) was calibrated. During the test drives, the participants had the opportunity to perform the NDRT (see Sect. 5.11.2.5). After the test block, a subsequent questionnaire was completed. Finally, the last test drive in the previous automation level was carried out. During this run, the automated driving system failed to respond to a stop sign. Shortly after this situation, the drive ended and the participants received a final questionnaire and their financial compensation. This was the end of the experiment. Each test run lasted about 3.5 h per participant.

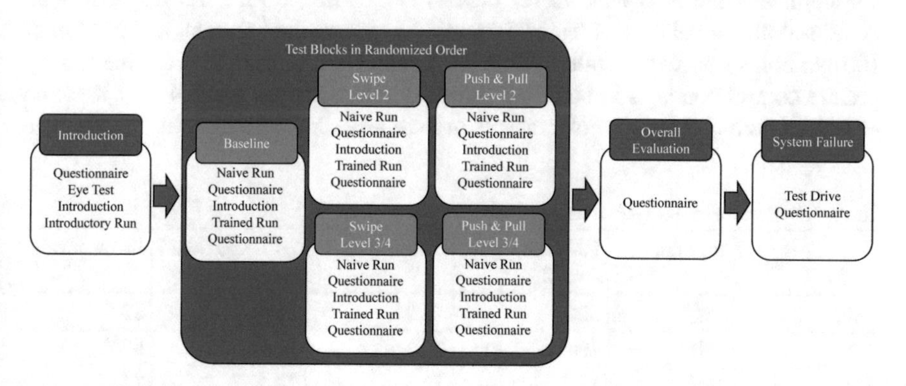

Fig. 5.33 Test procedure

5.11.2.7 Questionnaires

The study was implemented using the online survey tool Questback (Stangnes and Blekastad, 2015). It included an opt-in privacy and data protection statement and a socio-demographic part. Afterwards, the test block started. Here, after each naive run, a short questioning took place to find out how and if the participants had understood the gestures. After a short explanation and the subsequent trained run, the actual evaluation of the respective steering concept followed. In this part of the questionnaire, the dimensions of the extended Devil's Square (see Sect. 5.2.1) were recorded, mainly through validated instruments (such as NASA-TLX and system usability scale).

5.11.3 First Results

In the context of this publication, first results from the study are presented below. This includes the results of the system usability scale (SUS), the NASA-TLX and extracts from the overall evaluation. Further results from the overall evaluation, e.g. on intuitivity and learnability, as well as on the dimensions of the extended Devil's Square, will be published in further publications.

5.11.3.1 System Usability Scale (SUS)

The SUS is a technology-independent questionnaire for evaluating the usability of a system. A value of at least 72 (or higher) is a reference value for a system with good usability, a value of at least 85 represents an excellent system, and a value of 100 symbolizes perfect usability. Table 5.4 depicts the ratings of the baseline and the gesture control concepts in both SAE levels of automation 2 and 3/4. The baseline as well as both gesture control concepts in SAE level 3/4 were rated as "good". The

Table 5.4 Results of the system usability scale

	Baseline	Push and twist level 2	Swipe level 2	Push and twist level 3/4	Swipe level 3/4
Min	55.0	32.5	15.0	50.0	52.5
Max	100	100	82.5	100	97.2
SD	10.2	17.8	21.5	12.3	12.2
M	81.6	70.0	51.6	79.2	76.8
SUS-score *M*	=100	≥85.58	≥72.75	≥52.01	≥39.17
Results	Best imaginable	Excellent	Good	OK	Poor

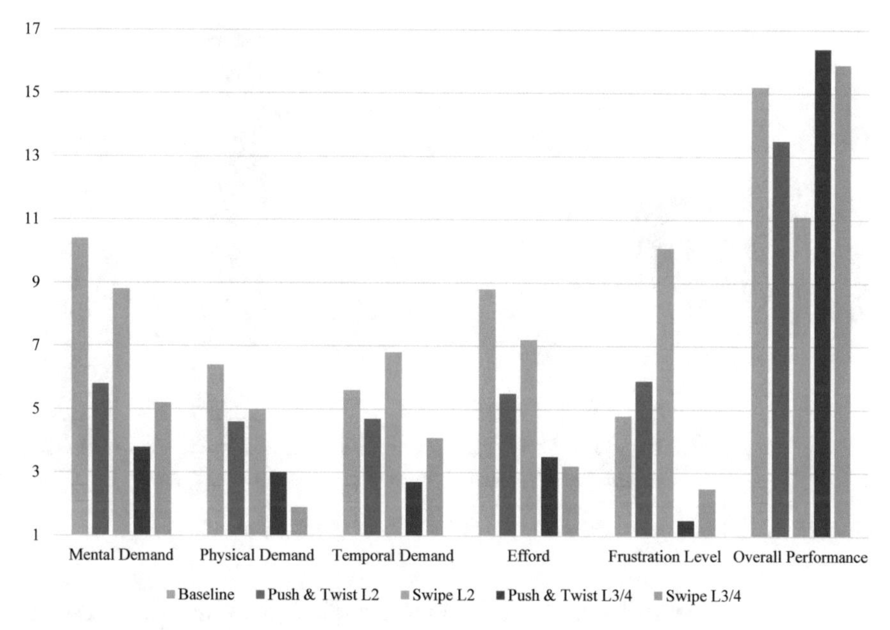

Fig. 5.34 Subjective ratings of the NASA-TLX

push/twist gestures in SAE level 2 are only just below the limit with a rating of 70. The swipe gestures in SAE level 2 are still rated OK with 51.6.

5.11.3.2 NASA-TLX

Figure 5.34 shows the results of the NASA-TLX. The scales "mental demand", "physical demand", "temporal demand", "effort" and "frustration level" represent negatively polarized items. High values in these scales also mean, for example, high frustration with the system. It shows that the baseline was rated highest in three of the five negative scales (mental demand, physical demand and effort). It is also noticeable that the gesture control concepts in SAE level 3/4 have lower values in all negative scales than in SAE level 2. Also notable are the ratings for the swipe gestures in SAE level 2, which are above those of the other gesture control concepts on all five negative scales. In the "overall performance", the SAE level 2 swipe gesture also achieves only the lowest rating.

5.11.3.3 Overall Preferences

Figure 5.35 depicts the subjective preferences and the legal admissibility of the steering gestures (see Sect. 5.7.4.2). As part of the overall evaluation (see Fig. 5.33), each participant could indicate which of the systems he or she preferred. Multiple

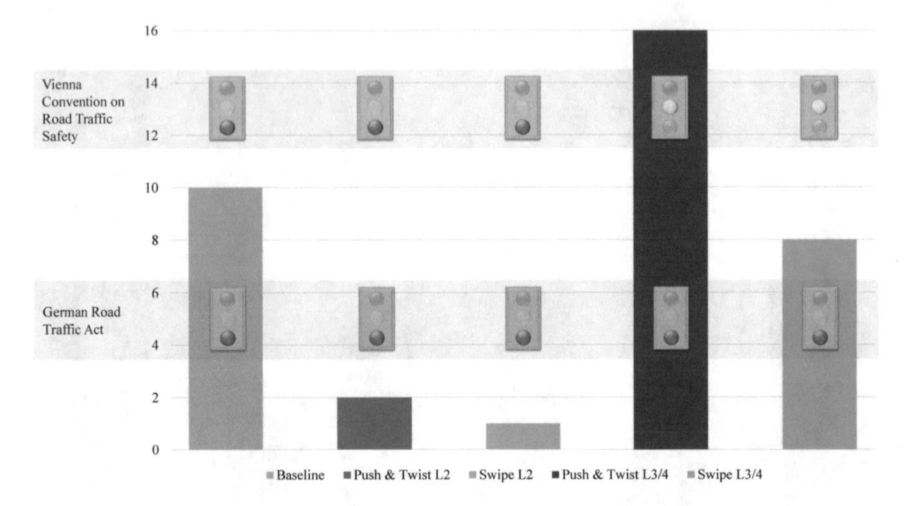

Fig. 5.35 Subjective preferences and legal admissibility of steering gestures

answers were also possible here. Similar to NASA-TLX, there are clear differences between the gesture control concepts regarding the level of automation. Regardless of the level of automation, push/twist gestures were preferred over swipe gestures. The push/twist as well as the swipe gestures in SAE level 3/4 were preferred as often or even more often than the conventional steering system in the baseline.

5.11.4 Discussion

The ratings in SUS showed that both level 3/4 steering concepts were rated as good and even only slightly lower than the baseline condition. These results demonstrate impressively that gesture control concepts and manoeuvre-based systems could be a real alternative to conventional steering systems in future. The systems in SAE level 2 still performed "OK" with values between 51.6 and 70, but still indicated clear difficulties. One challenge with the SAE level 2 systems was the fact that at least one hand on the steering wheel was required. This sometimes resulted in a mix up of the manoeuvre and the stabilization task. In the case of swipe gestures, this led to an error in parallel gesture input while holding the steering wheel. In this case, a two-handed gesture input was falsely detected, which often led to an error feedback (see Fig. 5.32). But also the push/twist gestures in SAE level 2 had to struggle with some difficulties: sometimes, a steering gesture input which was too gentle led to a situation where you stayed below the threshold value and thus changed lanes without actually triggering the corresponding gesture.

On the positive side, it should be emphasized that in highly automated driving, in the evaluation by NASA-TLX, the steering gestures scored even better than the baseline in the overall performance. However, the NASA-TLX also clearly reveals the difficulties with the swipe SAE level 2 steering system, which also received the lowest rating in the SUS. The results of the overall preferences reveal that the drivers' preference is either for manual or highly automated driving. These high preference values for manual driving and driving in SAE automation level 3/4 indicate that drivers still want to intervene in the driving task despite highly automated driving functions.

In summary, it can be said that manoeuvre-based driving using steering gestures was preferred by the test persons in highly automated driving. In contrast, the implementations of partially automated and coupled systems were rather rejected by the participant. As a limitation, it should be mentioned that the presented study is an experiment in a static simulator. For this reason, a number of driving parameters cannot be simulated and there is for example no feedback on road conditions or speed via the steering. Here, a real traffic study could provide important additional results, as the haptic feedback is highly relevant to the stabilization level.

As shown in the following section, the steering concepts were not only discussed with mass-market drivers but also with drivers with disabilities. These exploratory studies to derive further requirements are currently still ongoing.

5.11.5 Qualitative Study with Driver with Disability

To investigate if the developed gesture controls can actually apply to various users with different needs and limitations (design-for-all), we also performed a test run with a paraplegic person who uses a wheelchair. The participant was a 27-year-old male with a valid driving licence, an annual mileage of 6000–12,000 km and previous experiences with adaptive cruise control (ACC) and automated parking assistants. Due to the strict limitation of his leg movements, we could not test the push/twist gestures, neither on the steering wheel nor on the acceleration pedal. Therefore, we conducted test runs with the swipe gestures in SAE automation levels 2 and 3/4 and in the baseline condition. The subject reported small difficulties while driving, because the system did not recognize all of his gestures. Apart from that, he pointed out that the steer-by-wire concept of SAE level 3/4 is superior to the larger steering wheel movements in SAE level 2 and baseline since a decent feedback is enough to understand the actions of the automated driving system. In summary, he stated that the tested gesture controls could be a great relief for drivers with disabilities. Further studies with drivers with disabilities will be carried out in the near future to derive their requirements for automated driving and gesture control systems.

5.12 Outlook: Real Vehicle Testing Platform for and Towards a Standardized Catalogue of Cooperative Steering Gestures

The Vorreiter project was dedicated to developing input devices and input gestures to cooperatively interact with automated vehicles in different levels of automation. The explorative approach (see Sects. 5.2, 5.4 and 5.5; Kaschub et al. 2018; Sommer et al. 2019) resulted in a generic catalogue of steering gestures, in implementations of push and twist gesture interactions via the steering wheel (Sects. 5.8 and 5.11), swipe gestures on a specifically developed touch-sensitive steering wheel (Sects. 5.9–5.11), push gestures on a joystick input device and a virtually presented steering wheel (holowheel) (see Sects. 5.10). The driving concept and steering gestures on the steering wheel have also been assessed for legal admissibility under international, European and German law (Fig. 5.36).

The different interaction devices and interaction principles have been investigated within the Vorreiter project for different automation levels in driving simulators. With the described experiments, there are strong hints that push/twist gestures are suitable for SAE level 2 (SAE 2018) interactions and, if they do not immediately influence the vehicle position and are not a takeover, also in Level 3 or 4, while swipe gestures are relevant for interacting during SAE levels 3 and 4 driving. Joysticks apparently show high potential for manoeuvre-based driving when steering wheels become obsolete and the seating position in the car is not optimized for driving anymore. The holowheel (see Sect. 5.10) may become an option in vehicles that drive almost all times without the possibility for intervention by a passenger. It might be virtually projected on request.

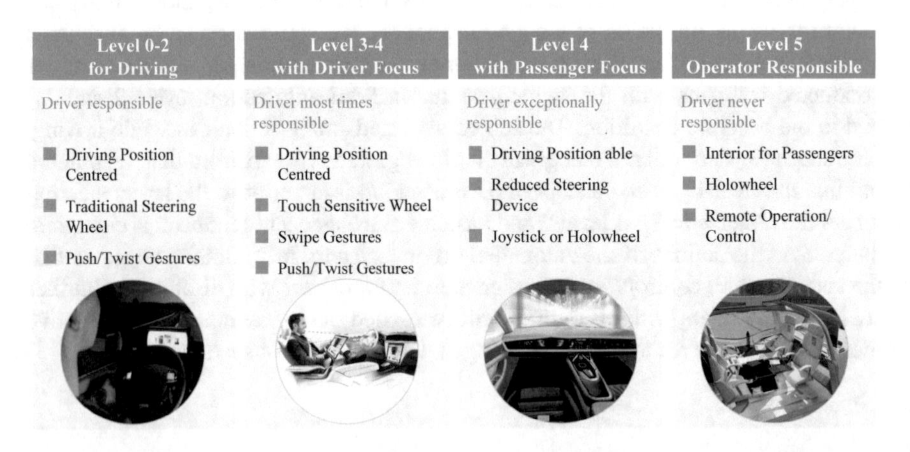

Fig. 5.36 Future passenger car interiors for different levels of automation and different priorities of steering as a principle task in the car. Adapted from Diederichs (2019), based on SAE (2016/2018ff), Gasser et al. (2012), Flemisch et al. (2003/2008); Sects. 5.5, 5.8 and 5.10

In Germany, swipe gestures on the steering wheel and sidesticks used for manoeuvre-based gestures in combination with automated driving are legally admissible as far as the technical equipment can be approved under Law of Approval (which still requires exemptions for new technology and concepts), and the use of automated systems is in line with Law of Conduct. Damages caused while using an automation system allowing for input of gestures are covered by German Liability Law according to the following principles. (i) The vehicle holder is liable in any case, which is covered by the mandatory insurance. (ii) The driver herself/himself is liable while driving partially automated (SAE level 2). Yet, with respect to highly automated driving (SAE level 4), he/she is regularly not liable, except when he/she entered a gesture at the system limit and when push/twist gestures change the situation of the vehicle in traffic (trace offset). On the other hand, the manufacturer is liable if the technology in use results in new safety risks. With regard to chance for people with special needs however, manoeuvre-based gestures in Germany so far do not augment the number of persons entitled to attain a driving licence, based upon the actual status of technology, the Law of Conduct and Driving License Law (Sect. 5.7). Thus, an implementation of admissibility and use of driving gestures in the legal setting would certainly enhance use of such technologies, which however does not depend alone on national law, given its interconnections with international as well as EU law.

All described conclusions rely on the fast implementation by the research partners in the available time; hence, these hypotheses need further investigation especially in more realistic environments and with more industrial development teams. While the push/twist gestures have a long history in research (e.g. Flemisch et al. 2005, Altendorf et al. 2015a, b), the swipe gestures have only been introduced in 2018 for the first time (Kaschub et al., 2018; Sommer et al. 2019). Meanwhile, such swipe gestures have also been demonstrated on the CES Conference 2020 by Honda in the Augmented Driving Concept (Honda 2020). Hence, Valeo strives for industrialization of the touch-sensitive steering wheel.

All developments and testing so far have been conducted in driving simulators. No investigations have been performed in real car driving so far, also due to the fact that safe SAE level 4 road vehicles are not available yet. As a consequence, a Wizard-of-Oz/theatre vehicle has been specified within the Vorreiter project by the consortium (Sect. 5.6). Based on these specifications, the vehicle has been equipped and delivered right at the end of the Vorreiter project by the former partner Paravan GmbH. This vehicle will allow for the first time real road testing of the Vorreiter cooperative gesture interaction (and other use cases).

In the Wizard-of-Oz/theatre vehicle, a Human Wizard (or confederate) is driving the car from the back seat, simulating the automation of the car (all levels from SAE 0-5). The participant on the left front seat perceives an automated car and is either not aware of the wizarded simulation (classical Wizard-of-Oz approach) or is told that there is a human controller in the loop (theatre method; Schieben et al. 2009a, b). A safety driver on the right front seat can intervene and take over control, similar to a driving school instructor. In the Vorreiter Wizard-of-Oz concept, the participant and the Wizard share the steering in a shared and cooperative control approach

Fig. 5.37 General set-up for the Wizard-of-Oz and theatre methods, and the Paravan implementation from the wizard's/confederates point of view (Diederichs et al. 2019a, b)

(Flemisch et al, 2019a, b; Schneemann and Diederichs, 2019). With this set-up and by integrating the different Vorreiter input devices, all Vorreiter cooperative control interaction principles can be tested on the road (Fig. 5.37).

The architecture of the Vorreiter Wizard-of-Oz car is explained in Sect. 5.6. The vehicle is available for further investigations at Fraunhofer IAO in Stuttgart, Germany, and planned to be used by the Vorreiter consortium in follow-up research and development projects.

Besides a further refinement of the gestures and experimental validation in real traffic, another important task of the future will be to further develop the catalogue of manoeuvre gestures in a way that it can be standardized across all vehicles. A standardized visualization concept with technical requirements and potential for personalized automated driving experiences are also important tasks for the future, especially towards the industrialization of hardware and software by Valeo.

References

Abbink DA (2006) Neuromuscular analysis of haptic gas pedal feedback during car following

Altendorf E, Baltzer M, Kienle M, Meier S, Weißgerber T, Heesen M, Flemisch F (2015) H-Mode 2D. In: Winner H, Hakuli S, Lotz F, Singer C (eds) Handbuch Fahrerassistenzsysteme. ATZ/MTZ-Fachbuch, 3rd edn. Springer Vieweg, Wiesbaden, pp 1123–1138

Altendorf E, Baltzer M, Heesen M, Kienle M, Weißgerber T, Flemisch F (2015) H-Mode, a haptic-multimodal interaction concept for cooperative guidance and control of partially and highly automated vehicles. In: Winner et al. (eds) Handbook of driver assistance systems. Springer, Berlin

Baltzer M, Flemisch F, Altendorf E, Meier S (2014) Mediating the interaction between human and automation during the arbitration processes in cooperative guidance and control of highly automated vehicles. In: Ahram T, Karwowski W, Marek T (eds) Proceedings of the 5th international conference on applied human factors and ergonomics AHFE 2014, Krakow, July 2014

Baltzer MCA (2020) Interaktionsmuster der Kooperativen Bewegungsführung von Fahrzeugen—Interaction Patterns for Cooperative Guidance and Control of Vehicles

Berndt S (2017) Der Gesetzentwurf zur Änderung des Straßenverkehrsgesetzes – ein Überblick. Straßenverkehrsrecht (SVR) 2017:121–127

Boren T, Ramey J (2000) Thinking aloud: reconciling theory and practice. IEEE Trans Profess Commun 43:261–278. https://doi.org/10.1109/47.867942

Brooke J (1996) SUS-A quick and dirty usability scale. Usability Eval Ind 189:4–7

Burmester M, Hassenzahl M, Koller F (2002) Usability ist nicht alles - Wege zu attraktiven interaktiven Produkten. I-Com 1:32–40

Cambridge University Press (2020) MASS MARKET | meaning in the Cambridge Business English Dictionary. https://dictionary.cambridge.org/dictionary/english/mass-market. Accessed 24 Jan 2020

Carmel E, Whitaker RD, George JF (1993) PD and joint application design: a transatlantic comparison. Commun ACM 36:40–48

Chihuri S, Mielenz TJ, DiMaggio CJ, Betz ME, DiGuiseppi C, JOnes VC, Li G (2015) Driving cessation and health outcomes in older adults. https://aaafoundation.org/driving-cessation-health-outcomes-older-adults/. Accessed 24 Jan 2020

Dauer P (2019) Vor FeV. In: Dauer P, König P (eds) Straßenverkehrsrecht, 45 edn, Munich

Deutscher Bundestag (2017) Entwurf eines ... Gesetzes zur Änderung des Straßenverkehrsgesetzes. Bundestag-Drucksache 18/11300

Dickmanns ED, Zapp A (1987) Autonomous high speed road vehicle guidance by computer vision. IFAC Proc Vol 20:221–226

Diederichs F (2019) New levels of automation from the user perspective. Advancements in human factors for highly automated cars. In: Autonomous vehicles interior design and technology symposium 2019, Stuttgart, Germany

Diederichs F, Döntgen B, Bopp-Bertenbreiter AV (2019) The Vorreiter/trailblazer project: maneuver-based driving—a steering paradigm change for automated vehicles. In: International symposium for intuitive partly and highly automated driving. Nov 27th, Aachen, Germany

Diederichs F, Marberger C, Jordan P, Melcher V (2010) Design and assessment of informative auditory warning signals for ADAS. In: Gepipari Tudomanyos Egyesulet (ed) FISITA World automotive congress 2010: Budapest, Hungary, 30 May–4 June 2010, Budapest, Hungary

Diederichs F, Döntgen B, Bopp-Bertenbreiter V (2019) The Vorreiter/trailblazer project: maneuver-based driving—a steering paradigm change for automated vehicles. In: Presentation on the international symposium for intuitive partly and highly automated driving. Nov 27th, Aachen, Germany

DIN EN ISO 9241-110:2008-09 (2008) DIN EN ISO 9241-110:2008-09, Ergonomie der Mensch-System-Interaktion_- Teil_110: Grundsätze der Dialoggestaltung (ISO_9241-110:2006); Deutsche Fassung EN_ISO_9241-110:2006. Deutsches Institut für Normung, Berlin

Economic Commission for Europe (ECE) (2018) UN-Regulation 79—uniform provisions concerning the approval of vehicles with regard to steering equipment. E/ECE/TRANS/505/Rev.1/Add.78/Rev.4

Economic Commission for Europe (ECE)/Global forum for road traffic safety (2017) Report of the global forum for road traffic safety on its seventy- fifth session. ECE/TRANS/WP.1/159

Economic Commission for Europe (ECE)/Global Forum for Road Traffic Safety (2018) Resolution on the deployment of highly and fully automated vehicles in road traffic. ECE/TRANS/WP.1/2018/4/Rev.3

Economic Commission for Europe (ECE)/World Forum for Harmonization of Vehicle Regulations (2019) Revised framework document on automated/autonomous vehicles. ECE/TRANS/WP.29/2019/34/Rev.1

Federal Court of Justice (Bundesgerichtshof - BGH) (2009a) Judgement of 03-17-2009—VI ZR 176/08. In: Neue Juristische Wochenschrift (NJW) 2009, pp 1669–1670

Federal Court of Justice (Bundesgerichtshof—BGH) (2009b) Judgement of 06-16-2009—VI ZR 107/08. In: Neue Juristische Wochenschrift (NJW) 2009, pp 2952–2953

Flemisch F, Abbink DA, Itoh M, Pacaux-Lemoine MP, Weßel G (2019) Joining the blunt and the pointy end of the spear: towards a common framework of joint action, human–machine cooperation, cooperative guidance and control, shared, traded and supervisory control. Cogn Technol Work 21(4):555–568

Flemisch F, Adams C, Conway S, Goodrich K, Palmer M, Schutte P (2003) The H-metaphor as guideline for vehicle automation and interaction. Langley Research Center, Hampton

Flemisch F, Goodrich K, Conway S (2005) At the crossroads of manually controlled and automated transport: the H-Metaphor and its first applications. In: Ertico (ed) 2005 intelligent transportation systems. The 5th european congress and exhibition on intelligent transport systems and services HITS, Hannover, June 2005

Flemisch F, Heesen M, Hesse T, Kelsch J, Schieben A, Beller J (2012) Towards a dynamic balance between humans and automation: authority, ability, responsibility and control in shared and cooperative control situations. Cogn Technol Work 14(1):3–18

Flemisch F, Kelsch J, Heesen M (2010) Skizze des Gestaltungsraumes haptisch-multimodaler Koppelung zwischen Mensch, Co-System und Regelstrecke als Teil einer kooperativen Bewegungsbeeinflussung. In: Neue Arbeits- und Lebenswelten gestalten. GfA-Frühjahrskonferenz, Darmstadt, March 2010

Flemisch F, Meyer R, Baltzer M, Sadeghian S (2019) Arbeitssysteme interdisziplinär analysieren, bewerten und gestalten am Beispiel der automatisierten Fahrzeugführung. GfA. Arbeit interdisziplinär analysieren - bewerten- gestalten. Dresden 2019

Flemisch F, Semling C, Heesen M, Meier S, Baltzer M, Krasni A, Schieben A (2013) Towards a balanced human systems integration beyond time and space: exploroscopes for a structured exploration of human–machine design spaces. Paper presented at the HFM-231 SYMPOSIUM On "Beyond Time and Space", Orlando, Florida, USA, Jan 2013

Flemisch F, Winner H, Bruder R, Bengler K (2015) Kooperative Fahrzeugführung. In: Winner H, Hakuli S, Lotz F, Singer C (eds) Handbuch Fahrerassistenzsysteme. ATZ/MTZ-Fachbuch, 3rd edn. Springer Vieweg, Wiesbaden, pp 1103–1110

Flemisch FO, Bengler K, Bubb H, Winner H, Bruder R (2014) Towards cooperative guidance and control of highly automated vehicles: H-mode and conduct-by-wire. Ergonomics 57:343–360. https://doi.org/10.1080/00140139.2013.869355

Flemisch F, Onken R (1997) Mensch-Maschine-Interaktion für ein kognitives Pilotenunterstützungssystem. DGLR BERICHT, pp 217–240.

Flemisch F, Altendorf E, Canpolat Y, Weßel G, Baltzer M, Lopez D, Herzberger ND, Voß GMI, Schwalm M, Schutte P (2017) Uncanny and unsafe valley of assistance and automation: first sketch and application to vehicle automation. In: Schlick CM, Duckwitz S, Flemisch F, Frenz M, Kuz S, Mertens A, Mütze-Niewöhner S (eds) Advances in ergonomic design of systems, products and processes. Springer, Berlin, pp 319–334

Flemisch F, Schieben A (eds) (2010) Highly automated vehicles for intelligent transport: validation of preliminary design of haveit systems by simulator tests. Brussels: EU-Commission. Public Deliverable D33.3.

Flemisch F, Schieben A, Kelsch J, Löper C (2008) Automation spectrum, inner/outer compatibility and other potentially useful human factors concepts for assistance and automation. Human factors for assistance and automation

Flemisch F, Schieben A, Kelsch J, Löper C (2008) Automation spectrum, inner/outer compatibility and other potentially useful human factors concepts for assistance and automation; In: Waard D, Flemisch F, Lorenz B, Oberheid H, Brookhuis K (eds) Human factors for assistance and automation. Shaker, Maastricht

Flemisch FO (2000) Pointillistische Analyse der visuellen und nicht-visuellen Interaktionsressourcen am Beispiel Pilot-Assistentensystem. Shaker

Flemisch FO, Kelsch J, Schieben A, Schindler J (2006) Stücke des Puzzles hochautomatisiertes Fahren: H-Metapher und H-Mode, Zwischenbericht 2006

Flemisch FO, Onken R (1998) The cognitive assistant system and its contribution to effective man/machine interaction. The Application of Information Technology (Computer Science) in Mission Systems

Foerste U, Graf von Westphalen F (2012) Produkthaftungshandbuch 3 24. 3. edition, Munich 2012

Franz B, Kauer M, Blanke A, Schreiber M, Bruder R, Geyer S (2012) Comparison of two human-machine interfaces for cooperative maneuver-based driving. Work 41(Supplement 1):4192–4199

Gasser TM, Arzt C, Ayoubi M, Bartels A, Bürkle L, Eier J, Flemisch F, Häcker H, Hesse T, Huber W, Lotz C, Maurer M, Ruth-Schumacher S, Schwarz J, Vogt W (2013) Legal consequences of an increase in vehicle automation—consolidated final report of the project group part 1, https://bast.opus.hbz-nrw.de/opus45-bast/frontdoor/deliver/index/docId/689/file/Legal_consequences_of_an_increase_in_vehicle_automation.pdf

Gasser TM, Seeck A, Smith BW (2015) Rahmenbedingungen für die Fahrerassistenzentwicklung. In: Winner H, Hakult S, Lotz F, Singer C (eds) Handbuch Fahrerassistenzsysteme, 3rd edn. Wiesbaden, pp 27–54

Gasser TM, Arzt C, Ayoubi M, Bartels A, Bürkle L, Eier J et al (2012) Rechtsfolgen zunehmender fahrzeugautomatisierung. Berichte der Bundesanstalt für Straßenwesen. Unterreihe Fahrzeugtechnik (83)

Geyer S, Baltzer M, Franz B, Hakuli S, Kauer M, Kienle M, Meier S, Weissgerber T, Bengler K, Bruder R, Flemisch F, Winner H (2014) Concept and development of a unified ontology for generating test and use-case catalogues for assisted and automated vehicle guidance. IET Intel Transport Syst 8(3):183–189

Green P, Wei-Haas L (1985) The rapid development of user interfaces: experience with the wizard of OZ method. In: Proceedings of the human factors society -29th annual meeting, vol 29. CA: SAGE Publications, Sage CA, Los Angeles, pp 470–474

Griffiths P Gillespie R (2004) Shared control between human and machine: haptic display of automation during manual control of vehicle heading. In: Proceedings of the 12th international symposium on haptic interfaces for virtual environment and teleoperator systems. IEEE, Chicago, IL

Haberfellner R, de Weck O, Fricke E, Vössner S (2015) Systems engineering-Grundlagen und Anwendung. Orell Füssli, Zürich

Harcourt AH, de Waal FB (eds) (1992) Coalitions and alliances in humans and other animals. Oxford University Press, Oxford, pp 445–471

Harper CD, Hendrickson CT, Mangones S, Samaras C (2016) Estimating potential increases in travel with autonomous vehicles for the non-driving, elderly and people with travel-restrictive medical conditions. Transp Res Part C: Emer Tech 72:1–9. https://doi.org/10.1016/j.trc.2016.09.003

Hoc J-M, Lemoine M-P (1998) Cognitive evaluation of human-human and human-machine cooperation modes in air traffic control. Int J Aviat Psychol 8:1–32

Holzinger A (2004) Application of rapid prototyping to the user interface development for a virtual medical campus. IEEE Softw 21(1):92–99

Honda (2020) Augmented driving steering wheel. Consumer Electronics Show CES 2020 https://global.honda/innovation/CES/2020/augmented_driving_concept.html

ISO 26022:2010(en) (2010) Road vehicles—ergonomic aspects of transport information and control systems. International Organization for Standardization, Geneva, Switzerland

Kaschub V, Sommer A, Bischoff S, Diederichs F, Widlroither H, Flemisch F, Döntgen B, Guenes EB, Graf R, Schmettow M (2018) Streich-und Lenkgesten für manöverbasiertes automatisiertes Fahren: Streicheln Sie schon oder Drücken Sie noch? In: 34. VDI/VW-Gemeinschaftstagung Fahrerassistenzsysteme und Automatisiertes Fahren 2018: Wolfsburg, 07. und 08. November 2018, vol 2335. VDI Verlag GmbH, Düsseldorf, pp 55–63

Kelley J (1984) An iterative design methodology for user-friendly natural language office information applications. ACM Trans Inf Syst 2(1):26–41

Kiss M, Schmidt G, Babbel E (2008) Das Wizard of Oz Fahrzeug: rapid prototyping und usability Testing von zukünftigen Fahrerassistenzsystemen. Presented at 3. Tagung Aktive Sicherheit durch Fahrerassistenz, Garching, 7–8 Apr 2008

Lee JD, See KA (2004) Trust in automation: designing for appropriate reliance. Hum Factors 46:50–80

Löper C, Kelsch J, Flemisch FO (2008) Kooperative, manöverbasierte Automation und Arbitrierung als Bausteine für hochautomatisiertes Fahren

Lutz LS (2014) Anforderungen an Fahrerassistenzsysteme nach dem Wiener Übereinkommen über den Straßenverkehr. Neue Zeitschrift für Verkehrsrecht (NZV) 2014:67–72

Lutz LS (2015) Autonome Fahrzeuge als rechtliche Herausforderung. Neue Juristische Wochenschrift (NJW) 2015:119–124

Lutz LS (2016) Automatisiertes Fahren: Änderung des Wiener Übereinkommens tritt im März 2016 in Kraft. Deutsches Autorecht (DAR) 2016:55–56

MacLean A, Young RM, Bellotti VME, Moran TP (1991) Questions, options, and criteria: elements of the design space analysis. Hum Comput Inter 6:201–250

MacLean A, Young RM, Moran TP (1989) Design rationale: the argument behind the artifact. SIGHCI Bull 20:247–252

Makrushin A, Dittmann J, Kiltz S, Hoppe T (2008) Exemplarische Mensch-Maschine-Interaktionsszenarien und deren Komfort-, Safety- und Security-Implikationen am Beispiel von Gesicht und Sprache. In: Alkassar A, Siekmann J (eds) SICHERHEIT 2008 – Sicherheit, Schutz und Zuverlässigkeit. Beiträge der 4. Jahrestagung des Fachbereichs Sicherheit der Gesellschaft für Informatik e.V. (GI). Bonn, 2004. Gesellschaft für Informatik e. V., pp 315–327

Marean CW (2015) The most invasive species of all. Sci Am 313(2):32–39

Maunder DAC, Venter CJ, Rickert T, Sentinella J (eds) (2004) Improving transport access and mobility for people with disabilities

Meyer R., von Spee RG, Altendorf E, Flemisch FO (2018) Gesture-based vehicle control in partially and highly automated driving for impaired and non-impaired vehicle operators: a pilot study. In: International conference on universal access in human-computer interaction. Springer, Cham, pp 216–227

Mourant RR, Rengarajan P, Cox D, Lin Y, Jaeger BK (2007) The effect of driving environments on simulator sickness. In: Proceedings of the human factors and ergonomics society 51st annual meeting, vol 51. Sage Publications Sage CA, Los Angeles, CA, pp 1232–1236

Müller A, Weinbeer V, Bengler K (2019) Using the wizard of Oz paradigm to prototype automated vehicles: methodological challenges. In: Proceedings of the 11th international conference on automotive user interfaces and interactive vehicular applications, Utrecht, Sep 2019. Association for Computing Machinery, New York, pp 181–186. https://doi.org/10.1145/3349263.3351526

Naumann A, Hurtienne J (2010) Benchmarks for intuitive interaction with mobile devices. In: Sá M de, Carriço L, Correia N (eds) Proceedings of the 12th international conference on Human computer interaction with mobile devices and services—MobileHCI '10. ACM Press, New York, USA, p 401

Nurmuliani N, Zowghi D, Williams SP (2004) Using card sorting technique to classify requirements change. In: Proceedings. 12th IEEE International requirements engineering conference. IEEE, pp 224–232

Oechsler J (2018) § 3 ProdHaftG. In: Hager J (ed) J. von Staudingers Kommentar zum Bürgerlichen Gesetzbuch - Buch 2: Recht der Schuldverhältnisse, §§ 826–829 - Unerlaubte Handlungen, Teilband 2: Produkthaftung, Neubearbeitung, Munich

Onken R, Schulte A (2010) System-ergonomic design of cognitive automation: dual-mode cognitive design of vehicle guidance and control work systems. Springer, Berlin
Pacaux-Lemoine MP, Flemisch F (2019) Layers of shared and cooperative control, assistance, and automation. Cogn Technol Work 21(4):579–591
Park GD, Allen RW, Fiorentino D, Rosenthal TJ, Cook ML (2006) Simulator sickness scores according to symptom susceptibility, age, and gender for an older driver assessment study. Proc Hum Factors Ergon Soc Ann Meet 50:2702–2706. https://doi.org/10.1177/154193120605002607
Pöhler G, Heine T, Deml B (2016) Itemanalyse und Faktorstruktur eines Fragebogens zur Messung von Vertrauen im Umgang mit automatischen Systemen. Zeitschrift für Arbeitswissenschaft 70(3):151–160
Rasmussen J, Pejtersen AM, Goodstein LP (1995) Cognitive systems engineering. Wiley, New York
Rettig M (1994) Prototyping for tiny fingers. Commun ACM 37:21–27. https://doi.org/10.1145/175276.175288
SAE (2016) Taxonomy and definitions for terms related to driving automation systems for on-road motor vehicles
SAE (2018) 3016: 2018 Taxonomy and definitions for terms related to on-road motor vehicle automated driving systems. Society of Automotive Engineers
Schieben A, Heesen M, Schindler J, Kelsch J, Flemisch F (2009) The theater-system technique: agile designing and testing of system behavior and interaction, applied to highly automated vehicles. In: Proceedings of the first international conference on automotive user interfaces and interactive vehicular applications (AutomotiveUI 2009), Essen, Germany, 21–22 Sept 2009
Schieben A, Heesen M, Schindler J, Kelsch J, Flemisch F (2009) The theater-system technique: agile designing and testing of system behavior and interaction, applied to highly automated vehicles. In: Proceedings of the 1st international conference on automotive user interfaces and interactive vehicular applications, Essen, Sep 2009. Association for Computing Machinery, New York, pp 43–46
Schieben A, Heesen M, Schindler J, Kelsch J, Flemisch F (2000) The theater-system technique: agile designing and testing of system behavior and interaction, applied to highly automated vehicles. Automotive user interfaces and interactive vehicular applications (Au-tomotiveUI), Essen
Schneemann F, Diederichs F (2019) Action prediction with the Jordan model of human intention: a contribution to cooperative control. Cogn Technol Work 21(4):711–721
Shaheen S, Niemeier D (2001) Integrating vehicle design and human factors: minimizing elderly driving constraints. Transp Res Part C: Emerg Technol 9:155–174. https://doi.org/10.1016/S0968-090X(99)00027-3
Shuttleworth J (2019) SAE standards news: J3016 automated-driving graphic update. SAE International
SILAB (2019) Würzburger Institut für Verkehrswissenschaften GmbH. https://wivw.de/de/silab
Siren A, Hakamies-Blomqvist L, Lindeman M (2004) Driving cessation and health in older women. J Appl Gerontol 23:58–69. https://doi.org/10.1177/0733464804263129
Siren A, Haustein S (2016) Driving cessation anno 2010: which older drivers give up their license and why? Evidence from Denmark. J Appl Gerontol 35:18–38. https://doi.org/10.1177/0733464814521690
Sneed H (1987) Software-management. Müller, Köln
Sommer A, Diederichs F, Bischoff S, Kaschub V, Graf R, Dierberger M (2019) SWIP-IT - Ein intuitives Streichgesteninteraktionskonzept zum Automatisierungslevel 4- ein iterativer nutzerzentrierter Designansatz zur Konzeptentwicklung. In: Binz H, Bertsche B, Bauer W, Riedel O, Spath D, Roth D (eds) Stuttgarter symposium für Produktentwicklung SSP 2019: Agilität und kognitives Engineering, pp 313–322
Stangnes N, Blekastad I (2015) questback ©. Retrieved from: https://www.questback.com/
Statistisches Bundesamt (2019) Unfälle von 18- bis 24-Jährigen im Straßenverkehr 2018
Thieme H (1997) Lower palaeolithic hunting spears from Germany. Nature 385(6619):807–810
Thrun SM, Montemerlo H et al (2006) Stanley: the robot that won the DARPA grand challenge. J Field Robot 23(9):661–692

Tomasello M (2014) A natural history of human thinking. Harvard University Press
Tsugawa S, Kato S, Matsui T, Naganawa H, Fujii H (2000) An architecture for cooperative driving of automated vehicles. In: Proceedings of intelligent transportation systems. IEEE, pp 422–427
Twisk D (1993) Improving safety in young drivers: in search of possible solutions. In: International symposium "young drivers", Lisbon
Vavoula GN, Sharples M (2007) Future technology workshop: a collaborative method for the design of new learning technologies and activities. Comput Support Learn 2:393–419. https://doi.org/10.1007/s11412-007-9026-0
von Bodungen B, Hoffmann M (2016) Das Wiener Übereinkommen über den Straßenverkehr und die Fahrzeugautomatisierung (Teil 1). Straßenverkehrsrecht (SVR) 2016:41–47
von Bodungen B, Hoffmann M (2017) Zur straßenverkehrsrechtlichen (Un-)Zulässigkeit automatisierter Fahrzeuge. Zeitschrift Innovations- und Technikrecht (InTeR) 2017:85–92
Wickens CD (1992) Engineering psychology and human performance, 2nd edn. Harper Collins, New York
Wickens CD (2008) Multiple resources and mental workload. Hum Factors 50:449–455
Wieland A, Wallenburg CM (2012) The influence of relational competencies on supply chain resilience: a relational view. Int J Phys Distrib Logist Manag 43(4):300–320
Winner H, Hakuli S (2006) Conduct-by-wire–following a new paradigm for driving into the future. In: Proceedings of FISITA world automotive congress (vol 22, p 27)
Wolfgang M (2020) Alterssimulationsanzug GERT: Unser Original, von einem Ergonomen entwickelt. https://www.produktundprojekt.de/alterssimulationsanzug/. Accessed 17 Feb 2020
World Health Organization (2002) Proposed working definition of an older person in Africa for the MDS Project. https://www.who.int/healthinfo/survey/ageingdefnolder/en/. Accessed 24 Jan 2020

Chapter 6
Light-Based Communication to Further Cooperation in Road Traffic

Matthias Powelleit, Susann Winkler, and Mark Vollrath

Abstract From the perspective of traffic safety, interactions between cars, bicyclists and pedestrians that lead to crashes comprise the largest part of all crashes. One major cause is missing or inadequate communication. This results in inadequate understanding and anticipation of what the other is going to do. Additionally, communication could further cooperation which leads to positive emotions and a better traffic climate. Within the project KOLA (Kooperativer Laserscheinwerfer; cooperative laser beam), we first examined needs in traffic which lead to positive emotions (positive arousal, being thrilled and feeling close to others) when fulfilled, and negative, if denied (being irritated, annoyed, horrified and feeling hostility). With regard to communication messages, explaining one's intentions and actions and thanking others was identified as the most relevant topics. Prototypical messages were designed and evaluated in two driving simulator studies using light which was projected from a car as the medium of the messages. We found that expressing one's intention in a specific situation furthers understanding and prosocial behavior of others, which leads to positive emotions on both sides. Explaining one's unusual behavior works similarly to increase understanding and further positive emotions but seems to require specific information about the causes of the behavior. Enabling drivers to thank others encourages prosocial behavior and leads to positive emotions. Thus, light-based communication is an excellent way to further communication, cooperation and positive emotions in traffic. The studies presented provide a sound starting point for further research and development.

M. Powelleit · S. Winkler · M. Vollrath (✉)
Technische Universität Braunschweig, Lehrstuhl für Ingenieur- und Verkehrspsychologie, Braunschweig, Germany
e-mail: mark.vollrath@tu-braunschweig.de

G. Meixner (ed.), *Smart Automotive Mobility*,
Human–Computer Interaction Series,
https://doi.org/10.1007/978-3-030-45131-8_6

6.1 Introduction

In Germany, only about 20% of all crashes in 2018 were single car crashes (Statistisches Bundesamt Destatis 2019). Thus, about 80% of all German crashes involve at least two traffic participants. While the term "interaction" is clearly inadequate for crashes this statistics clearly shows that adapting one's behavior to other traffic participants is one of the major tasks of drivers, bicyclists and pedestrians in traffic. From a psychological point of view, situation awareness (Endsley 1995) as a prerequisite for safe driving requires at the highest level an anticipation of what will happen in the near future. This enables a proactive planning of adequate behavior. However, with regard to the behavior of other traffic participants this requires to understand the goals and plans of these other humans. Communication is a very effective means to indicate one's goals and plans ("I am going to turn right", "I am looking for a parking place"). However, in traffic, verbal communication is hardly ever used.

Instead, explicit or implicit communication can be distinguished in traffic (Ceunynck et al. 2018). Consider the situation at an intersection with a main road with the right of way and a side road, where a driver is waiting to finally be able to enter the main road. He would like to communicate his wish to oncoming cars, which then could answer him and let him in. But how to do so in traffic? Explicit communication includes communicating by gestures, which is only possible in a very limited distance and is impeded by car windows that prevent a clear view into the interior. Turn signals, horns or use of the flashlight are either very limited in vocabulary (turn left, turn right) or very unspecific. Thus, our waiting car driver currently has no real means to explicitly communicate his wishes. Implicit communication is also unspecific and may be misunderstood. Our driver may begin so slowly move forward, but how can he be sure that the oncoming car will really let him in? This may slow down, but is this a reaction to the possible danger of a crash or really a signal to enter the main road? This kind of implicit communication may even be the direct cause of accidents. For example, a driver may slow down when approaching an intersection where he has to give way to other cars. However, a bicyclist may think that the driver is slowing down because he has seen him, may cross the road in front of the car and be hit by the car driver, who in fact was fully concentrated on the other car traffic (Räsänen and Summala 1998). Thus, at the moment, communication in traffic is very limited.

Besides preventing crashes, by better understanding what others are doing or going to do, communication would also contribute to cooperation in traffic. Missing information or misunderstandings will hinder cooperation and lead to frustration and anger (Dollard et al. 1939; Merten 1977). Anger while driving furthers an aggressive driving style which is found in conflict situations (Risser 1985), but also in driving in general (Zhang and Chan 2016). Furthermore, anger has negative effects on attention, information processing and decisions which may also increase crash risk (Blanchette and Richards 2010). Thus, in contrary, being able to communicate could further the common goal of arriving safely and fast by looking for behavioral solutions which ideally are advantageous for all traffic participants involved. Moreover, this could lead to positive emotions and thus a better traffic climate.

In contrast to the well-studied effects of negative emotions in traffic, there is a need for further research in the area of positive emotions and cooperative and prosocial behavior. Here, the actor intends to voluntarily improve the situation of another road user. For example, the exchange of clearer information about current and future behavior, the renunciation of one's own right of way and the appreciation of cooperative gestures by others could have a positive influence on the traffic climate and make the traffic flow more efficient. This may also promote traffic safety. Since such interactions between road users play an essential role in road traffic, it seems particularly useful to promote cooperative behavior through driver assistance. An important basis for the wellness-oriented design approach is the fact that prosocial behavior is not only positively experienced by the recipient, but also represents a source of joy and well-being for the actor. In order to facilitate prosocial behavior in road traffic, it is useful to provide road users with situational support for "better" behavior, for example, by clarifying their own options, recommending concrete actions or even enabling new ways of acting.

Measures with an indirect influence, such as traffic education and campaigns, aim to promote a general pattern of thinking, but this is often difficult to activate in a specific situation. Interestingly enough, the car as such (e.g., by means of appropriate driver assistance systems) has so far played little role in stimulating cooperative behavior. On the contrary, the compartmentalized design of the car even makes cooperative behavior more difficult. In addition, the available communication channels and signals are not standardized in their use, even though every vehicle has the same signaling equipment (e.g., headlamp flasher, indicators). As described above, a specific signal can convey different and even contradictory messages. A headlight flasher can mean "Please, after you" or "Watch out, here I come" (Merten 1977). Since signals currently available in the vehicle are often ambiguous, the perceived message depends strongly on the context and can only be understood correctly if the interacting persons share the same understanding of the situation.

A new communication channel designed for the transmission of specific messages could address the problems described and support clearer communication in traffic. New signals can be designed directly in terms of specific messages and are therefore more personal and unambiguous than, for example, a headlight flasher. At the same time, it is possible to send messages that were not possible before. The essential questions are the following: What are actually meaningful messages in this context? What distinguishes positively or negatively experienced traffic situations and what role does communication play here? How can positively experienced interactions be encouraged and negatively experienced interactions reduced?

The question of how a possible language of light could be designed and which aspects should generally be considered when designing interactions between road users is currently receiving increasing attention. Especially in the context of autonomous driving, questions concerning everyday communication between road users are being addressed. In mixed traffic of manually and autonomously driven vehicles, the understanding of implicit negotiation processes or declarations of intent

Fig. 6.1 A car is projecting a zebra-crossing pattern to communicate to the pedestrian that it will stop and let him cross the road. *Source* (https://www.experienceandinteraction.com/kola)

plays a major role for the concrete parameterization of behavior of autonomous vehicles. Communication with weaker road users plays an important role for the future feasibility of autonomous urban traffic.

In order to illustrate this line of thought, consider the following example of communication concepts, "the gentleman" (see Fig. 6.1).

A zebra crossing projected by the car communicates the intention of the driver to stop and invites the pedestrian to cross the road. It supports the driver in his action of letting pedestrians cross the road in darkness and thus promotes cooperative behavior. In addition to the outside light, the driver is illuminated inside the car so that the driver can be seen and eye-contact with the pedestrian is possible. The driver may additionally gesture the pedestrian to cross.

This and other similar concepts have already been taken up in various concept studies. An overview of the current state of the automotive industry regarding external human–machine interfaces (eHMIs) for communication between cars and vulnerable road users is given by Bazilinskyy, Dodou and de Winter (Bazilinskyy et al. 2019). Many of the technical concepts use vehicle-based displays at the front (bumper, radiator grill, windshield, roof) or side of the vehicle. In addition, some of the concepts use light projections to display content on the road. These are mostly directed forward, but occasionally also backward or to the side. The content of the displays or the projections varies quite a lot. Some displays imitate human features (eyes, smile), others use familiar symbols (arrow, crosswalk), others communicate purely textually, and finally there are also abstract luminous forms (surfaces, lines). Of the 28 interaction concepts reported, 23 are designed for the use of letting a pedestrian cross the street or signaling that he or she should wait. This clearly shows the focus of efforts to integrate autonomous vehicles into urban traffic. There are also concepts that visualize vehicle movement (forward, backward) or announce driving maneuvers (turning, lane change).

To summarize, while the need for more cooperation in traffic is being accepted, communication and cooperation concepts are still very limited. In order to address this problem, it is necessary to first examine where situations arise in traffic which would benefit from communication and cooperation, and, secondly, to develop concepts for communication for these situations. With regard to the first aim, an audio diary study and an online survey were conducted. The results are given in the next chapter. To the second aim, basic concepts were derived from these studies, developed and tested in two driving simulator studies. This results in a framework for light-based communication that furthers cooperation in traffic.

6.2 The Need for Cooperation in Traffic

Successful communication between road users may further safe and efficient traffic. But what does a successful communication mean? In which situations are communication and cooperation of particular importance for road traffic and to what extent do they influence the (emotional) evaluation of a situation? Which communication practices are already in use and where are potentials to improve these practices? An audio diary study and an online survey were carried out with the aim of recording initial scenarios and situation-specific communication needs.

With the audio diary method, very timely "live reports" were recorded immediately after experiencing the situation, so that memories were less distorted and the emotional component of the situation could be reproduced in detail. In addition, a wider range of situations was recorded, as the participants were generally expected to describe remarkable situations, even if they were only short interpersonal moments. In the online survey, on the other hand, only the situation that had been most significantly remembered from the previous month was reported. In addition, the audio recordings (as opposed to the free text comments in the online survey) made it possible to record the course of the situation in more detail and to better understand the emotional impact of the situation through the respondent's tone of voice. Finally, the personal contact with the drivers made it possible to directly address any ambiguities or queries in the evaluation process of the situation descriptions. For the online study, no further contact was possible. However, the online survey offers a greater range and larger sample size, so that the frequencies of different situations in real traffic can be better estimated.

6.2.1 Method of the Audio Study

The subjects kept a diary of their experiences in road traffic for one week and used a dictation machine to document everyday situations in which communication or cooperation played a role (e.g., particularly pleasant or unpleasant communication, understandable or misunderstandable, polite or impolite or missing communication).

Table 6.1 Key questions to be answered after the trip

Questions
How long ago was the situation?
What has happened (who participated, what was your role, where, when)?
Who did what and why? What were the intentions?
How did you feel, during and after the situation?
Why did you feel this way?
What did you like about this situation and what would be nice in other interactions?
What did you miss?
How could this situation be improved?

The focus was on the general questions in which situations road users communicate or cooperate with each other in road traffic and when these interactions were experienced as positive or negative. Of particular interest were the factors that influenced the emotional reaction of the test persons in a concrete situation. The participants documented their experiences for one week according to the following instruction.

"If a situation was experienced, then focus on the immediate impact of the situation. While driving (if situation allows) or as soon as possible, record your first relevant impressions and tell us why the situation is worth reporting (e.g., emotions, dangerousness of the situation). The focus on the intentions and needs of those involved in the situation. Finally, when stationary or after the end of the journey, record detailed impressions, in addition to the first impression and answer key questions." The key questions are shown in Table 6.1.

A total of 29 subjects (car drivers, cyclists and pedestrians) with an average age of 36 years (SD = 11 years, 16 males, 13 females) took part in the audio diary study. The average driving experience was 18 years. Twenty-two of these came from Braunschweig and seven from Siegen. For participating in the study, the test persons were rewarded with a total of €50 for the entire week.

During the study, the test persons were independently on the road and documented their impressions as car drivers, pedestrians or cyclists using a dictation machine. After one week, they returned the recording device including audio data and were compensated for their efforts. Optionally, voluntary follow-up interviews took place, to better reconstruct the course of the situations.

6.2.2 Method of the Online Survey

An online survey was used to further investigate communication and cooperation in a larger sample. Similar to the audio diary study, the subjects described a concrete, self-chosen situation in which communication played a role. The subjects were instructed to describe either a positive or a negative experience. The corresponding selection (positive or negative) was random. Particularly relevant were the emotions experienced, the perceived needs and intentions as well as the actual behavior. In addition, general motives, attitudes and everyday behavior with regard to respectful behavior in road traffic as well as currently used means of communication were recorded. The survey consisted of both open and closed questions.

A total of 165 subjects took part in the online survey, of which 92 completed the questionnaire in full and were thus included in the evaluation. The average age of the test persons was 27 years (SD = 9 years). 73% of the subjects were frequently or very frequently walking. The bicycle and the car were used by 52% and 45% of the subjects, respectively, and the bus or train by 35%. Furthermore, 64% of the subjects were frequently or very frequently on the move in the city. On country roads and motorways, the figures were 45% and 28%, respectively. The subjects described their driving style as cautious, comfortable, defensive, fast, not very risk-taking and not very anxious.

The questionnaire was created in cooperation with the University of Siegen and was available from March to June 2017, via the survey platform "Surveymonkey." Participants were acquired via various e-mail distribution lists of all project partners, including Facebook and eBay.

At the beginning of the online survey, the respondents described a situation from the previous month in which they had contact with another road user and which they remembered ("We are interested in your personally experienced situations in which communication or cooperation with other road users played a central role or the lack of these aspects. We are interested in an example from your everyday life in road traffic."). Here it did not matter whether the test person was walking, going by bike or driving, as long as at least one motor vehicle was involved. Similar to the audio diary study, the course of events in the situation should first be described by means of the following guiding questions:

- Where and when did the situation take place?
- Who was involved?
- What did you intend?
- What happened?
- How did you and the other person react?
- What were the reasons for your own behavior and that of the other person?
- How should the situation have proceeded ideally?

Afterward, the respondents stated how they had felt before, during and after the situation and which motives had been relevant for their actions. Then, they gave general attitudes with regard to mindfulness and respect in traffic and assessed their own respectful behavior (in different roles) as well as the behavior of other road users. The subjects also reported on their general motives for acting respectfully in traffic and how they would like to communicate if there were no limits or technical restrictions. Finally, the test persons answered some demographic questions about the sample description. Three €50 vouchers were raffled among all participants. The answering took about 20 min.

6.2.3 Subjective Needs in Road Traffic and Their Consequences

From the audio study as well as the open questions of the online survey, the answers were analyzed with regard to the needs that were relevant in the respective situations. These needs were then grouped into different subject areas and are shown in Fig. 6.4.

Overall, there is a basic need for the (physical) integrity of one's own person. With this in background, the major aim in traffic is to reach a destination safely (integrity), but also to reach it at all and, mostly, as fast as possible (mobility and transport of goods). This is specified in four sub-goals or needs which refer to this mobility aspect (see left of the figure), i.e., how effectively or efficiently the goal is achieved or how confident the subjects feel about it. At the right side, four social needs were identified that can be achieved in interactions in traffic.

With regard to the mobility needs, the first was a desire for a smooth traffic flow, rapid progress and fast arrival at the destination. If this need was fulfilled, this resulted in joy, for example, when the respondent was allowed to go ahead even though he had no right of way, or in pride in the fact that a negotiated situation could be resolved quickly and thus unnecessary braking could be avoided. If this was not fulfilled, subjects were frustrated, for example, that they were unable to drive at their desired speed ("Actually, it was green, but he drove totally slowly.") or they felt guilty themselves because they had blocked the following traffic and thus caused annoyance (Fig. 6.2).

Secondly, there was the need for autonomy and control. On the one hand, the subjects wanted to have the possibility to act in their usual way (e.g., to drive at their desired speed). On the other hand, they wanted to be the master of their own decisions. This need was particularly relevant in negative situations when other drivers had forced themselves on them (e.g., jostling on the motorway), restricted their own room to maneuver (e.g., suddenly turning into their own lane), generally behaved ruthlessly or felt compelled rather than voluntary to make their own decisions ("I decide whether I let someone else drive in front of me").

Third, there was the need for transparency of the behavior of other road users. The subjects, for example, found it unpleasant if they had to brake unnecessarily early or too hard from their point of view. In such situations, there was a lack of opportunities to adapt one's own actions to the requirements in good time, because the intentions or actions of other road users were not clear enough ("Why isn't he driving?"; "Is he talking to me?"; car and bicycle get in each other's way and cannot communicate who drives first). Interestingly, an understanding of the other's action seemed to be more important than the fact that one's own behavior was blocked. For example, subjects often reported their frustration and anger at situations where they were blocked by a slower moving vehicle. However, if the reasons for the slow drive were comprehensible, the subjects were able to accept the perspective of the other person and react more empathetically and understandingly, which resulted in less frustration and anger. In some cases, the evaluation of the situation was even reversed ("Why is he driving so slowly? - Ah, he wants to let me go first. Thanks!"; "I

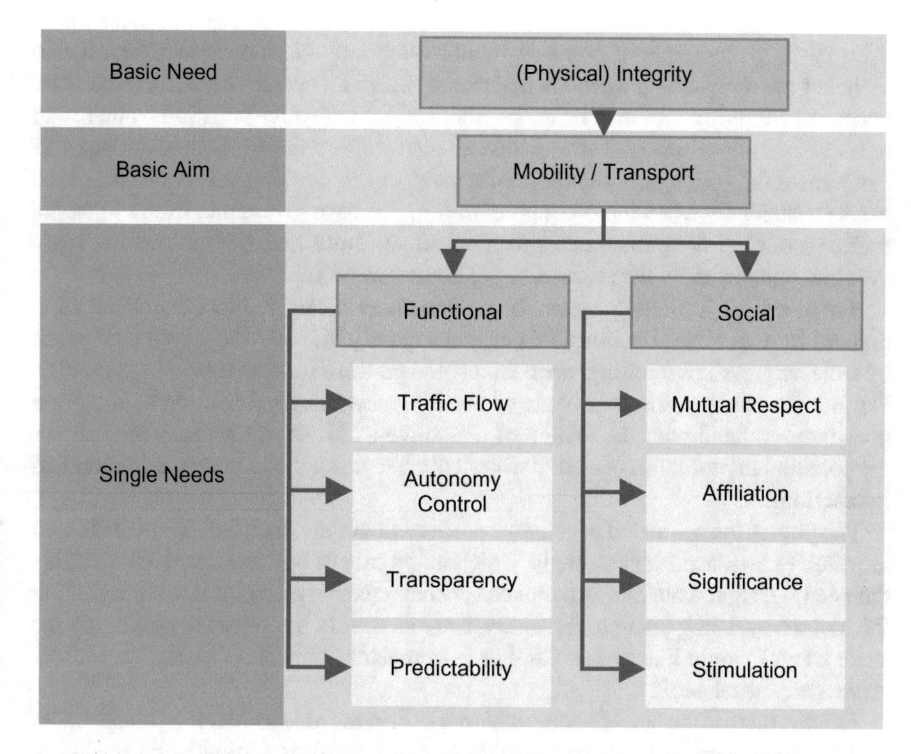

Fig. 6.2 Overview about the basic and single needs and aims within a traffic situation

was parked in the back. But the driver in the rear-view mirror apologized and smiled. All is well again."; "[On the motorway] Why is it driving so slow? Oh, an old woman. Cool that she's still so mobile"), which shows a great potential to improve the traffic climate through clearer communication and background information.

The fourth need of predictability is closely related to transparency, but focuses more on the safety aspect. For example, subjects felt insecure when other drivers drove up close or turned in very close to them, leaving little room for their own decisions. Similarly, unclear driving styles or signals (flashing lights of an oncoming car; truck honking on the motorway) led to distraction and the test persons could no longer concentrate fully on driving. Deviations from expectation- and rule-compliant behavior had the consequence that the basic requirement of safe mobility was called into question ("Somehow we all want to get home safely. [Commentary on a tailgater on the highway]").

With regard to the social needs, mutual respect and social appreciation were very important. In the broadest sense, this means manners and decency in road traffic. However, it also includes observing road traffic regulations and being a role model ("If I am met in a nice way, I myself am also more motivated in further contacts"; "I behave nicely in contact in order to be a role model and to promote nice contact also with others"). For example, the fulfillment of the need led to joy that other road users gave up their right of way and did not stubbornly insist on their rights. It also included mentally changing the perspective ("I also let cyclists cross the crosswalk because I myself would also be happy about it"). In negative situations, other road users were rude or unsocial if they disregarded traffic rules to their advantage ("If I'm almost out, other cars should let me park") or ignored the needs of other road users ("Especially when there is a lot of traffic, you have to communicate so that the traffic works"; "I drive too much already. And you come here boarded up on Friday evening, but somehow we all want to get home safely").

The need for affiliation means being somehow connected to other interaction partner. In respective situations, this happens mainly through direct communication by facial expressions (smiling, eye contact) and gestures (raising a hand in gratitude). Despite the slightly different focus of needs, a pleasant personal interaction often goes hand in hand with the feeling of a positive traffic environment, which shows the potential to promote a prosocial overall climate through more pleasant individual interactions.

Whether in the context of negative or positive situations, such feelings of affiliation are often experienced very strongly, which in the case of positive interactions fulfills the need for significance of experiences ("That gave me a beautiful morning"; "On the motorway, I sing along to the radio. The one next to me seems to be singing the same song. I would like to say: Cool taste in music! I hope you'll soon have it too. Have a nice weekend!").

Finally, there is the need for stimulation. Here, being able to do things independent of other road users is in the foreground. In some situations, stimulation is closely related to the need for autonomy and control ("I wanted to use the empty highway and drive fast, but he didn't see me and blocked the left lane").

Thus, two areas of needs were identified:

- Functional needs refer to being able to achieve the basic need of mobility, reaching one's destination safely and in time. Doing this fast and smooth and being in control focus on the subjective side, the transparency of what other persons do in traffic, the understanding why they do it and being able to predict what they are going to do. All show the relevance of interactions for a fulfillment of needs in traffic.
- Of, course, when people interact, social needs come into play. The need for mutual respect und closeness (affiliation) directly addresses these interactions. Experiencing significant moments and finally the need for stimulation independently of other supplement this by more subjective aspects which can be achieved in social situations or when not being hindered by others.

As the examples above show, positive emotions arise when these needs are fulfilled. Negative emotions are the consequences of these needs when impaired in a certain situation.

What emotions arise in these different situations? Fig. 6.3 shows the mean values of the different emotions experienced separately for the positive and negative situation examples.

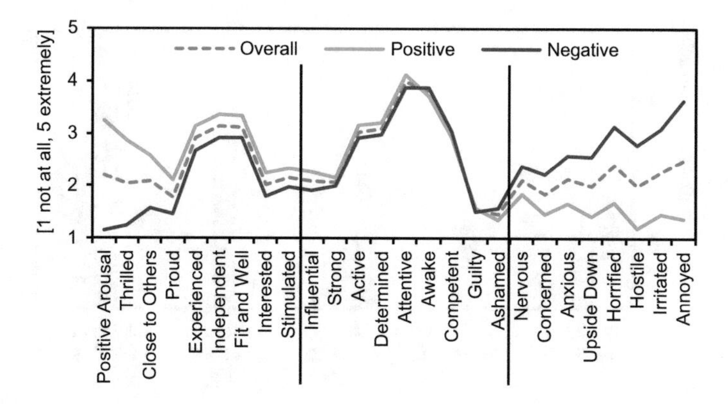

Fig. 6.3 Mean rating of emotions in different situations, also separated for positive and negative situations. The emotions were sorted with regard to the difference between positive and negative ratings

The emotions were sorted by the size of the difference between negative and positive descriptions. This shows that in positive descriptions of situations, as expected, some emotions with positive connotations dominate. These are a positive arousal, being thrilled, feeling close to others, proud, experienced, independent, fit and well, interested and stimulated (left part of the figure). At the right part, in negative situations negative emotions dominate such as being nervous, concerned, anxious, upside down, horrified, hostile, irritated and annoyed. In both positive and negative situations, people reported being attentive and awake, somewhat determined and active, and less often feeling strong and influential, but also not guilty or ashamed.

Thus, on the positive side emotions arise with a close relationship to the functional needs like the positive arousal, feeling independent, experience, fit and well, but also the social needs with feeling close to others, proud, interested and thrilled. Similarly, in the negative situations the functional needs especially related to safety may lead to negative emotions like feeling nervous, concerned, anxious or even horrified. The social needs are reflected in feelings like hostility, being irritated or annoyed.

- Overall, emotions are a frequent consequence of interactions in traffic.

Enabling communication in these situations may be able to shift these emotions to the positive side.

Furthermore, the subjects also described the motives for their actions in positive and negative situations and in different roles (car driver, pedestrian, cyclist). Figure 6.4 shows the ratings of the relevance of different motives. Intrinsic motives ("like to act like this") are most relevant for the drivers and even more so in positive situations, followed by something like norms ("it is important to act like this"). Of a lesser relevance is the evaluation by others. However, it is interesting to note that this motivation is stronger in car drivers than in pedestrians and least strong in cyclists. Not wanting to be a bad person and wanting to be liked play a lesser role as a motivation.

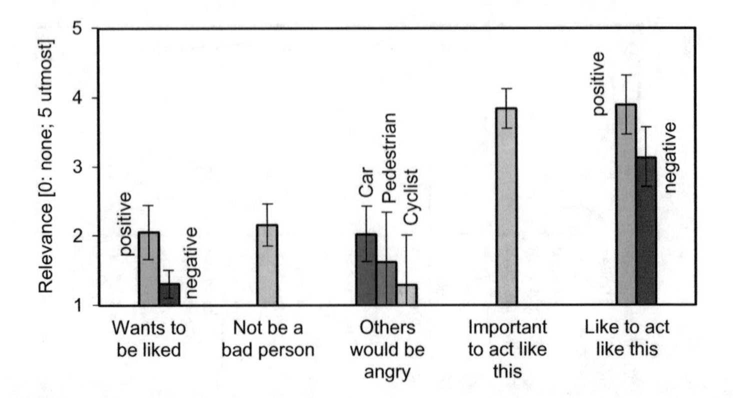

Fig. 6.4 Mean rating of relevance of the motives for positive and negative situations and different roles in traffic (if significant)

- Hedonistic aspects and norms seem to be the strongest motivator for actions in traffic.

Of course, why do people act in the way they do? Is it really because they like how doing this feels? It could also be that they like to do it because of the reaction of others to their behavior. In this manner, almost all subjects (98%) indicated that respectful behavior in traffic was important or very important to them and that they themselves often behaved respectfully. However, only 57% of the respondents reported that other road users also often behave respectfully. 21% of the subjects even reported that others would often behave disrespectfully.

Regarding the question of how respectful the test persons behaved in various roles in road traffic, the subjects rated themselves very respectful as drivers (94% as extremely or strongly respectful, see Fig. 6.5). If they were walking or cycling, only 72% and 68%, respectively, rated themselves as respectful. According to the subjects, the least respectful behavior was seen when they were traveling by truck or motorcycle. Only 40% of the subjects reported that they behaved respectfully in these roles. 30% of the motorcyclists and 40% of the truck drivers even reported not behaving respectfully at all or only a little bit. It should be noted here, however, that

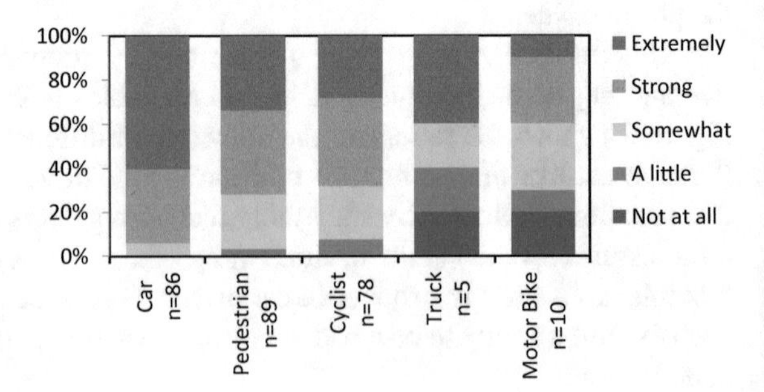

Fig. 6.5 How important is it to be respectful in traffic? Percentages of the answers in different road user groups

the case numbers of motorcyclists and truck drivers are significantly smaller than for the other means of transport and the estimates are therefore not as robust.

The main reasons for behaving respectfully in traffic are given in Fig. 6.6. Personal beliefs and responsibility, followed by norms and thinking about the greater good were the most relevant reasons. The law requiring that valuing interactions feelings of gratitude follow. The fear of punishment and wanting to be accepted by others was rarely given as a reason.

Overall, respect was found to be an important motivation in traffic:

- Most car drivers, but also pedestrians and cyclist, found it very important to behave respectfully in traffic.
- This motivated by personal beliefs and the own responsibility, thus a strong intrinsic motivation.

Respect can be furthered and impressed by communication. Thus, subjects were asked about their wishes with regard to communication in traffic. Above all, communication should be unambiguous, understandable and efficient ($n = 8$) and available at all times. In addition, it should be direct and immediate ($n = 2$) and also function over larger distances ($n = 2$). Finally, the way of communication should be pleasantly designed ($n = 5$), for example, by being friendly, considerate and reserved, by conveying general sympathy or by conveying small nice messages analogous to a smile.

The weaknesses of previous forms of communication in road traffic were ambiguity (flashing lights, gestures), unpleasantness (horn) or lack of availability (speaking from the car).

With regard to the modality, a large number of the subjects stated that they would like to talk to their counterpart ($n = 24$) or generally communicate in text form ($n = 6$). Gestures were said to have the potential to act as a positive sign, to transmit sympathy and to focus on what is important in the situation ($n = 8$). Furthermore, some wanted personal eye contact ($n = 3$) or the possibility to send an acoustic

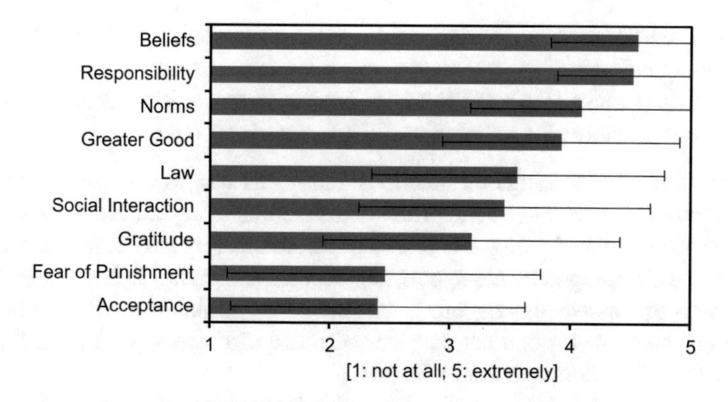

Fig. 6.6 What are the main reasons for behaving respectfully in traffic? Mean ratings of how frequent the different reasons were indicated

signal ($n = 2$), for example, a soft horn for cyclists and pedestrians or an acoustic background to a gesture of thanks.

In accordance with the desired modalities, technical implementations were requested that serve to transmit voice messages ($n = 15$), such as radio, telephone, microphone, headset, loudspeaker or voice messages. Other technical implementations included buttons on the steering wheel that send messages directly, functions for dictating and presenting messages, the conversion of gestures into speech or messages that can be sent to the on-board computer of another vehicle. Visual implementations were also suggested, such as an LED sign on the rear window, illuminated emojis or making one's own line of sight visible to others through moving headlights.

There was a great variety in the actual purpose of the communication. Many subjects wanted to provide other road users with information about their own intentions, behavior or moods ($n = 17$), for example, their own line of vision, the space they need, the limitations of their own vehicle or whether a certain area is free. In addition, there was also a large proportion of subjects who wanted to inform other drivers of their misconduct or to reprimand them ($n = 14$). Also frequently mentioned were warning of possible dangers ($n = 9$), coordinating behavior with other road users and negotiating priority ($n = 8$), and thanking others ($n = 8$). In addition, some wanted to send inquiries or requests ($n = 2$) or to apologize for their own mistakes ($n = 3$).

Some of the participants were satisfied with the current state of the communication possibilities ($n = 10$) and thought that gestures and previous means (headlights, horn) would be sufficient ($n = 6$) or additional technology would be unnecessary or distracting ($n = 3$). Some test persons had accordingly expressed their concerns ($n = 6$) and thought that increased or unmediated communication possibilities would lead to more misunderstandings, arguments and aggression or would endanger safety and order in road traffic and lead to confusion and disarray.

Some test persons ($n = 5$) referred to normative and legislative guidelines ("Socialize people as early as possible for social consideration"; "If everyone cared about the other road user and therefore would not be too selfish or hurried, there would be no problems"; "Everything is fine as long as everyone sticks to the rules").

- Overall, car drivers, pedestrians and cyclists see a large positive potential of improving communication in traffic.
- This could support respectful behavior, lead to more positive emotions in traffic and could possibly also increase traffic safety and efficiency.

To summarize, a wide range of situations and needs was found where interaction and communication would be important in traffic. Positively experienced situations were characterized by flowing, conflict-free and anticipatory traffic, which was made possible and encouraged by clear, polite and courteous driving styles. These driving styles were expressed, among other things, in the availability of clear information about the behavior and action plans of other road users, which confirms the importance of functioning communication.

To be understood and to understand the behavior of others characterized positively experienced interactions, as well as receiving gestures of appreciation, withdrawal or apology. This enabled the subjects to accept the perspective of the other person, which reduces the experience of frustration and anger, especially in situations with negative connotations.

The subjects also indicated that they moved respectfully in traffic because of their own personal beliefs. Thus, interventions do not necessarily have to address attitudes and values, but rather create new, easily accessible and understandable ways to act according to these convictions. Transferring the positive impressions from pleasant personal encounters to road traffic as a whole system opens up the possibility of promoting a prosocial overall traffic climate by improving the communication possibilities of individual road users.

Accordingly, in the second part of this chapter, different communication concepts are presented. These were developed in a human-centric manner using a driving simulator to demonstrate prototypes and evaluation studies to examine the reactions of car drivers, cyclists and pedestrians. In the first part, we were concerned about how to express appreciation, explaining one's intention and indicating one's progress as major messages which are relevant in many of the situations described above. In the second part, we examine how to express an intention and wish in a typical intersection situation where one car would like to enter the main road (although according to traffic rules one would have to wait until everything was free). Thus, this development will not provide a full catalogue and concept of communication for all situations in traffic. However, it will explore basic communication concepts.

6.3 Explaining Intentions and Expressing Gratitude

From the analyses of the two studies described above, two basic interaction goals were derived and explored in a first driving simulator study.

- Explaining intentions, e.g., when one is looking for a parking lot and has to go slowly, or simply showing where one is going to drive to.
- Expressing appreciation or gratitude, for example, if another car yields the right of way so that one can enter the main road.

In order to explore light-based communication concepts, typical situations were implemented in the driving simulator of TU Braunschweig, in which these interaction goals are relevant. Additionally, technical means were developed to project different light concepts in this driving simulation. Thus, for each situation and interaction goal, different concepts could be directly compared and be evaluated by users.

6.3.1 Method

Three different scenarios were used:

- (1) The driver of a vehicle that was allowed to enter the main road thanks for letting him enter the main road (thank you to the vehicle behind).

- (2) The car in front is looking for a parking space and wants to indicate that intention, thus explaining why the car is going so slow.
- (3) The own vehicle indicates its direction of travel to other road users so that these can better understand and anticipate where it will be going.

While the first situation is only of many possible, the focus of the study was on how to express gratitude, regardless of what the reason for this may be. Thus, the results may easily be transferred to a wide variety of other situations. Similarly, there may be many reasons to drive very slowly. The second situation examined a typical one. As reasons for driving slowly may differ, one basic question was if it was necessary to provide detailed information about the reasons for this behavior. Finally, the third situation was introduced as the idea to provide a higher transparency and enable a better anticipation by giving information about one's future driving behavior to the outside.

Each subject experienced each driving situation several times with different communication designs in randomized order. In the first two scenarios, there was a first control drive without a lighting concept before three different design variants were presented. In the third scenario, ten different design variants were compared, which were changed directly during the drive. The concepts will be described in the next section.

6.3.1.1 Scenarios and Communication Concepts

In the first scenario, subjects drove a car on a main road and a vehicle was trying to enter this road from the right from a side street. The subjects were instructed to behave cooperatively and let the other driver enter the main road. After the other vehicle had done this, it projected a message of "thanks" to the back. Three variants of the message were compared, varying both in their animation and their symbolism (see Fig. 6.9), with all symbols projected with green light (Fig. 6.7).

Fig. 6.7 Three design variants of the "thanks" message in the first scenario: **A** Check mark, at a fixed distance from the front vehicle, **B** smiley, firmly on the road and **c** thumbs-up, moving toward the driver (left). At the right, a screenshot is given from the simulation environment with implemented Variant B

Variant A consisted of a check mark symbol projected at a fixed distance from the front/transmitter vehicle. In variant B, a smiley symbol was projected onto the road at a fixed position. As soon as the driver reached this position, the symbol showed a "pop-up animation" before it finally disappeared (short waxing and then shrinking until it disappeared). In variant C, a thumbs-up symbol was used, which moved from the rear of the front vehicle toward the driver and slowly faded out before it reached the driver's vehicle.

In the second scenario, a vehicle directly in front drove very slowly for about 15 s before finally turning left into a parking space. In order not to frustrate the following traffic by an unclear situation, the driver in front projected a message to the back to explain his own behavior. Three different messages were compared, and their information content varied in three levels from generic to specific information. As described above, the main question here was whether a specific information is really required, which would lead to many different possible messages. All display variants were presented at a fixed distance from the vehicle in front. The variants are shown in Fig. 6.8.

Variant A was modeled on a loading animation, as it occurs, for example, when buffering video streams or with busy applications. This indicated that the driver is currently busy or very concentrated and therefore cannot drive at normal speed. Shortly before turning into the parking space, the front vehicle set the turn signal and the load symbol was faded out. At the same time, a green tick appeared to indicate that the process that had been going on before was now complete.

Variant B consisted of a fan that swung to the left and right, its movement imitating that of a searchlight or a head that turned back and forth during orientation. Shortly before the front vehicle turned into the parking space, the projection was faded out.

In variant C, a parking symbol was projected, whose blue filling faded in and out once per second. The symbol directly showed that the car in front was looking for a parking lot. The pulsating animation showed that the search was still in progress. When the car in front used the turn signal, the fill color pulsed three times slightly faster before the symbol was displayed continuously to signal that a free parking space had been found.

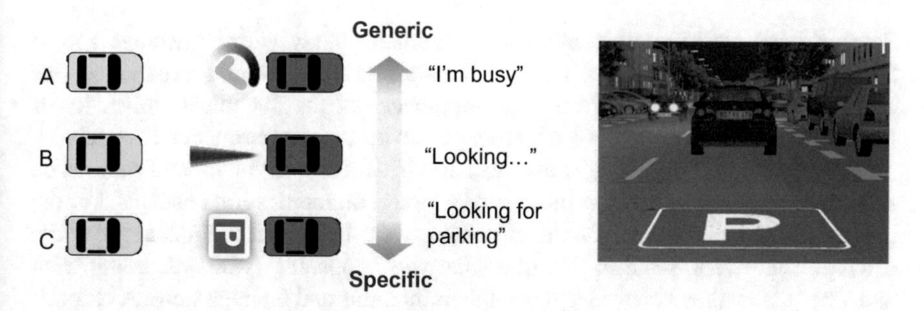

Fig. 6.8 Three design variants to explain one's own behavior, according to the level of detail from generic to specific in the second scenario (see right of the figure): **A** loading animation with a final check mark, **B** searchlight that swung back and forth and **C** pulsating parking symbol (left)

Fig. 6.9 Examples of the textures used (left) and screenshot of the simulation environment with implemented Variant B (right)

For the display of one's own driving intentions, drivers drove through a simulated town. The aim of this display was to communicate more clearly to other road users the own behavioral intentions. For this purpose, a "band of light" was projected in front of the vehicle, whose length and curvature provided information about speed and direction of travel. Based on the current speed and steering wheel position, the light band displayed the driving path within the next second. When the vehicle accelerated, the band became longer. When braking, the band shortened and disappeared completely as soon as the vehicle came to a halt. Ten variants were compared, which differed in terms of contrast, shape and pattern, and which had been designed by project partners from the University of Siegen (see Fig. 6.11 for examples) (Fig. 6.9).

This movement display was designed as a permanent display of the vehicle that reacted passively to the driving dynamics. For such a display, it is particularly relevant whether the driving environment is not overloaded by the permanent presence and whether it is understood by the other road users (e.g., when a vehicle brakes to let a pedestrian cross). However, before considering the effect of this display on other road users, this part of the study first examined how the drivers themselves reacted to this continuous display of one's own driving path.

6.3.1.2 Procedure

Seven drivers participated in the driving simulator study with an average age of 34 years (SD = 10 years, 2 male, 5 female). Average driving experience was eight years. Most of the drivers were employees of the Technical University of Braunschweig, half of them were themselves involved in experimental research.

The drivers went through an urban scenario in a driving simulator and were asked to evaluate the impact and significance of various communication concepts. For the evaluation of the concepts, the driver persons were instructed to think aloud while driving. They were asked to describe what was happening (who was doing what and why?), how they behaved and what their thoughts and feelings were. After each individual drive, a structured interview was additionally conducted.

The static driving simulator of the Technische Universität Braunschweig was used which is equipped with a steering wheel including pedals and sports seat. The driving

Table 6.2 Example of a category partitioning scale used for a two-step rating of different experiences (first step: verbal; second step "–", "0" or " + "), resulting in a 15-point scale

Very weak			Weak			Medium			Strong			Very strong		
–	0	+	–	0	+	–	0	+	–	0	+	–	0	+

environment and the light projections were displayed with Silab 5.0 (Krüger et al. 2005) on three screens (46 in. diagonal), which covered a field of view of about 180°. The programming of the light band (scenario 3) was done in Processing 3 (www.processing.org, Processing Foundation).

The entire structured interview was recorded in writing by one of the two test supervisors and recorded via an audio recorder and a camera, so that the spontaneous comments of the drivers on the individual driving experiences and respective light projections could be evaluated afterward.

In the interview and questionnaire, among other things, the clarity, comprehensibility and usefulness of the communication concepts were evaluated on a category partitioning scale (see Table 6.2).

In addition, the drivers were asked about their perception of the situation and the reasons for their behavior, as well as about their wishes for improvement with regard to the general course of the situation and the concrete implementation of the communication concepts. Finally, the best implementation of the individual concepts and design factors relevant for the evaluation were asked. In scenario 3, it was also of interest to what extent the permanent representation of one's own direction of movement had an influence on the perceived driving and viewing behavior.

6.3.2 *Results*

In the "Thank You" scenario, three different ways of implementing a thank you message were compared with each other. The variants differed both in the symbol used and in the animation.

With regard to the animations, the differences were mainly due to the extent to which the respondents felt addressed by the message. In variant A, the symbol was projected to the rear at a fixed distance from the front vehicle, which was understood less as a directed message from the driver and more as a kind of status display of the vehicle itself. In addition, the further away the front vehicle was, the smaller the symbol was. As a result, the symbol was sometimes harder to recognize than in the other variants B and C.

Variants B and C, on the other hand, appeared to be more clearly directed at the driver and differed in their presence. While the smiley symbol in variant B rested on the road until the driver reached his position, the thumb symbol in variant C was shorter because it moved toward the driver. As a result, the smiley symbol was more clearly visible and appeared calmer, but at the same time had the potential to attract more attention over a longer period of time. The thumb symbol was described as more subtle due to the shorter duration, but in return it could be overlooked more quickly and, compared to the static display of the smiley, appeared too hasty and less sincere ("looks like being thrown down"). The concluding "pop-up" animation in Variant B, which was triggered shortly before the drivers went over the static smiley symbol, was rated particularly positively. This gave the test persons the feeling that the message was intended specifically for them and that they had received something personal to take home ("Collect karma points").

With regard to the symbols used, there were clear differences in the meaning of the message and its undertone. The smiley symbol (variant B) conveyed a message that can be summarized as "Your actions have pleased me". The thumbs-up symbol (variant C) was described as "Your action was good". The check mark symbol (variant A) was interpreted as "Your action was correct".

All messages gave a positive feedback to the driver. However, the differently perceived undertone of the messages had an effect on the evaluation of the front driver. In variant B (smiley symbol), the front driver was described as open, communicative and very friendly ("He is simply happy and wants to tell me"). In contrast, the message in variant A (check mark symbol) was perceived as very impolite, as the drivers felt evaluated or instructed by the other driver. Although they had voluntarily given up their right of way in the situation, they had the impression that it would have been wrong to act differently. The driver in front was perceived here as know-all and arrogant. The driver in variant C (thumbs-up symbol) was between the other two variants in terms of the rating. He was perceived more politely than with the check mark symbol, but compared to the smiley symbol he was perceived a bit cooler and more distanced and less personal.

The results of the rating scales are in agreement with the free, verbal expressions. As Fig. 6.10 shows, the smiley symbol was rated most positively overall. It was very understandable, very distinct, useful, helpful and very pleasant. The difference to the other two concepts is especially pronounced for the design being pleasant. The

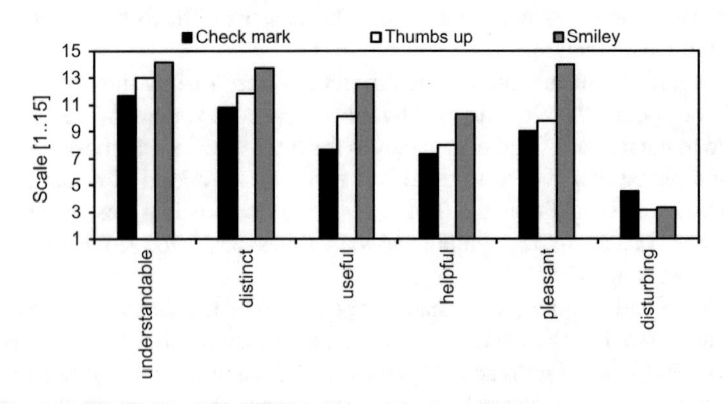

Fig. 6.10 Mean ratings of the three concepts for different dimensions

check mark and the thumbs-up were both rated as medium pleasant, while the smiley clearly was very pleasant. None of the symbols used was rated as disturbing.

Thus, to summarize the results of the "thank you" scenario:

- Displaying a smiley symbol from the back of a car provides a very efficient and positive feedback, resulting in positive emotions and a positive evaluation of the driver in the front car.

For the parking scenario, again three different variants of a message were compared. Here, the information content of the intended messages was varied, from a generic "I am busy" (variant A, "loading") to the specific message "I am looking for a parking space" (variant C, "parking"). Variant B ("searchlight") ranged in the middle.

Variant C (car park symbol) was rated most positively. The drivers were able to understand the symbol very quickly and clearly and found the display useful for understanding the intention of the driver in front. The drivers noted that the relevant information was immediately accessible and the display did not need to be observed any further, so that the journey could continue undisturbed. The front driver was accordingly experienced as very considerate and thoughtful, as he was aware of his behavior and did not leave the driver in uncertainty about the situation.

The two variants A and B performed very similarly on the evaluation scales, but both were rated significantly worse than variant C (see Fig. 6.11). However, the reasons for this were different.

The loading animation in variant A was recognized as such, but the drivers lacked a concrete reference to the situation in order to understand what the driver in front was occupied with ("Now the car has crashed…"). This led to uncertainty about how long the process would continue and how the drivers should behave. It was also noted that people with less PC experience might not know the load symbol. The final animation (loading animation fades out, green tick symbol fades in) of variant A was much more striking than in the other two variants. However, the intended message ("Finished", "Process completed") was not always received correctly, but

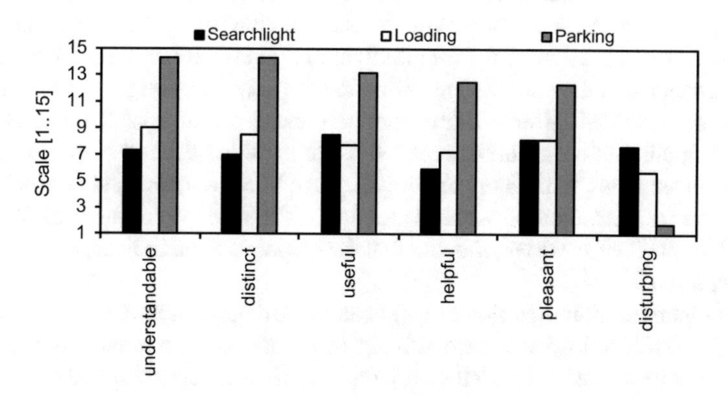

Fig. 6.11 Mean subjective evaluation on the category partitioning scale for the different dimensions, comparing the three concepts for this situation

was understood as a thank you or feedback ("Doing it right"). While the thanking was felt to be unnecessary, the feedback even led to an aversive perception of the driver in front ("know-it-all, arrogant"), similar to the Variant A from the thanking scenario.

The animation in Variant B (pendulum searchlight) was perceived as the most disturbing. On the one hand, the metaphor of the search movement was not understood. On the other hand, the animation was perceived as too aggressive, stressful and distracting. In comparison, the closed, rotating loading animation from variant A was rated more positively. Due to the sweeping animation of the searchlight beyond the width of the vehicle, the drivers were irritated and also stated that they had reservations about approaching or overtaking the vehicle in front. In summary, this variant was rated as "very much stress and distraction without getting information."

In the respective first condition, the control condition, in which the scenario was experienced by each test person without a projected message, almost all test persons expressed impatience and frustration with the slow, obstructive front driver. In contrast, all test persons experienced the drive with additional information as significantly more pleasant, also independent of the variants that were randomized in their order. In this respect, the light-based message provided more patience overall and even in some cases more understanding for the driver in front. In addition, a general detail of the concrete implementation of the projection proved to be advantageous: starting from the vehicle in front, a light beam could be seen on the road in the direction of the symbol (see Fig. 6.10, right), which made it clear who the sender of the message was. Especially when the symbol was not immediately understandable, this implementation helped in interpreting the message.

Thus, with regard to explaining an (annoying) behavior, the following resulted:

- Explaining the reasons for one's behavior clearly improves the evaluation of the other's behavior and leads to positive emotions. However, it seems to be necessary to explicitly specify the reasons for this behavior.

In the travel display scenario, the focus was to evaluate such a display first from the perspective of the driver providing this information to others. Overall, the driver liked the implementation of the longitudinal and lateral animation, although there were differences in the comprehensibility and the concrete design. During the ride, the drivers expressed different ideas about the exact meaning of the display while thinking aloud. Almost all drivers related the animation of the display to their own vehicle movements ("This is my braking distance", "Shows how fast I am driving"), but the correct interpretation was not mentioned ("Shows where the vehicle will be in a certain time"). One driver thought that the display was an indication to drive into a certain direction.

With regard to their direction of vision and distribution of attention, the majority of the drivers stated after a short driving time that they no longer looked at the vanishing point of the road near the horizon, but primarily concentrated on the top of the display. At the same time, this led to a reduced attention in the peripheral areas (e.g., pedestrians, side traffic, traffic lights and traffic signs).

The drivers justified this shift in attention by saying that they tried to avoid overlapping of the display with the edges of the road. Due to the concept, the projected display partially overlapped road markings at intersections and tight bends or protruded into the opposite lane or beyond the pavement (see Fig. 6.12). The drivers found that this superimposition significantly influenced their steering behavior. For example, drivers began to steer earlier than the actual road geometry required. In these situations, the display confused the drivers as they tried to stay in lane with both the vehicle and the projected light surface at the same time, which was not possible during normal driving due to the concept.

These distracting effects of the light beam were particularly strong when the texture of the motion indicator had strong contrasts and contours as well as high opacity (see Fig. 6.9, left, e.g., D and E). These variations were perceived as particularly intrusive and were difficult to ignore while driving. With regard to the design of the locomotion display, variants whose patterns or contours showed similarities to real road markings, e.g., a crosswalk or the centerline, also proved to be problematic (see Fig. 6.9, left, e.g., C to E). Of the displayed texture variants of the display, versions A and B were rated best (see Fig. 6.9, left).

The statements from the interview are also reflected in the scale ratings (see Fig. 6.13). Overall, the indicator was judged to be of little use or helpful. Here, the drivers distinguished between the outside and their own perspective. While the locomotion display could be useful for other road users, such as pedestrians or cyclists, in the immediate vicinity, the majority of the drivers found it rather unsuitable in the current version as a permanent display for the driver. The locomotion display did not provide any added value for the drivers, but rather led to distraction and sometimes disturbed the driving task. Overall, positive ratings are missing for this concept.

Thus, none of the design variants could really achieve the aim intended, not even from the inside of the car which projected this message.

Fig. 6.12 Light beam of the concept overlays road markings of the opposite lane (left) and leads onto the pavement (right)

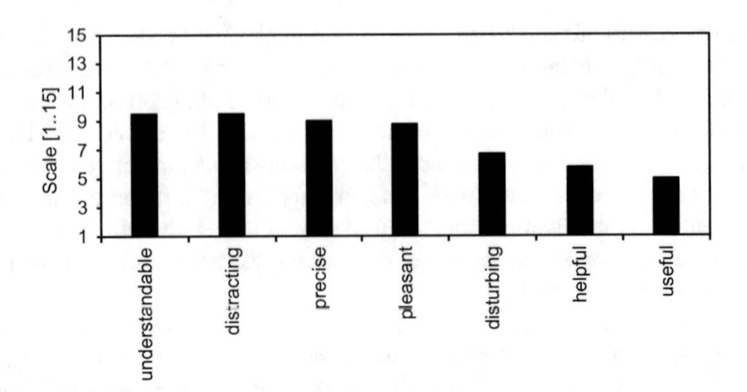

Fig. 6.13 Means of the subjective evaluation of all locomotion concepts taken together

- Displaying information to the outside about one's behavior in the very near future is distracting the drivers and does not really provide much information. It may be that this was due to the specific design variants. However, interviews indicate that this kind of information is not very helpful.

To summarize, this first exploration study investigated how a new communication channel (light-based projections onto the road) can be used to promote cooperative behavior in road traffic and thus positively influence the general traffic climate. The focus was on generic concepts (providing information about what one is doing from an outside and inside view, and thanking other drivers).

In the first scenario, "thanking," the drivers yielded the own right of way to another driver. This driver then provided a "thank you" to the subject. Three different implementations were compared in order to examine the influence of different symbols (check mark, smiley face, thumbs-up) and animations (fixed distance to the front vehicle, stationary on the road, moving toward the drivers) on the perceived content and the effect of the offerings.

Depending on the projected symbol, the front driver was perceived differently. The ratings ranged from "friendly" and "open" (smiley symbol) to "cool" and "reserved" (thumbs-up symbol) to "arrogant" and "know-it-all" (check mark symbol). The smiley symbol was generally preferred by the drivers, as it conveyed a simple, emotional reaction to the other driver ("I am happy"), similar to expressions of opinion in personal interactions. No explicit message is conveyed, but rather an emotional state implicit in the context of the situation. By the decision of the front driver to explicitly express this feeling, the front driver appears very open, communicative and like someone who consciously perceives and respects other road users.

Most of the drivers began to smile themselves and retained this positive basic mood even after the traffic situation they experienced. On the one hand, these findings show that light projections from the vehicle can actually convey emotional messages, even if the driver is not visible through the vehicle shell. On the other hand, it is clear that these messages can also have a positive influence on the emotional state of the recipient, which could have a positive effect on the traffic climate in general. Particularly with regard to the effects that frustration and anger have on driving behavior and the traffic climate (Bliersbach et al. 2002), potential safety-enhancing effects should be investigated more closely if drivers feel comfortable while driving.

In the second scenario, "Parking," the drivers followed a front driver who was driving much more slowly and who sent a message to the following driver to explain his behavior. Here, too, three implementations of the message were compared in order to investigate the influence of the specificity of the message ("I'm busy.", "I'm looking for something.", "I'm looking for a parking space.") on the comprehensibility and experience of the situation.

The most specific variant of the message used a car park symbol and was rated best by the drivers because it conveyed the greatest amount of information and thus explained the situation very quickly and clearly. The drivers showed significantly more understanding for the unusual behavior of the driver in front of them, even though the driver still hindered them. This increased understanding of the situation and the change in perspective that was made possible could have a positive overall impact on the traffic climate. This assumption is supported above all by the comparison with the first drive in the same scenario without light assistance, in which the test persons expressed considerably more impatience, incomprehension and aggression than when driving with light assistance.

For the scenario selected here, the drivers preferred the parking symbol. Due to its high specificity, the transferability of the symbol to other situations in which a slow vehicle also drives in front (e.g., address search, route selection) is limited by the concept. However, using a separate symbol for each situation did not appear to be a good solution for the drivers when asked. It has to be clarified whether different symbols would be necessary for different situations at all, or whether a substitute symbol would be sufficient to understand the situation of the driver in front and thus generate patience and forbearance in the following driver.

In the third scenario, the effects of a dynamic visualization of the driving path were examined. A flexible light band with different textures was projected in front of the drivers' vehicle for visualization. In the variants used here, the drivers evaluated the locomotion display as relatively intrusive and distracting, especially if the textures had a high contrast or opacity. At the same time, there was no added value for the drivers. On the contrary, the drivers reported an impairment of their driving and viewing behavior. These results highlighted above all a conceptual problem in the representation of the predicted driving path. The display was supposed to visualize the vehicle position within the next second and thus led to incongruities when drivers changed their behavioral intentions at short notice. In particular, these incongruities were disturbing during lateral movements, such as quick correction maneuvers when turning at intersections.

While according to the drivers, other road users could benefit from such a movement display, which improves the predictability of a vehicle, and the results of this study suggest that alternative forms of visualization should be developed. For example, steering movements could be neglected and only longitudinal movements could be displayed in order to avoid overlapping of the projection with road markings. In this case, the light band could follow the shape of the road geometry. Future studies should also take into account the perspective of other road users to verify their need for such a system.

- Being able to express gratitude and appreciation by means of light beams seems to be a very promising way to achieve cooperation and positive emotions in traffics.
- Explaining what one is doing at the moment and why one is driving very slowly has also a high potential, but also requires relatively specific messages. A parking sign was clearly evaluated best, but one would have to have a large library of different symbols which is not realistic for projections. Perhaps other channels of communication could be more useful here.
- Explaining where one is going to drive to is disturbing for the driver himself and does not provide additional benefit to him. However, it could be useful for outside people, including pedestrians and cyclists. Light beams could be a very good means to provide this kind of information. This should be examined more closely in future projects.

This last aspect of communicating what one is going to do was examined in more detail in a very specific situation in a second driver simulator study, which is presented in the following.

6.4 Communicating One's Intentions—"I Would like to Enter the Main Road!"

Providing information about where one is going to drive to as in the last scenario provided above may be not very useful as in most of the driving situations that is very clear and can be derived from where one is driving at the moment. However, there are a number of situations where the driver would like to do something which others have to consent to. Trying to enter a busy main road from a smaller road without the right of way is such a scenario that was examined in the second driving simulator study. How can one ask a favor, using light beams?

Ten variants of such a message were compared in their effect. Since the findings from the "thank you" scenario showed that different symbols can have a great influence on the basic tone of the message and thus also influence the perception of the sender, different representations were chosen for the message (symbolism, animation, spatial expansion), with the aim of clear but at the same time polite, unobtrusive communication. In addition, the influence of the visibility of the vehicle communicating was investigated, either by concealing it by vehicles parked at the roadside or by making it freely visible.

6.4.1 Method

In order to investigate the effect of the light message, the drivers experienced the driving situation several times with different variants of the communication concept in randomized sequence. The drivers first evaluated all concepts while the sending vehicle was hidden by parked vehicles. Afterward, the masking was removed so that the transmitter was clearly visible and a new evaluation was carried out. Figure 6.14 shows the scene from the perspective of the driver.

Since the test persons in the "thank you" scenario described above experienced the interaction with the driver in front in very different ways depending on the concrete interpretation of the message, ten message variants were selected here, which covered a broad spectrum for conveying a turning-in wish. The messages were varied, for example, in how active/initiative or passive/reactive they were. In addition, there were messages that had a communal, collegial or inquiring character as well as messages in which the own advantage was in the foreground, concrete actions were announced or instructions were given. Finally, the concepts varied in terms of symbolism, animation and the size of the projection screen. The first drafts of the concepts could finally be summarized to different core messages, which were conveyed individually or in combination:

- Make noticeable: In this case, only the own vehicle position is highlighted without giving any further information about the own intentions. A distinction can be made between a passive waiting ("I am here and standing.") and an active waiting ("I am here and rolling slowly forward.").
- Explain intention: Here one's intentions or desires are communicated without actually implying any concrete action. This can be a general expression of a wish to the environment ("I would like to turn in.") or a concrete request to a person ("Will you let me in?").
- Taking initiative: This is where the driver's wish is finally communicated clearly. This can be done implicitly in the form of an announcement ("I am going to turn now.[Stop there]") or as a direct appeal to the driver which is then explained ("Stop there.[I am going to turn here]").

Fig. 6.14 Screenshot of the scenario from the perspective of the driver. The car from the right is either not visible (left) or clearly visible. One of the concepts used can be seen as a projection on the road

From the various design variants, ten were finally selected by experts, representing a mixture of these possible core messages. Figure 6.15 gives an overview of these animations used.

In variant A, tire tracks spread out from the waiting vehicle into the driver's lane. These were used as a metaphor for an impending movement of the vehicle. The speed of propagation of the two parallel lines was modeled on the acceleration of an approaching vehicle. In the similar variant B, a filled surface was used instead of the two lines, which turned onto the roadway to create the impression of an object turning into the lane.

The variants C to E take up the metaphor of a vehicle carefully approaching. In Variant C, a light gradient was projected in front of the waiting vehicle, which faded out softly toward the road. The gradient expanded slowly so that it would cover the roadway after about ten seconds. The gradient pulsed at regular intervals of about one second to show the direction of expansion. In variant D, the vehicle emitted a kind of visible radio waves that followed each other increasingly closely and quickly, in order to suggest the approach of an object, similar to a direction finder. In variant E, a round shape moves back and forth from the waiting vehicle. The movement was once imitated that of a swinging pendulum and once that of a bouncing ball.

In variants F to I, the driver represented his waiting position. In variants F and G, a holding line was used to emphasize that the driver would not simply start to drive. At the same time, a smaller area pulsed on the right side of the holding line to communicate the desire to turn. The pulsation was in time with the flasher (which could not be seen from the test person's point of view) and was implemented either as a flashing (variant F) or forward expanding area (variant G). In variant H, the driver

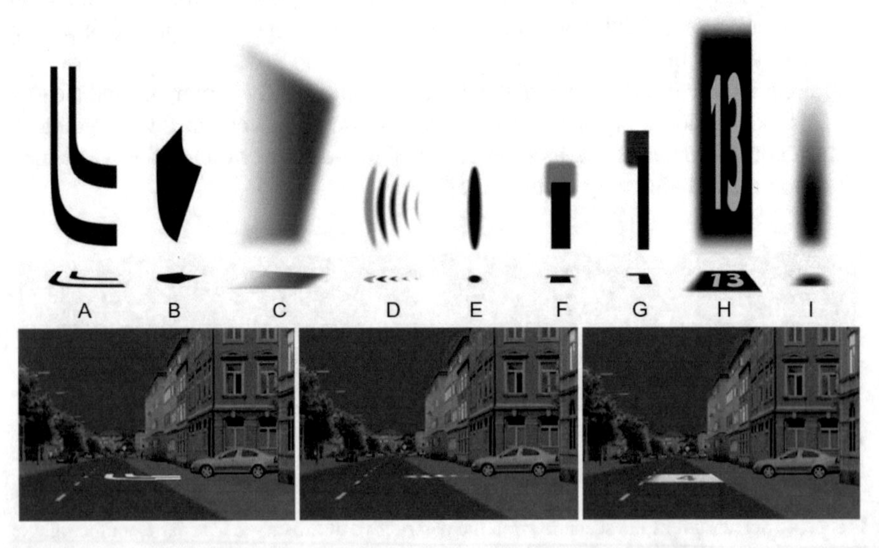

Fig. 6.15 Overview of the ten variants used to express one's intention to enter the main road (top). At the bottom, three of the variants are shown from the perspective of the driver

reported his previous waiting time. The numbers were counted up every second and projected onto the road. The message of variant I consisted merely of highlighting the exit from the property or the waiting vehicle without expressing a concrete intention or announcing any behavior. The vehicle projected a gentle gradient of light that swung back and forth in front of the car like a pendulum movement.

Six drivers with an average age of 34 years (SD = 11 years, 1 male, 5 females) took part in the driving simulator study. The average driving experience was 15 years. Most of the drivers were employees of the Technical University of Braunschweig.

These drivers experienced the urban scenario in the driving simulator and were asked to evaluate the different light projections with regard to their effect and significance. To evaluate the concepts, the drivers were instructed to think aloud while driving. They were asked to describe what they were seeing and how they understood the messages they were seeing. The experimental setup was identical to that of the first study.

During the interview, the drivers were also asked to evaluate each variant using category partitioning scales (see Table 6.2) and declare what they liked about it or what they would change and how. In addition, they were to indicate which of the following core statements were reflected to what extent in the perceived messages of the individual variants:

- "Someone is making himself noticed",
- "Someone is approaching cautiously on the street."
- "Someone is waiting (e.g., for an opportunity to act)."
- "Someone expresses his wish or intent."
- "Someone appeals to me."
- "Someone announces his action."
- "Somebody wants me to let me go first."

Finally, an evaluation of the most favored implementation of each subject was made with regard to relevant dimensions (see below).

6.4.2 Results

As in the previous studies, it was important to the drivers that their own progress was not hindered, i.e., that they did not have to brake unnecessarily hard and that traffic flow was not disturbed. To achieve that, messages should be clear, quick and unambiguous. Projections that did not have these characteristics were perceived as misleading and the drivers wanted to approach these situations much more cautiously because they felt uncomfortable.

Accordingly, projections with a very clear message about oncoming actions or behavior of the other car ("Someone is waiting", "Someone is announcing their action") were evaluated positively. Here, the drivers were easily able to anticipate the behavior of the sending driver and thus also plan their own behavior in good time. In contrast, when the intentions and actions were not so clear ("Someone approaches the road carefully", "Someone expresses his wish or intention") drivers tended to wait for further clear signals from the other vehicle, so that a moment of uncertainty arose and the drivers were annoyed because they did not know how to behave.

The messages were also to be communicated very directly and addressed to the driver. For example, concrete requests ("May I drive?") were perceived much more positively than implicit appeals ("I've been waiting for a very long time."). The latter, by implying norms and values, gave the driver the feeling that he or she was not allowed to make a free decision or to be seen as a bad person if they did not place the needs of the other person above their own. In the case of messages with a clear appeal ("Stay where you are, I'm driving the same way."), it was also important to the drivers that the applicable traffic regulations and thus their own right of way were not restricted. This was also expressed in the fact that some of the drivers already reacted negatively to the question about the perceived messages, although the message in question was not perceived at all in the projection ("Did you have the feeling that the other driver wanted to let you in?" "No, he can't let me go first, because I have the right of way. If I have the right, then I can let him go first.").

In general, projections that gave information about the sender ("Someone is making himself noticed.") were found useful, especially when the other car was blocked from view by parked cars. Overall, when the other car was blocked from view, most of the projected content was less meaningful (1–2 scale points) than when the sending vehicle was freely visible. In some variants, as expected, the visibility of the parked car made the intention to wait much clearer. In addition, by removing the occlusion, the drivers perceived an appeal and an announcement of action more strongly. Finally, concepts that clarified the waiting position or the intention to turn in were experienced much more positively with a not visible vehicle than with a visible car. If the other car was visible, the additional projection led to the negative emotions. The impression arose that the attention of the other car to its own needs was more important to it than the undisturbed flow of traffic, since the information conveyed (e.g., "I am standing here") was already obviously available (example: "I see that you are there, why do you impose yourself so much?).

With regard to general characteristics of the different variants, it was found that the comprehensibility and unambiguity of the message are essential. In order to plan one's own trajectory, it must become clear quickly (within a time window of 1–3 s) what the other person is planning to do, how and when he or she will act (stops vs. turns) and also what options the driver still has. The earlier drivers can plan their trajectory, the less they have to decelerate and accelerate again to be able to behave prosocial. Therefore, unclear symbols should be avoided as well as long, gradually building up animations, or longer pauses between repeated animations. These distract too long from the actual driving task.

Thus, moments of uncertainty and indecision should be avoided. Insufficiently clear lighting concepts caused the test subjects to delay out of caution in order to be able to better assess the situation. The waiting driver could, however, misunderstand this slowed-down driving style as an implicit invitation to enter the main road, which then creates the impression that the driver is just taking the right of way. In order to avoid such misunderstandings and their consequences, it is therefore important to have clear, unambiguous messages.

The majority of the drivers did not feel addressed by light projections until they were clearly visible on their own roadway (e.g., variants A–D, H in Fig. 6.15). Contrary to expectations, most of the drivers did not necessarily experience the large-area projections as an intrusion into their driving area or as too intrusive, but rather as situationally appropriate. Some drivers saw images that were not projected at all or only partially onto the road as not relevant for planning their own trajectory, as the other driver had not made his or her concerns sufficiently clear. This was especially the case when the other car was hidden.

The drivers preferred naturalistic movements that were based on the natural motion profile of a vehicle (e.g., variants A and B, see Fig. 6.15). These offered various advantages: They were associated with a vehicle very quickly, so that the sender could be traced even if the car was hidden. Furthermore, the direction of movement of the animation made it clear that this was a turning process, and the speed and acceleration of the animation also pointed to a turning vehicle. In general, animations should be smooth and flowing, as erratic, jumpy movements were more likely to distract or associated with critical situations. In addition, animations with a movement target in the middle of the roadway (e.g., variants C and D, see Fig. 6.15) or with changing movement directions (e.g., variants E and G, see Fig. 6.15) were more easily misunderstood.

Some drivers explicitly wished that the intended turning-in direction should be shown. In this case, the drivers were less willing to let drivers turn into their own lane. On the one hand, when the other car wanted to turn left and had to cross one's lane, drivers were reluctant to have a vehicle in front of them, which in turn emphasizes the relevance of fluent progress. In general, variants with a directional display were preferred, as they made the intention of the opposite party clearer, so that the drivers were able to decide and react more quickly. In addition to a clear movement target, it was also rated positively if the animation indicated the position of the sender, for example, by showing the origin of its movement.

With regard to the speed of the animations, the drivers wished that the message should be recognizable very early so that the traffic flow would be impaired as little as possible, even if they decided to let the other person turn. In other words, the message had to be recognizable from a certain distance and be present for a sufficiently long time, but without attracting too much attention. The message should be conveyed quickly so that the drivers could concentrate on the driving task as soon as possible. The animations of the C, D and H variants in particular developed very slowly, and it took several seconds to understand the message. In the variants E, F, G and I the animation repeated itself again and again. Here, too, the message did not appear immediately, but only after the test persons had seen 1–2 repetitions. Only in the variants A and B did the test persons succeed in understanding the message after the first run of the animation.

With regard to the message conveyed, short-lasting, erratic animations (flashing, pulsing) seemed more hesitant or inquiring (e.g., variants F and G, see Fig. 6.15) than longer lasting, even and slowly fading animations that conveyed more determination and a clearer announcement of action (e.g., variants A and B, see Fig. 6.15).

All variants were rated with regard to how well they reflected the following messages:

- Notice: "Someone is making himself noticed,"
- Approach: "Someone is approaching cautiously on the street."
- Waiting: "Someone is waiting (e.g., for an opportunity to act)."
- Wish: "Someone expresses his wish or intent."
- Appeal: "Someone appeals to me."
- Action: "Someone announces his action."
- Let me go: "Somebody wants me to let me go first."

Figure 6.16 shows the results for variant A (projection of wheel markings) and B (filled area). Both variants were consistently rated most positively. They were easy to understand and unambiguous. The animations were very naturalistic and reminded of tire tracks or a turning vehicle, especially in variant A. With variant A, it was very clear that another car wanted to get noticed and that it was approaching. The wish or appeal to enter the main road was very obvious, but it was not simply the announcement of the action. There was no misunderstanding that it was meant to let the driver go. Finally, it did not transmit the waiting, but more the wish the enter the road. Ratings were only somewhat clearer when the vehicle was visible as compared to when it was hidden.

The ratings of variant B made also clear that the other car wanted to get noticed. The approach was less clear, but it was also less of showing a waiting, but clearly a wish, somewhat of an appeal and less of an action. Again, there was a small and similar effect of the visibility of the other car with a better understanding when it was visible.

Variant C used a beam (see Fig. 6.17, left). The overall rating was in the middle range. With this variant, it was somewhat clear that another car wanted to get noticed. All the other messages lie in the middle range. Thus, some drivers thought more about an approach, others of a waiting, a wish, an appeal or an action. It did not matter much whether the other car was hidden or not. Thus, this was a very ambiguous variant.

Variant D is shown in Fig. 6.18. As the ratings show, drivers very clearly think that this variant shows that another car wants to get noticed. It does not convey the message that somebody is waiting, but wanting to take an action. So this is a very clear message that some other car is intending to enter the main road. However, as this other car does not have the right of way, this is not seen as a wish or appeal but more of a rude announcement of an action. This is reflected in the bad overall ratings.

Both variants C and D were understood only very slowly, which was partly due to the fact that the message only became clear after the test subjects had observed the animations for a few seconds. On the other hand, the movement was perpendicular to the direction of travel and was directed toward the center of the road. Therefore, the aim of the movement was not as clear as in variants A and B. Both variants were

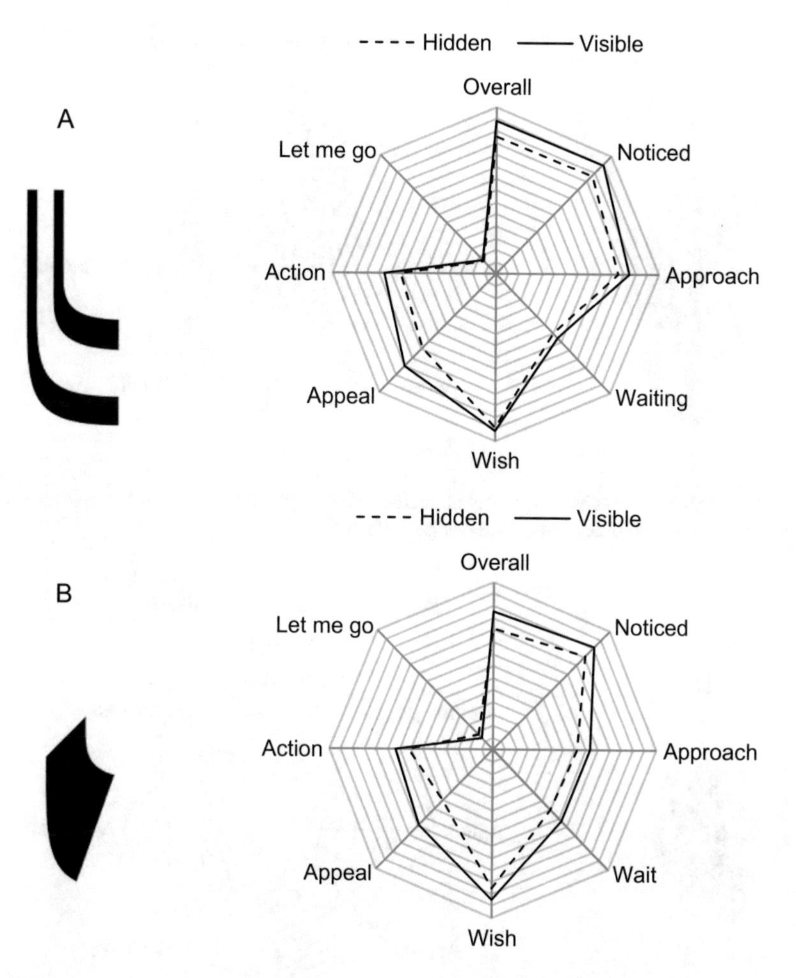

Fig. 6.16 Means of the subjective rating about the different possible messages in the situation with the hidden and visible other vehicle for design Variant A (top) and B (bottom)

also very dominant due to their surface, which in combination with the ambiguity of the message tended to result in a poor evaluation.

Variant E was once presented as a pendulum, once as a bouncing ball (see Fig. 6.19). Both are rated not very good. The pendulum makes it quite clear that somebody wants to get noticed but largely fails to convey the other intentions or actions. It might be that the other car is waiting (middle range of scale). The ratings for the second variant of E are even less clear.

Variants F and G projected a waiting line in front of the other car to make clear that this was waiting. A small area was pulsating in variant F or expanding in variant G to show the intention to turn into the main road. Both variants have a low overall rating. The most likely intention of variant F is that somebody wants to get noticed.

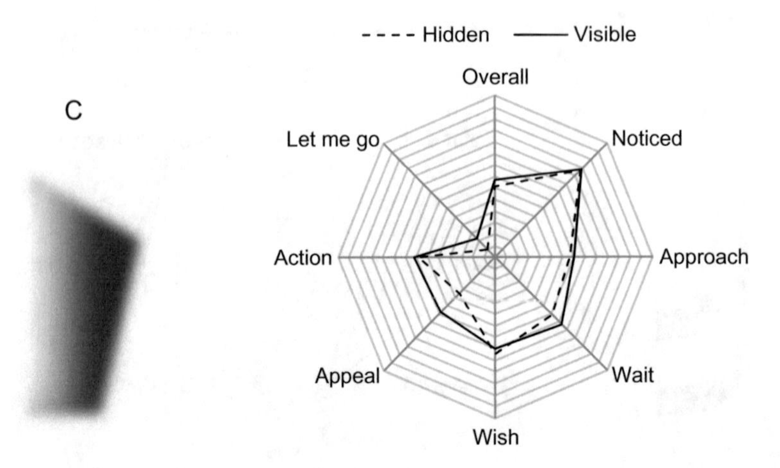

Fig. 6.17 Means of the subjective rating about the different possible messages in the situation with the hidden and visible other vehicle for design Variant C

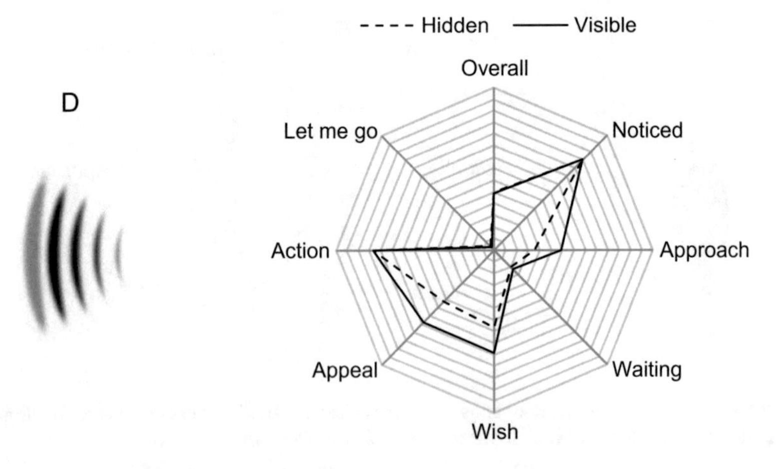

Fig. 6.18 Means of the subjective rating about the different possible messages in the situation with the hidden and visible other vehicle for design Variant D

The expansion of variant G strongly increases the message that somebody is waiting and wishes to enter the main road. However, this only works in an acceptable manner if the car is visible.

Variations E, F and G were generally perceived as very hesitant. Particularly when the other car was hidden, these representations were unable to convey a clear message. On the other hand, when the other car was visible, the messages ("I'll wait here." "I'd like to turn in.") were perceived as unnecessary or too hesitant or even as intrusive, since the driver's attention was demanded without generating any gain in information ("Yes, I see that you are standing there. Either you drive now or you

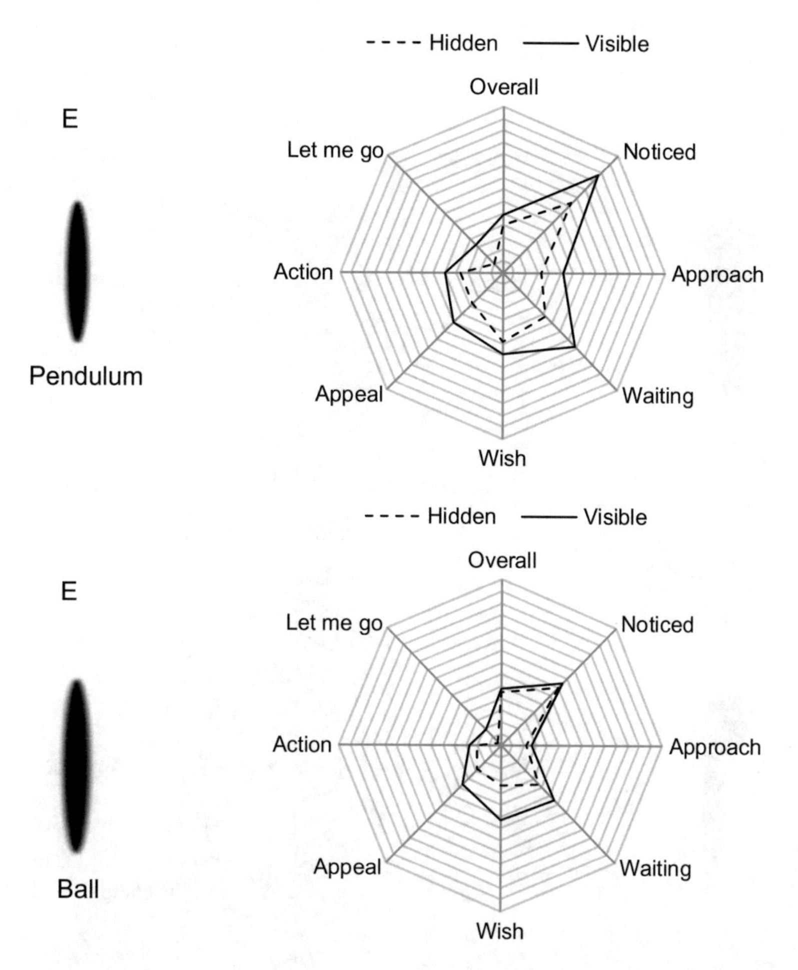

Fig. 6.19 Means of the subjective rating about the different possible messages in the situation with the hidden and visible other vehicle for design Variant E (pendulum, top, and ball, bottom)

leave it. Do I have to decide that for you now?"). In the case of variants E and G, the changing animation direction (forward and backward) was also criticized, as this led to uncertainty as to whether the transmitter wanted to drive up or stay and wait. Depending on when a respondent looked at the projection, the forward or backward movement was first perceived as more dominant (Fig. 6.20).

The countdown of variant H (see Fig. 6.21) has a very low overall rating. However, it is quite clear that this is some kind of an appeal to let the other car in. But it seems that drivers do not really like to be approached in such a manner.

This variant H also consistently led to the greatest confusion. Most of the drivers understood the numbers as a countdown (although the numbers were still counting up), but did not know what for. Even if they noticed that the numbers actually counted up, they were not able to deduce the correct message at first. Some drivers finally

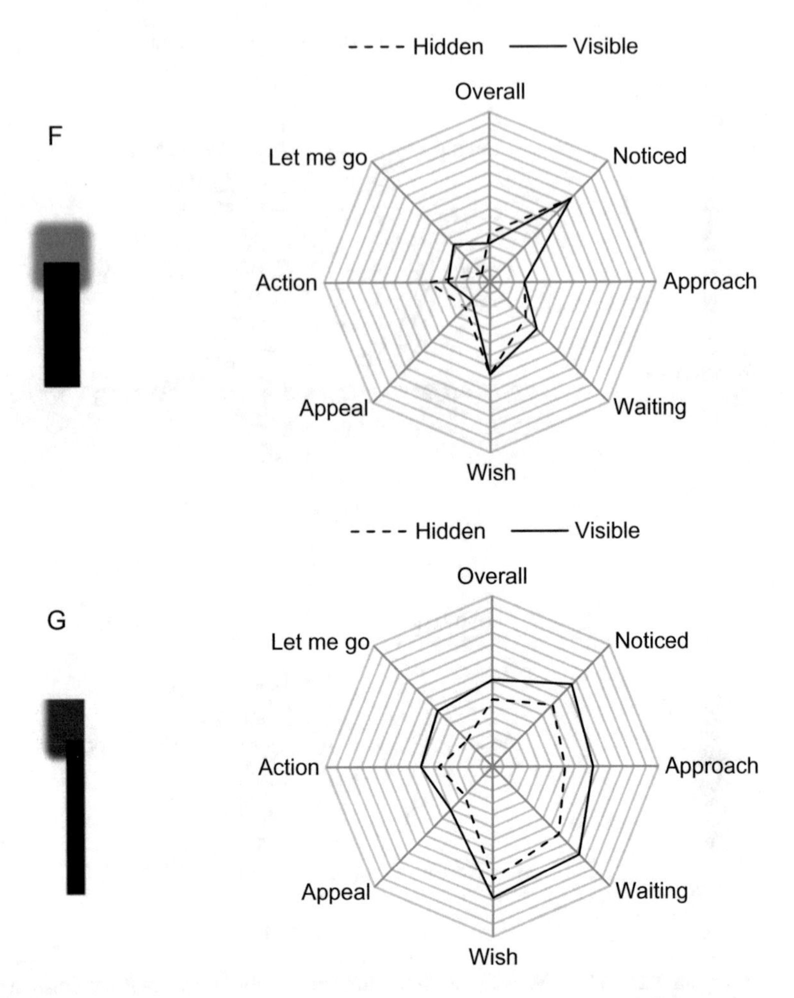

Fig. 6.20 Means of the subjective rating about the different possible messages in the situation with the hidden and visible other vehicle for design Variants F (top) and G (bottom)

understood the statement ("I've been waiting here for XX seconds."), but felt that the waiting driver was appealing to a guilty conscience and demanding an implicit right of way. All in all, this variant was assessed most negatively, since the message—if at all—only became accessible very late and the dominant projection surface additionally acted as a barrier.

Finally, variant I was meant to clarify the situation by projecting a pendulum in front of the car without really entering the main road. This gives a medium overall rating, and the message becomes quite clear that somebody is waiting. However, it is not seen as a wish or announcement of action. Although variant I was rated as pleasant, it was not considered to be very informative. When the other car was concealed, the display was primarily understood as an indication that there was

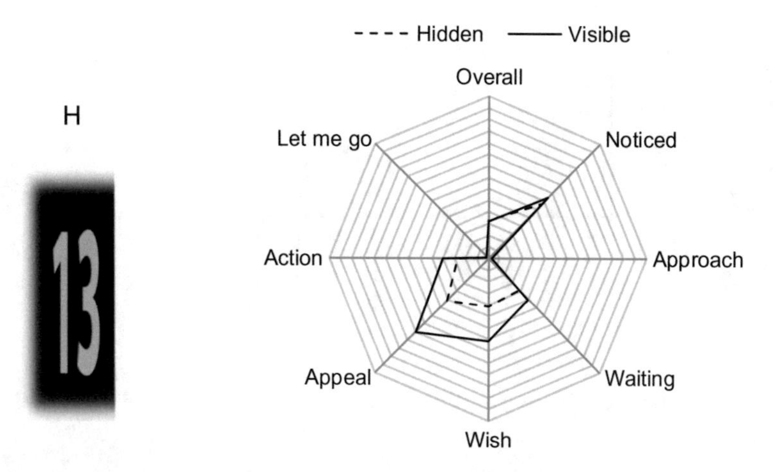

Fig. 6.21 Means of the subjective rating about the different possible messages in the situation with the hidden and visible other vehicle for design Variant H

something at the edge of the road. When the vehicle was visible, the message changed more toward a waiting position. By moving along their own lane, some drivers also had the impression that they were explicitly let in (Fig. 6.22).

Figure 6.23 shows the final evaluation of the two variants that were most frequently chosen as favorites (variants A and B). Variant A was perceived as better, more pleasant, more aesthetic and more useful than variant B. Among other things, the evaluation was based on the very calm animation and the quick association with an actual vehicle. Variant A, however, was also rated as somewhat more disturbing and

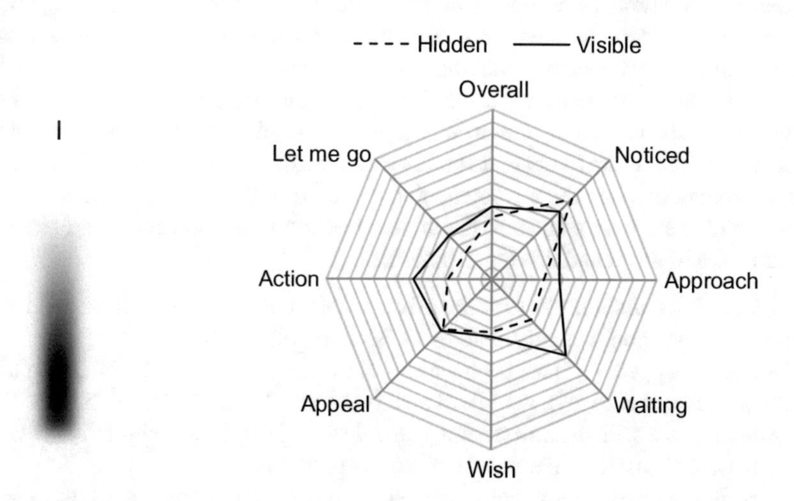

Fig. 6.22 Means of the subjective rating about the different possible messages in the situation with the hidden and visible other vehicle for design Variant I

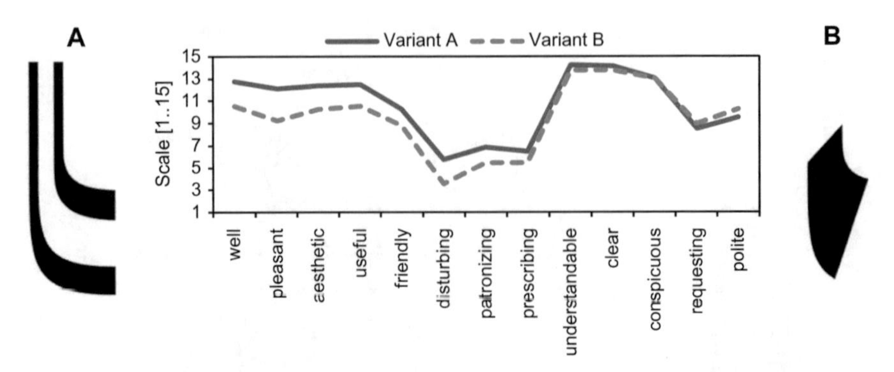

Fig. 6.23 Mean ratings of the two most favored variants **A** (tire tracks) and **B** (filled area)

patronizing. Here, too, the animation was used as a reason, since the slow fading out of variant A left a less hesitant impression and thus had a more pronounced announcement character, which at the same time implied that the test persons had less freedom of decision.

To summarize, in this second driving simulator study, a scenario was used in which the drivers had the possibility to let a waiting vehicle turn into the road. The waiting vehicle projected its intention to turn onto the road as a message to the approaching test driver. Ten variants of the projected intention were compared in their effect. In addition, the influence of the visibility of the sending vehicle was also examined, either by obscuring the waiting vehicle from vehicles parked at the side of the road or by providing a clear view of the waiting vehicle.

It was shown that clear and unambiguously visible representations on the own roadway were preferred to those which were discreetly displayed at the edge of the roadway to indicate the intention to turn off. Such an unambiguous representation better conveyed to the drivers that they were addressed by the light projection. This was particularly important when the waiting vehicle could not be seen directly. In addition, animations were preferred which, compared to more abstract, symbolic animations, were more likely to resemble a naturalistic movement pattern. The naturalistic animations had the advantage that the wish of the waiting vehicle could be understood at an early stage and the drivers had more time to decide and react and could return to the actual driving task faster.

- In a driving situation where some driver wants to ask a favor from other drivers, the message should clearly state this wish in a polite manner.
- Neither an appeal ("I have been waiting for a long time") nor the intention to act ("I will enter the road") were seen positively and should thus be avoided.
- In accordance with general design principles, projections and animations that are realistic and easy to understand are clearly preferable.
- Overall, the possibility to communicate by means of light messages was seen as a way to achieve better cooperation in traffic.

6.5 Light-Based Communication in Traffic

To further explicit communication in traffic would not only increase traffic safety, but make positive interactions and cooperation possible, thus promoting a well-being and joy in traffic which leads to a better traffic climate. We have examined light-based communication as one means of explicit communication of cars with other cars, but also pedestrians and cyclists. This could be used by human drivers, but in the future also by automatic cars.

The first question was about the messages, which would be useful and beneficial in this context. More specifically, what kind of information would lead to positive interactions (joy, fun, well-being) and would prevent negative interactions (frustration, anger, fear)?

Using a diary study and an online survey, basic needs in traffic were identified that lead to positive emotions, if fulfilled. On the contrary, if these needs are denied, negative emotions results. Four needs were connected to mobility, like wanting to have a smooth flow in traffic, being autonomous and in control, and wanting other traffic participants to act in a manner that is transparent and predictable. Secondly, four needs were directly related to social aspects which one could expect to be directly influenced by new communication means in traffic. These comprise the need for mutual respect, for affiliation, but also wanting to experience significant moments and stimulation by interactions. Overall, these needs present the theoretical background of how to further interactions in traffic. The more these needs are fulfilled by enabling communication and cooperation, the more positive will the resulting emotions be.

When directly asking about the resulting positive and negative emotions, a positive arousal, being thrilled and feeling close to others are dominant at the positive side. Being irritated, annoyed or even horrified are the most relevant negative effects, supplemented by feelings of hostility in the social area. These are the direct consequences if the needs described are not fulfilled. On the contrary, if better social interactions are enabled in traffic, this could not only further social emotions like feeling close and avoid negative emotions like hostility. It would also lead to positive, significant moments in traffic, avoiding feelings of insecurity, annoyance and fear.

While these emotions directly results from positive social interactions in traffic drivers indicated that there was mainly an intrinsic motivation for cooperation and respectful behavior in traffic. Just liking to act like that and the feeling that it is important to act like this seems to be a major motivation to cooperate in traffic. Thus, people do not have to be extrinsically motivated to cooperate in traffic, but they have to be enabled to do so and follow their own intrinsic motivation. This, in turn, will results in positive emotions of actors and co-actors and thus a more positive climate in traffic.

Accordingly, we found that people wish for clear, easy, efficient and available communication in traffic, and being focused on positive interactions. Talking or using text would be most natural and preferred. Most drivers wanted to communicate their intentions, behavior or mood, but also tell others what these did wrong.

But how to communicate explicitly in traffic? We examined in which manner light-based communication could achieve this kind of communication, focusing on very general outlines of communication concepts.

The first aspect examined was how to thank others in traffic, directly relating to the need for positive interactions in traffic. Projecting a smiley symbol from the car in front proved to be very effective and positive. It was very interesting to see that even subtle changes in the symbols lead to negative impressions about the car driver projecting the symbols. Again, a close relation to the needs could be shown. If the symbol indicated that the driver in front gave a kind of "moral" feedback, this went against the need for autonomy and control and was disliked. This was especially strong with a checkmark and less so with a thumbs-up symbol. Thus, the selection of the adequate symbols seems quite important to achieve the positive social emotions one would like to have.

The second aspect examined was how to explicitly express one's intentions. As a typical example, the search for a parking lot was taken where the driver followed another car which was going very slowly (because looking for a parking lot). First of all, experiencing this kind of behavior without any information was very frustrating, and any of the design variants clearly improved this situation. Moreover, results show that it is necessary to provide specific information about the causes of the unusual behavior. Thus, on the one hand the approach to further positive interactions in traffic by communication is clearly supported by this research. On the other hand, explicit communication seems to be necessary to fully achieve this potential. This would need to a larger number of symbols which would then have to be matched to the current situation. This might impede the implementation of such concepts.

As a third aspect, we looked at the projection of one's intention to the outside, in the first study from the inside view of the driver. The trajectory during the next second was projected on the road. Independent of the design variant, drivers found this distracting and unnecessary. It was also thought to not provide too much new information, as in most cases during normal driving one can quite well anticipate where the car will be in the next second.

Thus, a second driver simulator study looked at the projection of one's intention in a very specific situation where a driver likes to enter a main road, although usually he should simply wait until all is clear. Additionally, we looked at this situation from the outside that means from another car which was supposed to let this driver in. In this situation, the information provided proved very useful and helpful. However, in accordance to the needs described above and similar to the "thank-you" study, the specific design of the information proved to be very important. Again, drivers liked it most when the other driver clearly communicated their wishes, but did not act without being asked to and when they did not appeal to pity ("I have been waiting so long!"). However, it seemed important that wishes were clearly communicated as just showing that one is waiting was not found to be very helpful.

Overall, enabling communication in traffic even in such a specific way using light seems to be very promising:

- Expressing one's intention in specific situation furthers understanding and prosocial behavior of others, which leads to positive emotions on both side.
- Explaining one's unusual behavior works similarly to increase understanding and further positive emotions, but seems to require specific information about the causes of the behavior.
- Enabling drivers to thank others encourages prosocial behavior and leads to positive emotions.

Thus, light-based communication is an excellent way to further communication, cooperation and positive emotions in traffic. The studies presented provide a sound starting point for further research and development.

Acknowledgements The project KOLA (Kooperativer Laserscheinwerfer; cooperative laser beam; see www.experienceandinteraction.com/kola) was funded by the Federal Ministry of Education and Research under the funding code 16SV7614. The ideas and concepts were developed in close cooperation with Prof. Marc Hassenzahl and Kai Eckoldt from Universität Siegen.

The translation was done with www.DeepL.com/Translator (free version).

References

Bazilinskyy P, Dodou D, de Winter J (2019) Survey on eHMI concepts: the effect of text, color, and perspective. Transp Res Part F: Traffic Psychol Behav 67:175–194. https://doi.org/10.1016/j.trf.2019.10.013

Blanchette I, Richards A (2010) The influence of affect on higher level cognition: a review of research on interpretation, judgement, decision making and reasoning. Cogn Emot 24(4):561–595. https://doi.org/10.1080/02699930903132496

Bliersbach G, Culp W, Geiler M, Heß M, Schlag B, Schuh K (2002) Gefühlswelten im Straßenverkehr. Emotionen, Motive, Einstellungen, Verhalten

de Ceunynck T, Polders E, Daniels S, Hermans E, Brijs T, Wets G (2018) Road Safety Differences between Priority-Controlled Intersections and Right-Hand Priority Intersections. Transp Res Rec J Transp Res Board 2365(1):39–48. https://doi.org/10.3141/2365-06

Dollard J, Miller NE, Doob LW, Mowrer OH, Sears RR (1939) Frustration and aggression. Retrieved from https://search.ebscohost.com/direct.asp?db=pzh&jid=%22200416227%22&scope=site

Endsley MR (1995) Toward a theory of situation awareness in dynamic systems. Hum Factors 37(1):32–64. https://doi.org/10.1518/001872095779049543

Krüger H-P, Grein M, Kaussner A, Mark C (2005) SILAB—a task-oriented driving simulation. In: Proceedings of the driving simulator conference (DSC) North America, Orlando, Florida, Nov 30 - Dec 2, 2005.

Merten K (1977) Kommunikationsprozesse im Straßenverkehr. In: Unfall- und Sicherheitsforschung Straßenverkehr. Referate, Ergebnisse und Folgerungen des Symposions "Unfallforschung und Verkehrssicherheit" der Bundesanstalt für Straßenwesen. Symposium '77 (Heft 14, S. 115–126).

Räsänen M, Summala H (1998) Attention and expectation problems in bicycle-car collisions: an in-depth study. Accid Anal Prev 30(5):657–666. https://doi.org/10.1016/S0001-4575(98)00007-4

Risser R (1985) Behavior in traffic conflict situations. Accid Anal Prev 17(2):179–197. https://doi.org/10.1016/0001-4575(85)90020-X

Statistisches Bundesamt Destatis (2019) Verkehrsunfälle 2018, Statistisches Bundesamt, Wiesbaden
Zhang T, Chan AHS (2016) The association between driving anger and driving outcomes: a meta-analysis of evidence from the past twenty years. Accid Anal Prev 90:50–62. https://doi.org/10.1016/j.aap.2016.02.009

Printed in the United States
by Baker & Taylor Publisher Services